Process Modelling and Simulation

Process Modelling and Simulation

Special Issue Editors

César de Prada
Constantinos Pantelides
José Luis Pitarch

MDPI • Basel • Beijing • Wuhan • Barcelona • Belgrade

Special Issue Editors
César de Prada
University of Valladolid
Spain

Constantinos Pantelides
Imperial College London
UK

José Luis Pitarch
University of Valladolid
Spain

Editorial Office
MDPI
St. Alban-Anlage 66
4052 Basel, Switzerland

This is a reprint of articles from the Special Issue published online in the open access journal *Processes* (ISSN 2227-9717) from 2018 to 2019 (available at: https://www.mdpi.com/journal/processes/special_issues/process_model)

For citation purposes, cite each article independently as indicated on the article page online and as indicated below:

LastName, A.A.; LastName, B.B.; LastName, C.C. Article Title. *Journal Name* **Year**, *Article Number*, Page Range.

ISBN 978-3-03921-455-6 (Pbk)
ISBN 978-3-03921-456-3 (PDF)

Cover image courtesy of José Luis Pitarch

Contents

About the Special Issue Editors . . . . **vii**

César de Prada, Constantinos C. Pantelides and José Luis Pitarch
Special Issue on "Process Modelling and Simulation"
Reprinted from: *Processes* **2019**, *7*, 511, doi:10.3390/pr7080511 . . . . **1**

Hao Li, Zhien Zhang and Zhe-Ze Zhao
Data-Mining for Processes in Chemistry, Materials, and Engineering
Reprinted from: *Processes* **2019**, *7*, 151, doi:10.3390/pr7030151 . . . . **4**

Kris Villez, Julien Billeter, and Dominique Bonvin
Incremental Parameter Estimation under Rank-Deficient Measurement Conditions
Reprinted from: *Processes* **2019**, *7*, 75, doi:10.3390/pr7020075 . . . . **15**

Zhenyu Wang, Hana Sheikh, Kyongbum Lee and Christos Georgakis
Sequential Parameter Estimation for Mammalian Cell Model Based on In Silico Design of Experiments
Reprinted from: *Processes* **2018**, *6*, 100, doi:10.3390/pr6080100 . . . . **48**

Xiangzhong Xie, René Schenkendorf, and Ulrike Krewer
Toward a Comprehensive and Efficient Robust Optimization Framework for (Bio)chemical Processes
Reprinted from: *Processes* **2018**, *6*, 183, doi:10.3390/pr6100183 . . . . **60**

Jose Luis Pitarch, Antonio Sala and Cesar de Prada
A Systematic Grey-Box Modeling Methodology via Data Reconciliation and SOS Constrained Regression
Reprinted from: *Processes* **2019**, *7*, 170, doi:10.3390/pr7030170 . . . . **86**

Maximilian Sixt, Lukas Uhlenbrock and Jochen Strube
Toward a Distinct and Quantitative Validation Method for Predictive Process Modelling—On the Example of Solid-Liquid Extraction Processes of Complex Plant Extracts
Reprinted from: *Processes* **2018**, *6*, 66, doi:10.3390/pr6060066 . . . . **109**

Logan D. R. Beal, Daniel C. Hill, R. Abraham Martin and John D. Hedengren
GEKKO Optimization Suite
Reprinted from: *Processes* **2018**, *6*, 106, doi:10.3390/pr6080106 . . . . **136**

Der-Sheng Chan, Jun-Sheng Chan and Meng-I Kuo
Modelling Condensation and Simulation for Wheat Germ Drying in Fluidized Bed Dryer
Reprinted from: *Processes* **2018**, *6*, 71, doi:10.3390/pr6060071 . . . . **162**

Shashank Muddu, Ashutosh Tamrakar, Preetanshu Pandey and Rohit Ramachandran
Model Development and Validation of Fluid Bed Wet Granulation with Dry Binder Addition Using a Population Balance Model Methodology
Reprinted from: *Processes* **2018**, *6*, 154, doi:10.3390/pr6090154 . . . . **180**

Cristian Pablos, Alejandro Merino and L. Felipe Acebes
Modeling On-Site Combined Heat and Power Systems Coupled to Main Process Operation
Reprinted from: *Processes* **2019**, *7*, 218, doi:10.3390/pr7040218 . . . . **205**

Xiuli Wang, Yajie Xie, Yonggang Lu, Rongsheng Zhu, Qiang Fu, Zheng Cai and Ce An
Mathematical Modelling Forecast on the Idling Transient Characteristic of Reactor Coolant Pump
Reprinted from: *Processes* **2019**, *7*, 452, doi:10.3390/pr7070452 . . . . **231**

Lei Wang, Mengting Wang, Mingming Guo, Xingqian Ye, Tian Ding and Donghong Liu
Numerical Simulation of Water Absorption and Swelling in Dehulled Barley Grains during Canned Porridge Cooking
Reprinted from: *Processes* **2018**, *6*, 230, doi:10.3390/pr6110230 . . . . **245**

Son Ich Ngo, Young-Il Lim and Soo-Chan Kim
Wave Characteristics of Coagulation Bath in Dry-Jet Wet-Spinning Process for Polyacrylonitrile Fiber Production Using Computational Fluid Dynamics
Reprinted from: *Processes* **2019**, *7*, 314, doi:10.3390/pr7050314 . . . . **258**

Florian Markus Penz, Johannes Schenk, Rainer Ammer, Gerald Klösch and Krzysztof Pastucha
Evaluation of the Influences of Scrap Melting and Dissolution during Dynamic Linz–Donawitz (LD) Converter Modelling
Reprinted from: *Processes* **2019**, *7*, 186, doi:10.3390/pr7040186 . . . . **273**

About the Special Issue Editors

César de Prada (Prof.) is with the Department of Systems Engineering and Automatic Control in the School of Industrial Engineering, University of Valladolid, Spain. He graduated from the university's Physics (Electronics) program in 1972. After getting his Ph.D., he became full professor in 1987 with the Department of Computer Science at the Autonomous University of Barcelona. His fields of interest center on the control and dynamic optimization of process systems as well as in modelling and simulation. His research topics cover model predictive control and optimal management of large scale systems, considering aspects such as uncertainty, the presence of hybrid continuous-discrete elements, and non-linear physical modelling. He has published 145 journal papers and book chapters and has made 259 contributions to international conferences by combining research on methods and algorithms with the development of software systems and industrial applications. For his contributions and relevant trajectory, he has been recognized both by the Spanish scientific community, which gave him the CEA award in 2016, and by the International Society of Automation with the ISA-Spain award in 2008.

Constantinos Pantelides (Prof.) is currently the Managing Director of Process Systems Enterprise, a position he has held for the past 14 years. He is also a professor of Chemical Engineering at Imperial College London. He holds B.Sc. and Ph.D. degrees from Imperial College, and an MS degree from the Massachusetts Institute of Technology. He has been working in the area of process modelling technology for more than three decades, and played a leading role in the development of the gPROMS and SPEEDUP software. A key focus of his current activities is the role that deep knowledge, captured and encoded in mathematical models, can play in the ongoing digital transformation of the process industries, and the architecture and design of general digital application platforms that can support that role. His contributions have been honoured by several awards including the 2007 Royal Academy of Engineering MacRobert Award, the UK's highest prize for engineering innovation, and the 2016 Sargent Medal of the UK Institution of Chemical Engineers. He recently received a Doctor Honoris Causa degree from the Technical University of Dortmund, Germany, and the 2019 Computing Practice award of the American Institute of Chemical Engineers. He is a Fellow of both the Institution of Chemical Engineers and of the Royal Academy of Engineering.

José Luis Pitarch (Dr.) received an M.Sc. degree in Industrial Engineering with honors from the Universitat Jaume I (Castellón, Spain) in 2008. After working with BP Oil Refinery of Castellón as a process control engineer in 2009, he moved to the Universitat Politècnica de Valencia (Spain), where he received an MS degree in Control and Industrial Informatics in 2010 and a Ph.D. degree in Control Engineering in 2013. Currently, he is a postdoc at the Universidad de Valladolid (Spain), where he is working in process modelling, control, and real-time optimization. He has coauthored 30 conference papers, 14 journal papers indexed in JCR, a book, and three book chapters. His research interests are in machine learning, grey-box and fuzzy modelling, stability analysis, dynamic optimization, nonlinear MPC, invariance-based control, and production-maintenance scheduling, among others.

processes

MDPI

Editorial

Special Issue on "Process Modelling and Simulation"

César de Prada [1,2,*], Constantinos C. Pantelides [3,4] and José Luis Pitarch [1]

[1] Systems Engineering and Automatic Control DPT, Universidad de Valladolid, 47011 Valladolid, Spain
[2] Institute of Sustainable Processes, Universidad de Valladolid, 47011 Valladolid, Spain
[3] Process Systems Enterprise Ltd., London W6 7HA, UK
[4] Centre for Process Systems Engineering, Imperial College London, London SW7 2AZ, UK
* Correspondence: prada@autom.uva.es; Tel.: +34-98342-3164

Received: 2 August 2019; Accepted: 2 August 2019; Published: 5 August 2019

Collecting and highlighting novel developments that address existing as well as forthcoming challenges in the field of process modelling and simulation was the motivation for proposing this special issue on "Process Modelling and Simulation" in the journal *Processes*. Our objective was to provide interested readers with an overview of the current state of research, tools and applications on the use of models for simulation and decision support in the process industry. The special issue brings together fourteen contributions on topics ranging from the process systems [1–3] and (bio)chemical engineering [4,5] fields, to software development [6] and applications in heat and power systems [7,8]. Moreover, the hot topic of data mining and machine learning is also discussed from a process engineering perspective in [9,10]. This conveys the broadness of use and impact that models will have (and already have) for industrial decision support in the approaching digital era.

Process models are the foundation that other applications (sensitivity analysis, predictive simulation, real-time optimization, etc.) build upon. Accordingly, half of the published articles in this special issue focus on model building and parameter estimation and validation. From the chemical and process systems engineering field, we received two contributions [11,12] that model the underlying physical phenomena beyond the classical macro scale, with the aim of having a reliable simulation for predicting the effects of different process operation regimes on product quality, and hence reducing experimentation costs. Also related to this goal, two contributions brought heat and power systems into the scope: [7] proposed a grey-box model of limited complexity that couples the production process with the plant's combined heat and power system in order to reduce operation costs, whereas [8] modeled the hydraulic dynamics in a nuclear reactor cooling pump with respect to different vane structures to ensure safe operation in case of power failures.

Models for decision support must be tailored to the actual process, or the underlying equations should allow the transfer of the lab-scale data to any desired scale. In this sense, [3,4] proposed iterative methods for parameter estimation to progressively improve the plant-model match under realistic conditions, and [5] considered uncertainty in the estimation via robust optimization. Furthermore, a methodology for obtaining physically coherent grey-box models (or plant surrogate ones) from fundamental principles and plant data was proposed in [10], while [1] presented a quantitative validation method based on partial least squares to devise the suitable modelling depth according to the quality of the available experimental data.

Once reliable prediction models are available, they can be used in numerical simulations to analyze the main features of the process or to evaluate the influence of the operating conditions as well as of the external disturbances. Three examples of different applications were published in this regard: [2] developed a 3D simulation that describes the hydration behavior of cereals during cooking; [13] presented a dynamic simulation of a hot-metal steel converter based on thermodynamic and kinetic equations, used to evaluate the influences of different scrap features on the process; and [14] built a 3D model to simulate the fluid dynamics inside the coagulation bath of a spinning process for synthetic fiber production. Nevertheless, the use of models is not limited to offline or real-time predictive simulation, but is likely to extend to process (dynamic and real time) optimization in the near future. Although model-based optimization was not directly within the scope of this special issue, the authors of [5,6] proposed steps in this direction from the application and software viewpoints, respectively.

Although there are almost as many types of models as processes/applications, as well as multiple modelling methodologies to choose from, some key conclusions can be extracted from the received contributions. Plant models in the process industry are no longer just built from very detailed first-principles equations, and their applications often go beyond their classical use in process design to strongly influence the process operation in real time. Therefore, the tradeoff between model complexity and accuracy needs to take account of the decision level where the model is to be used. The increasing computational power, availability of big datasets and improved machine learning algorithms will facilitate model building in the materials, (bio)chemical and process engineering fields [9]. However, the big data that are already available in the process industry are not always complete and informative, and performing further experimental tests on demand may be expensive. Thus, as models are often required to provide reliable predictions outside the plant's current or usual region of operation, data-driven modelling methodologies need to be combined with process physical knowledge derived from first principles, resulting in a hybrid or grey-box model. The characterization of uncertainty from available plant data and its incorporation in process modeling are also important topics that require further research, as they directly affect the quality and reliability of model predictions and the inherent risk in making use of these predictions for decision support.

Finally, the full realization of the benefits of process modeling will depend on being able to deploy detailed first-principle or hybrid models throughout the process lifecycle. Of particular interest in this context is the use of such models, and the calculations based on them, in online decision support and control systems for process operations. This includes many important applications, from equipment condition monitoring, to real-time optimization and nonlinear model-predictive control, all of which would constitute major steps towards the digitalization of the process industries. Achieving this objective on a large scale, however, poses several significant technical challenges. Some of these are computational, arising from the need to perform complex calculations robustly and efficiently in real time. Other challenges are related to devising general software architectures that can support the development of complex digital applications involving multiple model-based computations communicating with each other and with external data servers. Successful advances in these areas will provide process engineers with a complete suite to implement advanced process management systems, boosting the development of virtual plants or digital twins that integrate plant information updated in real time.

We would like to end this editorial note with expressing our sincere gratitude to all the scientific contributors of the papers submitted to this special issue, as well as to the editor-in-chief of *Processes*, Michael A. Henson, the managing editor, Jamie Li, and the rest of the editorial staff for their effort and endless support.

Prof. Dr. Cesar de Prada
Prof. Dr. Constantinos Pantelides
Dr. Jose Luis Pitarch
Guest Editors

References

1. Sixt, M.; Uhlenbrock, L.; Strube, J. Toward a Distinct and Quantitative Validation Method for Predictive Process Modelling—On the Example of Solid-Liquid Extraction Processes of Complex Plant Extracts. *Processes* **2018**, *6*, 66. [CrossRef]
2. Wang, L.; Wang, M.; Guo, M.; Ye, X.; Ding, T.; Liu, D. Numerical Simulation of Water Absorption and Swelling in Dehulled Barley Grains during Canned Porridge Cooking. *Processes* **2018**, *6*, 230. [CrossRef]

3. Villez, K.; Billeter, J.; Bonvin, D. Incremental Parameter Estimation under Rank-Deficient Measurement Conditions. *Processes* **2019**, *7*, 75. [CrossRef]
4. Wang, Z.; Sheikh, H.; Lee, K.; Georgakis, C. Sequential Parameter Estimation for Mammalian Cell Model Based on In Silico Design of Experiments. *Processes* **2018**, *6*, 100. [CrossRef]
5. Xie, X.; Schenkendorf, R.; Krewer, U. Toward a Comprehensive and Efficient Robust Optimization Framework for (Bio)chemical Processes. *Processes* **2018**, *6*, 183. [CrossRef]
6. Beal, L.; Hill, D.; Martin, R.; Hedengren, J. GEKKO Optimization Suite. *Processes* **2018**, *6*, 106. [CrossRef]
7. Pablos, C.; Merino, A.; Acebes, L.F. Modeling On-Site Combined Heat and Power Systems Coupled to Main Process Operation. *Processes* **2019**, *7*, 218. [CrossRef]
8. Wang, X.; Xie, Y.; Lu, Y.; Zhu, R.; Fu, Q.; Cai, Z.; An, C. Mathematical Modelling Forecast on the Idling Transient Characteristic of Reactor Coolant Pump. *Processes* **2019**, *7*, 452. [CrossRef]
9. Li, H.; Zhang, Z.; Zhao, Z.Z. Data-Mining for Processes in Chemistry, Materials, and Engineering. *Processes* **2019**, *7*, 151. [CrossRef]
10. Pitarch, J.; Sala, A.; de Prada, C. A Systematic Grey-Box Modeling Methodology via Data Reconciliation and SOS Constrained Regression. *Processes* **2019**, *7*, 170. [CrossRef]
11. Muddu, S.; Tamrakar, A.; Pandey, P.; Ramachandran, R. Model Development and Validation of Fluid Bed Wet Granulation with Dry Binder Addition Using a Population Balance Model Methodology. *Processes* **2018**, *6*, 154. [CrossRef]
12. Chan, D.S.; Chan, J.S.; Kuo, M.I. Modelling Condensation and Simulation for Wheat Germ Drying in Fluidized Bed Dryer. *Processes* **2018**, *6*, 71. [CrossRef]
13. Penz, F.; Schenk, J.; Ammer, R.; Klösch, G.; Pastucha, K. Evaluation of the Influences of Scrap Melting and Dissolution during Dynamic Linz–Donawitz (LD) Converter Modelling. *Processes* **2019**, *7*, 186. [CrossRef]
14. Ngo, S.I.; Lim, Y.I.; Kim, S.C. Wave Characteristics of Coagulation Bath in Dry-Jet Wet-Spinning Process for Polyacrylonitrile Fiber Production Using Computational Fluid Dynamics. *Processes* **2019**, *7*, 314. [CrossRef]

processes

MDPI

Discussion

Data-Mining for Processes in Chemistry, Materials, and Engineering

Hao Li [1,*,†], Zhien Zhang [2,*] and Zhe-Ze Zhao [3]

1 College of Chemistry, Sichuan University, Chengdu 610064, China
2 William G. Lowrie Department of Chemical and Biomolecular Engineering, The Ohio State University, 151 West Woodruff Avenue, Columbus, OH 43210, USA
3 School of Life Sciences and State Key Laboratory of Agrobiotechnology, The Chinese University of Hong Kong, Shatin, New Territories, Hong Kong, China; zheze.zhao@hotmail.com
* Correspondence: lihao@utexas.edu (H.L.); zhang.4528@osu.edu (Z.Z.)
† Current Address: Department of Chemistry and the Institute for Computational and Engineering Sciences, The University of Texas at Austin, 105 E. 24th Street, Stop A5300, Austin, TX 78712, USA.

Received: 11 February 2019; Accepted: 4 March 2019; Published: 11 March 2019

Abstract: With the rapid development of machine learning techniques, data-mining for processes in chemistry, materials, and engineering has been widely reported in recent years. In this discussion, we summarize some typical applications for process optimization, design, and evaluation of chemistry, materials, and engineering. Although the research and application targets are various, many important common points still exist in their data-mining. We then propose a generalized strategy based on the philosophy of data-mining, which should be applicable for the design and optimization targets for processes in various fields with both scientific and industrial purposes.

Keywords: data-mining; machine learning; neural networks; chemistry; materials; engineering; energy

1. Introduction

Data-mining is a strategy for discovering intrinsic relationships and making proper predictions based on statistics from scientifically-collected data [1]. With the rapid progress in machine learning techniques and methodologies in the recent decade [2–7], data-mining has become a popular study since machine learning provides an efficient technique for non-linearly fitting the intrinsic relationships between the independent and dependent variables in a mathematical form. Therefore, without knowing the exact physical or empirical form of the relationships among data, machine learning can come up with a non-linear form of math that could precisely predict the trends of data, including interpolation and extrapolation [8–10]. Although those non-linear forms do not contain the exact correlation knowledge, a general approximation of data-based machine learning (with both supervised and unsupervised processes [11–13]) always shows precise prediction and could address the problem in an easier way.

In recent years, data-mining has been widely applied for solving problems in chemical, materials, and engineering processes, based on the data collected from either experiments or simulations [14–17]. In many worldwide pressing issues, such as greenhouse gas capture [18,19], catalytic materials design and optimization [20–31], and renewable energy studies [32–39], data-mining has shown predictive power for mining the relationships between the intrinsic and extrinsic properties [40–45]. Usually, the mission of a data-mining process is to predict (or output) those variables that are difficult to acquire from experiments/simulations by using the easy variables which can be acquired as the inputs. Through a well-fitted non-linear form, the predicted variables can be rapidly outputted with the inputs of those independent variables. In other words, a machine learning assisted data-mining process is able to expedite the (i) optimization of engineering processes, (ii) discovery of new functional materials, and (iii) understanding of chemical processes.

Despite a number of studies that have been published in the recent decade, there is no well-established philosophy that provides a standard guideline for doing data-mining. Therefore, in this discussion paper, we are motivated to summarize some recent typical studies of data-mining in the processes of chemistry, materials, and engineering. Based on the brief review, comments, and discussions, we then generalize a simple but useful data-mining strategy for these scientific and application processes, which should ultimately benefit to the standard development of knowledge-based data-mining through a machine learning modeling process.

2. Typical Studies

Due to the high-dimensional variables, trends in the chemical processes are sometimes difficult to understand and predict. For example, a chemical process usually depends on multiple factors, including temperature, pressure, as well as the component and composition of reactants. Previously, to capture the relationships between these independent and dependent factors, a response surface methodology (RSM) was usually applied to fit the trends between the independent and dependent variables with multiple 3-D plots [46]. This method is useful for the design and optimization of chemical and materials processes. However, RSM is only able to deal with very limited independent variables in one model, which is not applicable for higher dimension problems in a big-data scale. To address this issue, artificial neural networks (ANNs), as the most widely used machine learning algorithms, have been applied for the same target, replacing RSM [8,47]. People have found that not only being able to deal with high-dimension problems ANNs also have a generalized approximation capacity and tunable algorithmic architectures, which guarantees that they can exhaustively capture the potential relationships between inputs and output(s) after a proper data training and validation process.

Mining the Trends and Properties in Chemistry and Materials

A typical application for mining the trends and properties in a chemical process is the greenhouse gas capture and utilization. In our recent study, it was found that a kernel-based ANN, the general regression neural network (GRNN), is able to properly fit the relationships between the solution properties (temperature, operating gas pressure, component, and concentration of the blended solutions) and the solubility of CO_2, based on the literature-extracted experimental data [48]. Afterwards, the trends of CO_2 solubility can be predicted with the function of temperature, operating CO_2 pressure, concentration, and type of blended solutions (Figure 1). It can be seen from Figure 1 that though the trends are non-linear and usually difficult to be predicted with regular non-linear mathematical forms, a GRNN model trained from representative experimental data is able to capture these trends and provide proper understandings for CO_2 capture in solutions. A similar study on predicting CO_2 thermodynamic properties is shown in Reference [49], where the inputs of blend concentration, temperature, and CO_2 operating partial pressure can be used as inputs and specifically predict the CO_2 solubility, density, and viscosity of a solution. Similar studies for mining the gas capture and separation can be found in References [50,51]. In addition to the use of ANNs, Günay et al. used a decision tree model to evaluate the important factors of the reaction activity and selectivity of catalysts during CO_2 electro-reduction process (Figure 2) [52]. By extracting a large number of experimental literatures, they classified the catalysts with the best Faradaic efficiency, max activity, or most selective pathway. Other catalytic applications through data-mining can be found in References [53,54]. Since most of the chemical and reaction-related processes are based on temperature, pressure, component, composition, and energetic values, it is expected that the data-mining strategy shown here is general and should be applicable for addressing other similar chemical issues through machine learning.

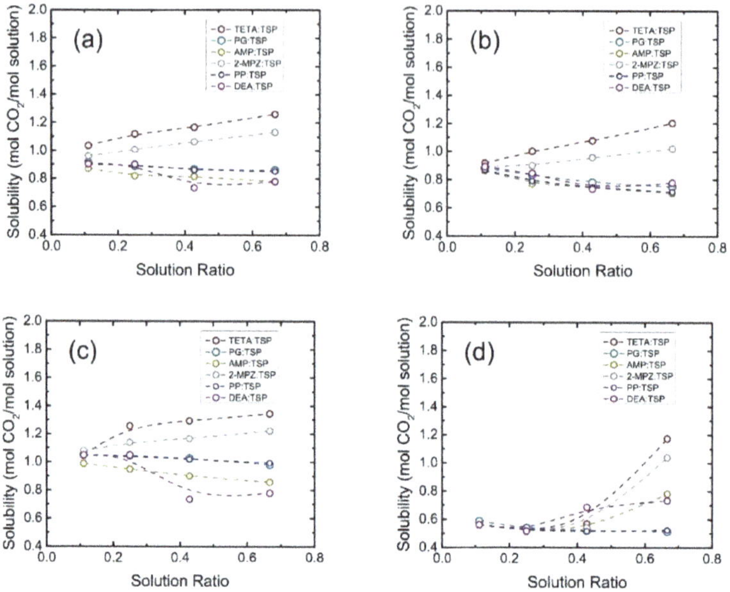

Figure 1. Trends in the CO_2 capture in blended solutions, predicted by a well-trained general regression neural network model. (**a**) T = 303 K, P = 14 kPa, C = 2.5 M; (**b**) T = 323 K, P = 14 kPa, C = 2.5 M; (**c**) T = 303 K, P = 42 kPa, C = 2.5 M; (**d**) T = 303 K, P = 14 kPa, C = 1.5 M. T, P, and C represent temperature, CO_2 partial pressure, and concentration, respectively. Reproduced with permission from *J. CO2 Util.*; published by Elsevier, 2018 [48].

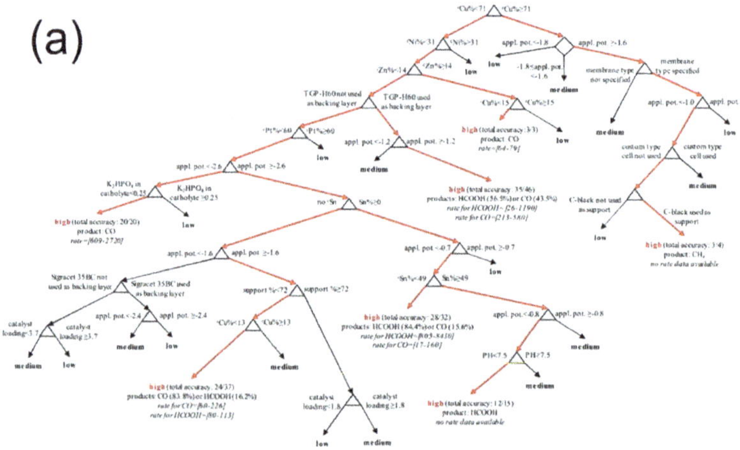

Figure 2. *Cont.*

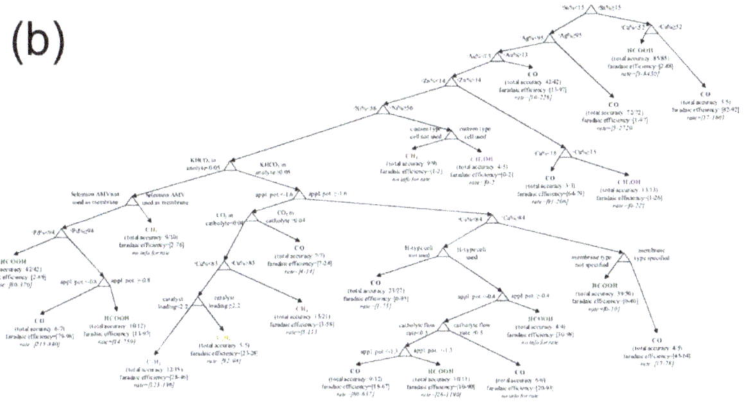

Figure 2. Decision tree analysis for (**a**) catalysts with maximum faradaic efficiency and (**b**) catalysts with the highest selective product, for CO_2 reduction. Reproduced with permission from *J. CO2 Util.*; published by Elsevier, 2018 [52].

In terms of mining the materials properties, one of the most typical works is the discovery of nature's missing ternary oxide compounds, as described by Ceder et al. [55]. They developed a machine learning model based on the crystal structure database and suggested new compositions and structures through a data-mining process. Then, using density function theory (DFT) as the quantum mechanical computation method [56,57], they calculated and confirmed the stability of those suggested ternary oxides (Figure 3). Similar studies can be found in recent References [58–61]. Due to the complexity of the structural information and the electronic structures of the periodic table elements [62–71], a challenge of their data-mining is the definition of suitable descriptors as the model inputs. In the past decades, there was a large number of descriptors that have been applied for the machine learning process of chemical and materials systems, such as bond length, bond angle, and group contribution analysis [72]. However, since the structural information is usually dependent on the coordination and reference, it was hard to generalize the methods for more complicated systems. To address these issues and provide a generalized machine learning representation, Behler and Parrinello developed a set of new symmetry functions that converts all the atomistic environments into the terms of pair and angular interactions [73]. Together with an architecture of conventional ANN, the relationship between the atomistic structures and the materials properties (e.g., energy) can be efficiently mined. So far, this Behler-Parrinello representation has proven to be highly effective for capturing the structural information of materials during machine learning, which especially benefits to the data-mining in theoretical chemistry and computational materials based on quantum mechanical calculated data.

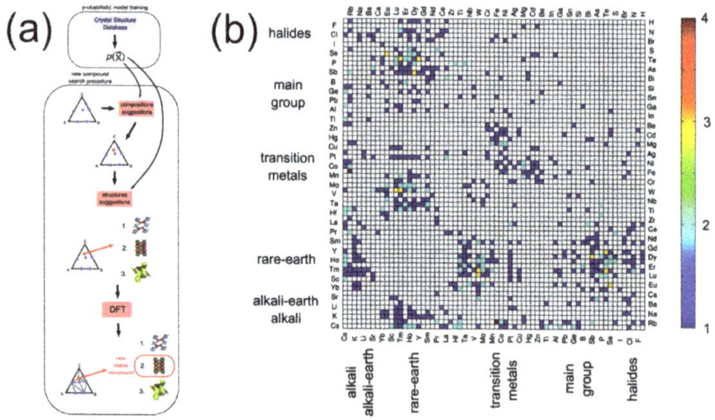

Figure 3. (**a**) A data-mining compound searching procedure proposed by Ceder et al. (**b**) Distribution of the newly discovered compounds. Reproduced with permission from *Chem. Mater.*; published by American Chemical Society, 2010 [55].

3. Processes in Engineering

3.1. *Engineering Optimization and Design*

Engineering process is somewhat different from the processes of chemistry and materials discussed above. The main reason is that most of the knowledge in engineering are based on various empirical equations, due to the complexity of the systems. Therefore, mining the intrinsic relationships during engineering processes are particularly challenging but also important. A typical study using data-mining method for the optimization and design of engineering applications is proposed by Kalogirou [74], where an ANN was applied to train a small number of data from TRNSYS simulations on a typical solar energy system for industrial engineering. Then, a genetic algorithm (GA) [75–77] was employed to estimate the optimum size of parameters based on the results from ANN. Interestingly, the use of GA has shown a promising process that could generate reliable data combinations in a short time (Figure 4). Instead of listing the interpolated trends as discussed above, the GA method is a fast way that could expedites the industrial decision on the processes.

```
                         Genetic Algorithm
Begin (1)
    t = 0 [start with an initial time]
    Initialize Population P(t) [initialize a usually random population of individuals]
    Evaluate fitness of Population P(t) [evaluate fitness of all individuals in population]
    While (Generations < Total Number) do begin (2)
        t = t + 1 [increase the time counter]
        Select Population P(t) out of Population P(t-1) [select sub-population for
                                                          offspring production]
        Apply Crossover on Population P(t)
        Apply Mutation on Population P(t)
        Evaluate fitness of Population P(t) [evaluate new fitness of population]
    end (2)
end (1)
```

Figure 4. A genetic algorithm procedure for optimizing the solar energy systems together with a well-trained artificial neural network model. Reproduced with permission from *Appl. Energy*; published by Elsevier, 2004 [74].

3.2. *A Computational High-Throughput Screenig Method*

Though a GA method is sufficient for generating a limited amount of data, its strategy sometimes would omit the important possible parameters during design. In addition, being different from materials design (as shown in Figure 3), engineering applications require to operate a larger size of data since the materials types are limited by the finite number of elements. And thus, there are many more different possibilities exist in the design and optimization of engineering processes. To overcome these problems, in very recent years, a high-throughput screening (HTS) method was developed for optimizing the engineering devices and processes (Figure 5) [78,79]. As illustrated in Figure 5, it can be seen that an HTS method can generate a large number of possible combination of inputs at the beginning, then a well-trained ANN can rapidly output the performance of all these possible input combinations. Then all those combinations which predicted with good performance would be recorded in a database as future candidates. Then the experimental process can pick a few of these candidates for testing. In previous studies, it has been shown that a regular ANN (trained with 1~2 hidden layers, respectively, with less than 50 hidden neurons) is able to quickly output thousands of predictions in a relatively short period [78]. More importantly, an HTS method is able to fully mine the trends between input and output variables for engineering processes.

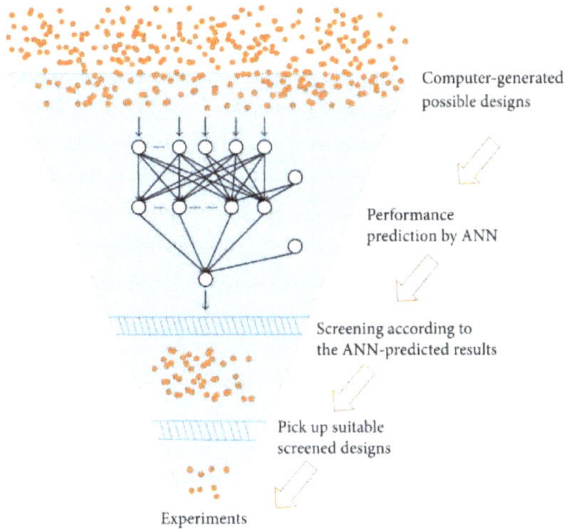

Figure 5. A high-throughput screening process for engineering system optimization. Reproduced with permission from *Int. J. Photoenergy*; published by Hindawi, 2017 [78].

4. Discussions

With the case analysis discussed above, we can see that a machine learning assisted data-mining is a powerful technique for fitting the intrinsic relationships in the processes of chemistry, materials, and engineering. In addition, it is clear that there are a couple of important steps for these data-mining. First, the choice of model inputs is important since it should be the independent variables that have potential relationships with the output variable(s). Therefore, the use of descriptors should be carefully selected. Second, since the predictions are usually for interpolation, the database used for machine learning model training should be sufficiently representative and diverse. Otherwise, the model might easily get over-fitted [80]. Finally, for prediction, optimization, and/or design applications, the way to generate new combined input data could be carefully chosen: for new materials design, the combination of different types of elements from the periodic table is a good way to screen all the possible materials which are predicted with high-performances; for targeting a good design with less computational cost, a GA method could help to rationally generate new input combinations; to exhaustively screen all the possible optimization in engineering, an HTS method could be a good strategy since the prediction through an already-trained machine learning (e.g., ANN) model is usually computationally costless [78].

Overall, the general data-mining process remains similar regardless of its applications, as summarized in Figure 6. After data collection, a statistical analysis would evaluate whether the data scale is diverse and representative. Then the most reasonable independent variables can be chosen as the descriptors in the model inputs. By training and validation of the machine learning model, we can evaluate whether the descriptors are suitable for capturing the potential relationships with the output(s). If the model is well-trained, it can be used for further mining of the new properties by performing its predictive power. Those new input combinations generated by GA or HTS can be set as the input of the trained model, and the predictions can be rapidly outputted. Finally, a new database can be constructed by having the original experimental data as well as the predicted data from the well-trained machine learning model.

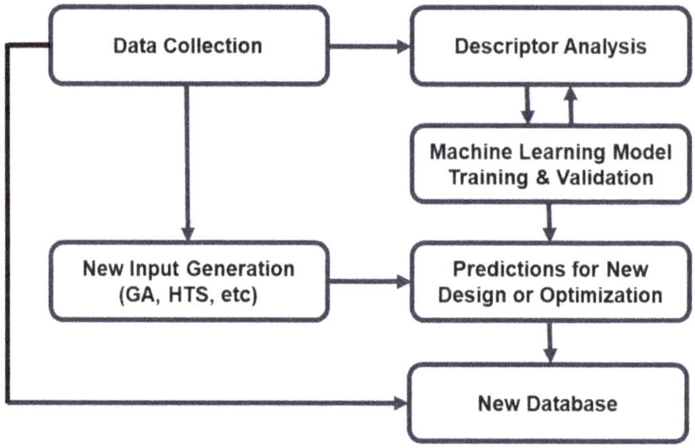

Figure 6. Flow chart of the data-mining for processes in natural science and engineering applications.

5. Conclusions

In the new era of machine learning development, data-mining for processes in chemistry, materials, and engineering has become a popular way to promote efficiency in both scientific and industrial research. In this discussion, we have summarized several typical cases for the optimization and design of chemistry, materials, engineering, and other related applications. We found that though there is a variety of research and application fields, the basic strategy, process, and philosophy of data-mining are highly similar. We then have proposed a generalized strategy for the basic philosophy of data-mining, which should be applicable for the design and optimization targets for the processes in various fields. We also expect that in future studies with larger data-scale in science and industry, some more advanced machine learning (e.g., deep learning) techniques could fulfill the future requirement of data-mining, leading to faster and more efficient scientific development.

Author Contributions: Both H.L. and Z.Z. wrote this discussion paper. Z.-Z.Z. provided important insights in the discussion of the paper.

Funding: This research received no external funding.

Acknowledgments: We are grateful for all the editorial works from the *Processes* editorial office.

Conflicts of Interest: The authors declare no conflict of interest.

References

1. Wu, X.; Kumar, V.; Ross Quinlan, J.; Ghosh, J.; Yang, Q.; Motoda, H.; McLachlan, G.J.; Ng, A.; Liu, B.; Yu, P.S.; et al. Top 10 algorithms in data mining. *Knowl. Inf. Syst.* **2008**, *14*, 1–37. [CrossRef]
2. Goh, K.L.; Singh, A.K. Comprehensive Literature Review on Machine Learning Structures for Web Spam Classification. *Procedia Comput. Sci.* **2015**, *70*, 434–441. [CrossRef]
3. Sattlecker, M.; Stone, N.; Bessant, C. Current trends in machine-learning methods applied to spectroscopic cancer diagnosis. *TrAC Trends Anal. Chem.* **2014**, *59*, 17–25. [CrossRef]

4. Schmidhuber, J. Deep Learning in neural networks: An overview. *Neural Netw.* **2015**, *61*, 85–117. [CrossRef] [PubMed]
5. Kotsiantis, S.B. Supervised Machine Learning: A Review of Classification Techniques. *Informatica* **2007**, *31*, 249–268. [CrossRef]
6. Lin, J.; Yuan, J.-S. Analysis and Simulation of Capacitor-Less ReRAM-Based Stochastic Neurons for the in-Memory Spiking Neural Network. *IEEE Trans. Biomed. Circuits Syst.* **2018**, *12*, 1004–1017. [CrossRef] [PubMed]
7. Lin, J.; Yuan, J. Capacitor-less RRAM-Based Stochastic Neuron for Event-Based Unsupervised Learning. In Proceedings of the 2017 IEEE Biomedical Circuits and Systems Conference (BioCAS), Turin, Italy, 19–21 October 2017.
8. Li, H.; Zhang, Z.; Liu, Z. Application of Artificial Neural Networks for Catalysis: A Review. *Catalysts* **2017**, *7*, 306. [CrossRef]
9. Li, H.; Chen, F.; Cheng, K.; Zhao, Z.; Yang, D. Prediction of Zeta Potential of Decomposed Peat via Machine Learning: Comparative Study of Support Vector Machine and Artificial Neural Networks. *Int. J. Electrochem. Sci.* **2015**, *10*, 6044–6056.
10. Li, H.; Tang, X.; Wang, R.; Lin, F.; Liu, Z.; Cheng, K. Comparative Study on Theoretical and Machine Learning Methods for Acquiring Compressed Liquid Densities of 1,1,1,2,3,3,3-Heptafluoropropane (R227ea) via Song and Mason Equation, Support Vector Machine, and Artificial Neural Networks. *Appl. Sci.* **2016**, *6*, 25. [CrossRef]
11. Kawamoto, Y.; Takagi, H.; Nishiyama, H.; Kato, N. Efficient Resource Allocation Utilizing Q-Learning in Multiple UA Communications. *IEEE Trans. Netw. Sci. Eng.* **2018**. [CrossRef]
12. Siniscalchi, S.M.; Salerno, V.M. Adaptation to new microphones using artificial neural networks with trainable activation functions. *IEEE Trans. Neural Netw. Learn. Syst.* **2017**, *28*, 1959–1965. [CrossRef] [PubMed]
13. Hofmann, T. Unsupervised learning by probabilistic Latent Semantic Analysis. *Mach. Learn.* **2001**, *42*, 77–196. [CrossRef]
14. Witten, I.H.; Frank, E. *Data Mining: Practical Machine Learning Tools and Techniques*; Morgan Kaufmann: Boston, MA, USA, 2005.
15. Wu, H.; Yu, Y.; Fu, H.; Zhang, L. On the prediction of chemical exergy of organic substances using least square support vector machine. *Energy Sources Part A Recover. Util. Environ. Eff.* **2017**, *39*, 2210–2215. [CrossRef]
16. Wu, K.; Liu, D.; Tang, Y. In-situ single-step chemical synthesis of graphene-decorated $CoFe_2O_4$composite with enhanced Li ion storage behaviors. *Electrochim. Acta* **2018**, *263*, 515–523. [CrossRef]
17. Wu, K.; Du, K.; Hu, G. Red-blood-cell-like $(NH_4)[Fe_2(OH)(PO_4)_2]•2H_2O$ particles: Fabrication and application in high-performance $LiFePO_4$ cathode materials. *J. Mater. Chem. A* **2018**, *6*, 1057–1066. [CrossRef]
18. Zhang, Z.; Li, Y.; Zhang, W.; Wang, J.; Soltanian, M.R.; Olabi, A.G. Effectiveness of amino acid salt solutions in capturing CO_2: A review. *Renew. Sustain. Energy Rev.* **2018**, *98*, 179–188. [CrossRef]
19. Song, J.; Feng, Q.; Wang, X.; Fu, H.; Jiang, W.; Chen, B.; Song, J.; Feng, Q.; Wang, X.; Fu, H.; et al. Spatial Association and Effect Evaluation of CO_2 Emission in the Chengdu-Chongqing Urban Agglomeration: Quantitative Evidence from Social Network Analysis. *Sustainability* **2018**, *11*, 1. [CrossRef]
20. Li, H.; Henkelman, G. Dehydrogenation Selectivity of Ethanol on Close-Packed Transition Metal Surfaces: A Computational Study of Monometallic, Pd/Au, and Rh/Au Catalysts. *J. Phys. Chem. C* **2017**, *121*, 27504–27510. [CrossRef]
21. Li, H.; Evans, E.J.; Mullins, C.B.; Henkelman, G. Ethanol Decomposition on Pd–Au Alloy Catalysts. *J. Phys. Chem. C* **2018**, *122*, 22024–22032. [CrossRef]
22. Wu, K.; Yang, H.; Jia, L.; Pan, Y.; Hao, Y.; Zhang, K.; Du, K.; Hu, G. Smart construction of 3D N-doped graphene honeycombs with $(NH_4)_2SO_4$ as a multifunctional template for Li-ion battery anode: "A choice that serves three purposes". *Green Chem.* **2019**. [CrossRef]
23. Sun, Q.; Yang, Y.; Zhao, Z.; Zhang, Q.; Zhao, X.; Nie, G.; Jiao, T.; Peng, Q. Elaborate design of polymeric nanocomposites with Mg(ii)-buffering nanochannels for highly efficient and selective removal of heavy metals from water: Case study for Cu(ii). *Environ. Sci. Nano* **2018**, *5*, 2440–2451. [CrossRef]
24. Li, H.; Zhang, Z.; Liu, Y.; Cen, W.; Luo, X. Functional Group Effects on the HOMO–LUMO Gap of g-C3N. *Nanomaterials* **2018**, *8*, 589. [CrossRef] [PubMed]

25. Li, H.; Zhang, Z.; Liu, Z. Non-Monotonic Trends of Hydrogen Adsorption on Single Atom Doped g-C3N. *Catalysts* **2019**, *9*, 84. [CrossRef]
26. Yang, L.; Chen, Z.; Cui, D.; Luo, X.; Liang, B.; Yang, L.; Liu, T.; Wang, A.; Luo, S. Ultrafine palladium nanoparticles supported on 3D self-supported Ni foam for cathodic dechlorination of florfenicol. *Chem. Eng. J.* **2018**, *359*, 894–901. [CrossRef]
27. Shi, C.; He, Y.; Ding, M.; Wang, Y.; Zhong, J. Nanoimaging of food proteins by atomic force microscopy. Part II: Components, imaging modes, observation ways, and research types. *Trends Food Sci. Technol.* **2018**. [CrossRef]
28. Shi, C.; He, Y.; Ding, M.; Wang, Y.; Zhong, J. Nanoimaging of food proteins by atomic force microscopy. Part I: Components, imaging modes, observation ways, and research types. *Trends Food Sci. Technol.* **2018**, 0–1. [CrossRef]
29. Li, N.; Tang, S.; Rao, Y.; Qi, J.; Zhang, Q.; Yuan, D. Peroxymonosulfate enhanced antibiotic removal and synchronous electricity generation in a photocatalytic fuel cell. *Electrochim. Acta* **2019**, *298*, 59–69. [CrossRef]
30. Tang, S.; Yuan, D.; Rao, Y.; Li, M.; Shi, G.; Gu, J.; Zhang, T. Percarbonate promoted antibiotic decomposition in dielectric barrier discharge plasma. *J. Hazard. Mater.* **2019**, *366*, 669–676. [CrossRef] [PubMed]
31. Kang, L.; Du, H.L.; Du, X.; Wang, H.T.; Ma, W.L.; Wang, M.L.; Zhang, F.B. Study on dye wastewater treatment of tunable conductivity solid-waste-based composite cementitious material catalyst. *Desalin. Water Treat.* **2018**, *125*, 296–301. [CrossRef]
32. Voyant, C.; Notton, G.; Kalogirou, S.; Nivet, M.L.; Paoli, C.; Motte, F.; Fouilloy, A. Machine learning methods for solar radiation forecasting: A review. *Renew. Energy* **2017**, *105*, 569–582. [CrossRef]
33. Liu, X.; He, Y.; Fu, H.; Chen, B.; Wang, M.; Wang, Z. How Environmental Protection Motivation Influences on Residents' Recycled Water Reuse Behaviors: A Case Study in Xi'an City. *Water* **2018**, *10*, 1282. [CrossRef]
34. Liu, G.; Chen, B.; Jiang, S.; Fu, H.; Wang, L.; Jiang, W.; Liu, G.; Chen, B.; Jiang, S.; Fu, H.; et al. Double Entropy Joint Distribution Function and Its Application in Calculation of Design Wave Height. *Entropy* **2019**, *21*, 64. [CrossRef]
35. Wang, J.; Zhou, S.; Zhang, Z.; Yurchenko, D. High-performance piezoelectric wind energy harvester with Y-shaped attachments. *Energy Convers. Manag.* **2019**, *181*, 645–652. [CrossRef]
36. Wang, J.; Tang, L.; Zhao, L.; Zhang, Z. Efficiency investigation on energy harvesting from airflows in HVAC system based on galloping of isosceles triangle sectioned bluff bodies. *Energy* **2019**, *172*, 1066–1078. [CrossRef]
37. Yang, G.; Wang, J.; Zhang, H.; Jia, H.; Zhang, Y.; Gao, F. Applying bio-electric field of microbial fuel cell-upflow anaerobic sludge blanket reactor catalyzed blast furnace dusting ash for promoting anaerobic digestion. *Water Res.* **2019**, *149*, 215–224. [CrossRef] [PubMed]
38. Yu, L.; Li, Y.P. A flexible-possibilistic stochastic programming method for planning municipal-scale energy system through introducing renewable energies and electric vehicles. *J. Clean. Prod.* **2019**, *207*, 772–787. [CrossRef]
39. Yu, L.; Li, Y.P.; Huang, G.H. Planning municipal-scale mixed energy system for stimulating renewable energy under multiple uncertainties—The City of Qingdao in Shandong Province, China. *Energy* **2019**, *166*, 1120–1133. [CrossRef]
40. Eichler, U.; Brändle, M.; Sauer, J. Predicting Absolute and Site Specific Acidities for Zeolite Catalysts by a Combined Quantum Mechanics/Interatomic Potential Function Approach. *J. Phys. Chem. B* **2002**, *101*, 10035–10050. [CrossRef]
41. Zurek, E.; Grochala, W. Predicting crystal structures and properties of matter under extreme conditions via quantum mechanics: The pressure is on. *Phys. Chem. Chem. Phys.* **2015**, *17*, 2917–2934. [CrossRef] [PubMed]
42. Fischer, C.C.; Tibbetts, K.J.; Morgan, D.; Ceder, G. Predicting crystal structure by merging data mining with quantum mechanics. *Nat. Mater.* **2006**, *5*, 641. [CrossRef] [PubMed]
43. Ceder, G.; Morgan, D.; Fischer, C.; Tibbetts, K.; Curtarolo, S. Data-mining-driven quantum mechanics for the prediction of structure. *MRS Bull.* **2006**, *31*, 981–985. [CrossRef]
44. Kim, H.; Stumpf, A.; Kim, W. Analysis of an energy efficient building design through data mining approach. *Autom. Constr.* **2011**, *20*, 37–43. [CrossRef]
45. Fan, C.; Xiao, F.; Wang, S. Development of prediction models for next-day building energy consumption and peak power demand using data mining techniques. *Appl. Energy* **2014**, *127*, 1–10. [CrossRef]

46. Prakash Maran, J.; Priya, B. Comparison of response surface methodology and artificial neural network approach towards efficient ultrasound-assisted biodiesel production from muskmelon oil. *Ultrason. Sonochem.* **2015**, *23*, 192–200. [CrossRef] [PubMed]
47. Huang, S.-M.; Hung, T.-H.; Liu, Y.-C.; Kuo, C.-H.; Shieh, C.-J. Green Synthesis of Ultraviolet Absorber 2-Ethylhexyl Salicylate: Experimental Design and Artificial Neural Network Modeling. *Catalysts* **2017**, *7*, 342. [CrossRef]
48. Li, H.; Zhang, Z. Mining the intrinsic trends of CO_2 solubility in blended solutions. *J. CO2 Util.* **2018**, *26*, 496–502. [CrossRef]
49. Zhang, Z.; Li, H.; Chang, H.; Pan, Z.; Luo, X. Machine Learning Predictive Framework for CO_2 Thermodynamic Properties in Solution. *J. CO2 Util.* **2018**, *26*, 152–159. [CrossRef]
50. Abdi-Khanghah, M.; Bemani, A.; Naserzadeh, Z.; Zhang, Z. Prediction of solubility of N-alkanes in supercritical CO_2 using RBF-ANN and MLP-ANN. *J. CO2 Util.* **2018**. [CrossRef]
51. Soroush, E.; Shahsavari, S.; Mesbah, M.; Rezakazemi, M.; Zhang, Z. A robust predictive tool for estimating CO_2 solubility in potassium based amino acid salt solutions. *Chin. J. Chem. Eng.* **2018**, *26*, 740–746. [CrossRef]
52. Günay, M.E.; Türker, L.; Tapan, N.A. Decision tree analysis for efficient CO_2 utilization in electrochemical systems. *J. CO2 Util.* **2018**, *28*, 83–95. [CrossRef]
53. Günay, M.E.; Yildirim, R. Neural network analysis of selective CO oxidation over copper-based catalysts for knowledge extraction from published data in the literature. *Ind. Eng. Chem. Res.* **2011**, *50*, 12488–12500. [CrossRef]
54. Davran-Candan, T.; Günay, M.E.; Yildirim, R. Structure and activity relationship for CO and O_2 adsorption over gold nanoparticles using density functional theory and artificial neural networks. *J. Chem. Phys.* **2010**, *132*, 174113. [CrossRef] [PubMed]
55. Hautier, G.; Fischer, C.C.; Jain, A.; Mueller, T.; Ceder, G. Finding natures missing ternary oxide compounds using machine learning and density functional theory. *Chem. Mater.* **2010**, *22*, 3762–3767. [CrossRef]
56. Li, H.; Shin, K.; Henkelman, G. Effects of Ensembles, Ligand, and Strain on Adsorbate Binding to Alloy Surfaces. *J. Chem. Phys.* **2018**, *149*, 174705. [CrossRef] [PubMed]
57. Li, H.; Luo, L.; Kunal, P.; Bonifacio, C.S.; Duan, Z.; Yang, J.C.; Humphrey, S.M.; Crooks, R.M.; Henkelman, G. Oxygen Reduction Reaction on Classically Immiscible Bimetallics: A Case Study of RhAu. *J. Phys. Chem. C* **2018**, *122*, 2712–2716. [CrossRef]
58. Kim, C.; Chandrasekaran, A.; Huan, T.D.; Das, D.; Ramprasad, R. Polymer Genome: A Data-Powered Polymer Informatics Platform for Property Predictions. *J. Phys. Chem. C* **2018**, *122*, 17575–17585. [CrossRef]
59. Graser, J.; Kauwe, S.K.; Sparks, T.D. Machine Learning and Energy Minimization Approaches for Crystal Structure Predictions: A Review and New Horizons. *Chem. Mater.* **2018**, *30*, 3601–3612. [CrossRef]
60. Mansouri Tehrani, A.; Oliynyk, A.O.; Parry, M.; Rizvi, Z.; Couper, S.; Lin, F.; Miyagi, L.; Sparks, T.D.; Brgoch, J. Machine Learning Directed Search for Ultraincompressible, Superhard Materials. *J. Am. Chem. Soc.* **2018**, *140*, 9844–9853. [CrossRef] [PubMed]
61. Yang, Y.; Kang, L.; Li, H. Enhancement of photocatalytic hydrogen production of BiFeO3 by Gd3+ doping. *Ceram. Int.* **2018**, *45*, 8017–8022. [CrossRef]
62. Duan, C.; Huo, J.; Li, F.; Yang, M.; Xi, H. Ultrafast room-temperature synthesis of hierarchically porous metal–organic frameworks by a versatile cooperative template strategy. *J. Mater. Sci.* **2018**, *53*, 16276–16287. [CrossRef]
63. Duan, C.; Li, F.; Xiao, J.; Liu, Z.; Li, C.; Xi, H. Rapid room-temperature synthesis of hierarchical porous zeolitic imidazolate frameworks with high space-time yield. *Sci. China Mater.* **2017**, *60*, 1205–1214. [CrossRef]
64. Yin, K.; Chu, D.; Dong, X.; Wang, C.; Duan, J.A.; He, J. Femtosecond laser induced robust periodic nanoripple structured mesh for highly efficient oil-water separation. *Nanoscale* **2017**, *9*, 14229–14235. [CrossRef] [PubMed]
65. Wang, K.; Pang, J.; Li, L.; Zhou, S.; Li, Y.; Zhang, T. Synthesis of hydrophobic carbon nanotubes/reduced graphene oxide composite films by flash light irradiation. *Front. Chem. Sci. Eng.* **2018**, *12*, 376–382. [CrossRef]
66. Kai, W.; Li, L.; Lan, Y.; Dong, P.; Xia, G. Application Research of Chaotic Carrier Frequency Modulation Technology in Two-Stage Matrix Converter. *Math. Probl. Eng.* **2019**, *2019*, 8. [CrossRef]
67. Kai, W.; Shengzhe, Z.; Yanting, Z.; Jun, R.; Liwei, L.; Yong, L. Synthesis of Porous Carbon by Activation Method and its Electrochemical Performance. *Int. J. Electrochem. Sci.* **2018**, *13*, 10766–10773. [CrossRef]

68. Duan, C.; Cao, Y.; Hu, L.; Fu, D.; Ma, J. Synergistic effect of TiF_3 on the dehydriding property of α-AlH3 nano-composite. *Mater. Lett.* **2019**, *238*, 254–257. [CrossRef]
69. Duan, C.W.; Hu, L.X.; Ma, J.L. Ionic liquids as an efficient medium for the mechanochemical synthesis of α-AlH_3 nano-composites. *J. Mater. Chem. A* **2018**, *6*, 6309–6318. [CrossRef]
70. Yin, K.; Yang, S.; Dong, X.; Chu, D.; Gong, X.; Duan, J.-A. Femtosecond laser fabrication of shape-gradient platform: Underwater bubbles continuous self-driven and unidirectional transportion. *Appl. Surf. Sci.* **2018**, *471*, 999–1004. [CrossRef]
71. Yin, K.; Yang, S.; Dong, X.; Chu, D.; Duan, J.A.; He, J. Robust laser-structured asymmetrical PTFE mesh for underwater directional transportation and continuous collection of gas bubbles. *Appl. Phys. Lett.* **2018**, *112*, 243701. [CrossRef]
72. Constantinou, L.; Gani, R. New group contribution method for estimating properties of pure compounds. *AIChE J.* **1994**, *40*, 1697–1710. [CrossRef]
73. Behler, J.; Parrinello, M. Generalized neural-network representation of high-dimensional potential-energy surfaces. *Phys. Rev. Lett.* **2007**, *98*, 146401. [CrossRef] [PubMed]
74. Kalogirou, S.A. Optimization of solar systems using artificial neural-networks and genetic algorithms. *Appl. Energy* **2004**, *77*, 383–405. [CrossRef]
75. Kumar, M.; Husian, M.; Upreti, N.; Gupta, D. Genetic Algorithm: Review and Application. *Int. J. Inf. Technol. Knowl. Manag.* **2010**, *2*, 451–454.
76. Iba, H.; Aranha, C.C. *Adaptation, Learning, and Optimization*; Springer: Berlin/Heidelberg, Germany, 2012.
77. Whitley, D. A genetic algorithm tutorial. *Stat. Comput.* **1994**, *4*, 65–85. [CrossRef]
78. Li, H.; Liu, Z.; Liu, K.; Zhang, Z. Predictive Power of Machine Learning for Optimizing Solar Water Heater Performance: The Potential Application of High-Throughput Screening. *Int. J. Photoenergy* **2017**, *2017*, 4194251. [CrossRef]
79. Li, H.; Liu, Z. Performance Prediction and Optimization of Solar Water Heater via a Knowledge-Based Machine Learning Method. In *Handbook of Research on Power and Energy System Optimization*; IGI Global: Hershey, PA, USA, 2018; pp. 55–74.
80. Tetko, I.V.; Livingstone, D.J.; Luik, A.I. Neural Network Studies. 1. Comparison of Overfitting and Overtraining. *J. Chem. Inf. Comput. Sci.* **1995**, *35*, 826–833. [CrossRef]

processes MDPI

Article

Incremental Parameter Estimation under Rank-Deficient Measurement Conditions

Kris Villez [1,2,*], Julien Billeter [3] and Dominique Bonvin [3]

[1] Eawag, Swiss Federal Institute of Aquatic Science and Technology, 8600 Dübendorf, Switzerland
[2] ETH Zürich, Institute of Environmental Engineering, 8093 Zürich, Switzerland
[3] Laboratoire d'Automatique, École Polytechnique Fédérale de Lausanne, CH-1015 Lausanne, Switzerland; julien.billeter@alumni.epfl.ch (J.B.); dominique.bonvin@epfl.ch (D.B.)
* Correspondence: kris.villez@eawag.ch; Tel.: +41-58-765-5280

Received: 13 December 2018; Accepted: 27 January 2019; Published: 2 February 2019

Abstract: The computation and modeling of extents has been proposed to handle the complexity of large-scale model identification tasks. Unfortunately, the existing extent-based framework only applies when certain conditions apply. Most typically, it is required that a unique value for each extent can be computed. This severely limits the applicability of this approach. In this work, we propose a novel procedure for parameter estimation inspired by the existing extent-based framework. A key difference with prior work is that the proposed procedure combines structural observability labeling, matrix factorization, and graph-based system partitioning to split the original model parameter estimation problem into parameter estimation problems with the least number of parameters. The value of the proposed method is demonstrated with an extensive simulation study and a study based on a historical data set collected to characterize the isomerization of α-pinene. Most importantly, the obtained results indicate that an important barrier to the application of extent-based frameworks for process modeling and monitoring tasks has been lifted.

Keywords: extents; graph theory; model identification; observability; optimal clustering; parameter estimation; state decoupling

---

1. Introduction

Despite advances in model identification theory, parameter estimation can still be very challenging in practice. Such challenges include the lack of identifiability, large computational cost, the need to formulate appropriate experimental designs, and the fact that many methods, such as those for uncertainty analysis, are still being investigated and therefore not standardized. In this work, we focus on a novel method to tackle the computational challenge associated with the identification of kinetic parameters in large dynamic models. Hence, the other challenges associated with parameter identifiability, experimental design, and uncertainty analysis, though relevant, are not investigated here.

To handle the computational challenge, it is typical to devise a protocol for model fitting and model validation. Such protocols can be divided into two broad classes. In the first class, the protocols are domain specific. For instance, several protocols for the identification of activated sludge models are discussed in [1]. Similarly, a protocol for environmental system models is proposed in [2]. These protocols incorporate significant expertise specific to the particular application domain, thereby leading to protocols that are fine-tuned for that domain. While they tend to be similar on a conceptual level, it is rather difficult to apply these protocols universally.

The second class includes protocols that are more general and thus—in principle—broadly applicable. The incremental model identification framework studied in [3–7] is a good example. This method is grounded on the computation of *extents*, which are—loosely speaking—linear combinations of the original model states and capture the progress of individual dynamic phenomena, such as chemical reactions. A more precise definition will be given below.

The applicability of the extent-based framework is limited to cases where all extents can be computed based on measurements and invariant relationships. However, since this is rarely the case for biological process models, the extent-based framework has been extended in [8] to the case where each extent is either observable or non-sensed—see below for precise definitions. While this recent work provides a meaningful improvement, extent-based incremental model identification remains inapplicable for a wide range of biological scenarios found in practice. For example, this method cannot deal with the frequent case where an extent is sensed but unobservable.

The goal of this study is to present a novel method for incremental parameter identification that is more universally applicable. This method is based on the formulation of a generalized framework for extent computation and the use of a graph-based clustering algorithm. In what follows, we present the method and demonstrate its applicability to cases that could not be handled in an extent-based incremental model identification framework before. The expected impact of the newly developed tools is discussed at the end of this study.

2. Notation and Symbols

The matrix composed of the columns c of the matrix $\mathbf{M}$ is denoted $\mathbf{M}_{\bullet,c}$, while the matrix composed of the rows r of the matrix $\mathbf{M}$ is denoted $\mathbf{M}_{r,\bullet}$. All vectors are column vectors unless mentioned otherwise. Table 1 lists all symbols used in this study.

Table 1. List of symbols.

Symbol	Description	Dimensions
ϵ_h	Measurement error at time $t = t_h$	$M \times 1$
$\boldsymbol{\theta}\left(\boldsymbol{\theta}^{(j)}\right)$	Kinetic parameters (in subsystem j)	$T \times 1\left(T^{(j)} \times 1\right)$
$\boldsymbol{\Lambda}_{\mathrm{o}}^{(j)}, \boldsymbol{\Lambda}_{\theta}^{(j)}$	Selection matrix	$\rho_{\mathrm{o}}^{(j)} \times \rho_{\mathrm{o}}, T^{(j)} \times T$
$\rho_{\mathrm{o}}\left(\rho_{\mathrm{o}}^{(j)}, \rho_{\mathrm{o,i}}^{(j)}\right)$	Number of observable extent directions (in system j, interpolated in system j)	1×1
ρ_{u}	Number of unobservable extent directions	1×1
ρ_{aug}	Number of extents and observable extent directions	1×1
$\boldsymbol{\Sigma}_{\epsilon}$	Measurement error variance-covariance matrix	$M \times M$
τ	Time (integrand)	1×1
$\boldsymbol{\Sigma}_{\overline{\chi}}$	Estimation error variance-covariance matrix	$A \times A$
χ_{aug}	Extents and observable extent directions	$\rho_{\mathrm{aug}} \times 1$
$\chi_{\mathrm{o}}\left(\chi_{\mathrm{o}}^{(j)}\right)$	Observable extent directions (in subsystem j)	$\rho_{\mathrm{o}} \times 1\left(\rho_{\mathrm{o}}^{(j)} \times 1\right)$
$\overline{\chi}_{\mathrm{o}}\left(\overline{\chi}_{\mathrm{o}}^{(j)}\right)$	Observable extents and extent directions (in subsystem j)	$A \times 1\left(A^{(j)} \times 1\right)$
$\tilde{\overline{\chi}}_{\mathrm{o},h}\left(\tilde{\overline{\chi}}_{\mathrm{o},h}^{(j)}\right)$	Computed observable extents and extent directions (in subsystem j)	$A \times 1\left(A^{(j)} \times 1\right)$
$\tilde{\overline{\chi}}_{\mathrm{o,i}}\left(\tilde{\overline{\chi}}_{\mathrm{o,i}}^{(j)}\right)$	Interpolated observable extents and extent directions (in subsystem j)	$A \times 1\left(A^{(j)} \times 1\right)$
χ_{u}	Unobservable extent directions	$\rho_{\mathrm{u}} \times 1$
$\overline{\chi}_{\mathrm{u}}$	Unobservable extents and extent directions	$(R-A) \times 1$
$A\left(A^{(j)}\right)$	Number of observable extents and extent directions (in subsystem j)	1×1
$\mathbf{a}$	Indices of ambiguous extents	$R_{\mathrm{a}} \times 1$
$\mathbf{B}$	Reduced row echelon form of $\mathbf{G}$	$M \times R$
$\boldsymbol{d}_h\left(\boldsymbol{d}_h^{(j)}\right)$	Model prediction residuals at t_h (in subsystem j)	$A \times 1\left(A^{(j)} \times 1\right)$
$\mathbf{e}^{(j)}$	Indices of computed extents and extent directions simulated by subsystem j	$A^{(j)} \times 1$
$\mathcal{F}\left(\mathcal{F}^{(j)}\right)$	Information flow graph (for subsystem j)	$--$
$f\left(f^{(j)}\right)$	Rate expressions (in subsystem j)	$R \times 1\left(R^{(j)} \times 1\right)$

Table 1. *Cont.*

Symbol	Description	Dimensions
$\mathbf{G}$	Extent-based measurement matrix	$M \times R$
$\mathcal{G}_o, \overline{\mathcal{G}}_o$	Measurement matrix for the observable extent directions/observable extents and extent directions	$M \times \rho_o / A$
H	Number of measurements	1×1
h	Measurement sample index	1×1
i	Subsystem index	1×1
J	Number of subsystems	1×1
$\mathbf{j}$	Indices of extents in subsystem j	$R^{(j)} \times 1$
j	Subsystem index	1×1
$\mathbf{M}$	Species-based measurement matrix	$M \times S$
M	Number of measurement samples	1×1
$\mathbf{N}$	Stoichiometric matrix	$R \times S$
$\overline{\mathcal{N}}_o, \overline{\mathcal{N}}_u$	Extent and extent direction-based stoichiometric matrices	$A/(R-A) \times S$
$\mathbf{n}$	Indices of non-sensed extents	$R_n \times 1$
$\boldsymbol{n}$ ($\boldsymbol{n}_0$)	Number of moles (at time $t = 0$)	$S \times 1$
$\mathbf{o}$ $\left(\mathbf{o}^{(j)}\right)$	Indices of observable extents (in subsystem j)	$R_o \times 1$ $\left(R_o^{(j)} \times 1\right)$
$\mathbf{P}$	Projection matrix	$A \times M$
$Q_0, Q_0^*, Q_1^{(j)}$	Objective function	1×1
R $\left(R^{(j)}, R_a, R_n, R_o, R_o^{(j)}, R_{o,i}^{(j)}, R_s\right)$	Number of reactions (in subsystem j, ambiguous/non-sensed/ observable/observable in subsystem j/interpolated in subsystem j/sensed)	1×1
$\boldsymbol{r}$	Reaction rates	$R \times 1$
r	Reaction index	1×1
S	Number of species	1×1
$\mathbf{s}$	Indices of sensed extents	$R_s \times 1$
T $(T^{(j)})$	Number of parameters (in subsystem j)	1×1
t (t_h)	Time (of measurement sample h)	1×1
$\mathbf{U}^{(j)}, \overline{\mathbf{U}}^{(j)}$	Mixing matrix	$\rho_o^{(j)} / A^{(j)} \times R^{(j)}$
$\mathbf{V}_o, \mathbf{V}_u$	Direction matrix for observable/unobservable direction directions	$R \times \rho_o, R \times \rho_u$
V	Reactor volume	1×1
$\mathbf{W}$ $\left(\mathbf{W}^{(j)}\right)$	Weight matrix (for subsystem j)	$A \times A$ $\left(A^{(j)} \times A^{(j)}\right)$
$\boldsymbol{x}$ $\left(\boldsymbol{x}^{(j)}, x_r\right)$	Extents (in subsystem j, of reaction r)	$R \times 1$ $\left(R^{(j)} \times 1, 1 \times 1\right)$
$\boldsymbol{y}$ $(\boldsymbol{y}_0, \boldsymbol{y}(t_h))$	Noise-free measurements (at time $t = 0, t = h$)	$M \times 1$
$\tilde{\boldsymbol{y}}_h$	Measurements in sample h	$M \times 1$

3. Methods

3.1. System Representation and Extents

3.1.1. Dynamic Model in Terms of Numbers of Moles

We study batch process systems whose dynamic behavior is described by a set of differential equations of the following form:

$$\dot{\boldsymbol{n}}(t) = \mathbf{N}^{\mathrm{T}}\, V\, \boldsymbol{r}(t), \qquad \boldsymbol{n}(0) = \boldsymbol{n}_0\,. \tag{1}$$

In the present work, the above equations represent a single-phase reaction system in a vessel with constant volume V. $\boldsymbol{n}(t)$ is the S-dimensional vector of numbers of moles; $\boldsymbol{n}_0$ specifies the initial conditions; $\boldsymbol{r}(t)$ is the R-dimensional vector of reaction rates; $\mathbf{N}$ is the $(R \times S)$-dimensional stoichiometric matrix.

We assume that M noisy measurements, $\tilde{\boldsymbol{y}}_h$, are available at the H sampling times t_h ($h = 1, \ldots, H$), according to the following equations:

$$\tilde{y}_h := y(t_h) + \epsilon_h = \frac{1}{V}\,\mathbf{M}\,n(t_h) + \epsilon_h\,, \qquad \epsilon_h \sim \mathcal{N}\left(\mathbf{0}, \mathbf{\Sigma}_\epsilon\right), \tag{2}$$

where $y(t_h)$ is the M-dimensional vector of noise-free measurements, ϵ_h the M-dimensional vector of measurement errors, $\mathbf{\Sigma}_\epsilon$ the M-dimensional measurement error variance-covariance matrix, and $\mathbf{M}$ is the $(M \times S)$-dimensional *species-based measurement matrix*.

We further assume that kinetic laws for the reaction rates are available. These laws are functions of the component masses, $n(t)$, and a T-dimensional vector of kinetic parameters, θ:

$$r(t) := f\left(n(t), \theta\right). \tag{3}$$

In the present study, we assume that the rate laws are known, except for the values of the parameters θ that need to be estimated.

3.1.2. Dynamic Model in Terms of Extents

We now adopt the definition of the *extents of reaction*, hereafter simply called *extents*, as the number of times each reaction has occurred since $t = 0$. These extents are measured in moles and defined mathematically as:

$$x(t) := V \int_0^t r(\tau)\, d\tau. \tag{4}$$

It follows that

$$n(t) = n_0 + \mathbf{N}^{\mathrm{T}}\, x(t). \tag{5}$$

Accordingly, (1)–(3) can be represented equivalently in the following form:

$$\dot{x}(t) = V\, f\left(n(t), \theta\right), \qquad x(0) = \mathbf{0} \tag{6}$$

$$\begin{aligned} \tilde{y}_h &= \frac{1}{V}\,\mathbf{M}\,\left(n_0 + \mathbf{N}^{\mathrm{T}}\, x(t_h)\right) + \epsilon_h \\ &= y_0 + \frac{1}{V}\,\mathbf{G}\,x(t_h) + \epsilon_h \end{aligned} \tag{7}$$

$$n(t) = n_0 + \mathbf{N}^{\mathrm{T}}\, x(t), \tag{8}$$

where

$$y_0 := \frac{1}{V}\,\mathbf{M}\,n_0 \tag{9}$$

$$\mathbf{G} := \mathbf{M}\,\mathbf{N}^{\mathrm{T}}, \tag{10}$$

with the $(M \times R)$-dimensional matrix $\mathbf{G}$ being labeled the *extent-based measurement matrix*.

Let A denote the rank of $\mathbf{G}$, that is, $A := \mathsf{rank}\,(\mathbf{G}) \le \mathsf{min}\,(M, R)$.

3.2. *Labeling Extents*

3.2.1. Definitions

To label the extents, the following definitions are proposed:

- The rth extent $x_r(t)$ is labeled *sensed* if the measurements $\tilde{y}_h$ are affected by $x_r(t)$ through the measurement Equation (7), that is, if $\mathbf{G}_{\bullet,r}$ has at least one non-zero element.
- The rth extent is labeled *observable* if (i) it is sensed ($\mathbf{G}_{\bullet,r}$ not null); and (ii) the change in the measurements $\tilde{y}_h$ caused by a change in $x_r(t)$ can be unambiguously attributed to the change in that extent, that is, $\mathbf{G}_{\bullet,r}$ is independent of all other column vectors in $\mathbf{G}$.

The above labeling is based on the structure of **G** and does not depend on the temporal resolution or quality of the recorded measurements ($\tilde{y}_h$). As such, this means only *structural* observability is considered. This terminology is similar to the one used in data reconciliation [9,10] as the labels are produced without a dynamic model.

3.2.2. Labeling Procedure

To label the extents, one first computes the $(M \times R)$-dimensional reduced row echelon form of **G** denoted **B**:

$$\mathbf{B} := \mathrm{rref}\left(\mathbf{G}\right). \tag{11}$$

This $(M \times R)$-dimensional matrix is composed of A non-zero rows and $M - A$ zero rows. Using **B**, the extents are labeled with the following procedure:

(a) Label the R_n extents corresponding to zero columns in **B** as *non-sensed* and use the vector **n** to identify their positions in x. Label the R_s remaining extents as *sensed* and use the vector **s** to identify their positions in x.

(b) Find all rows in **B** with a *single* non-zero element and find the column positions of these non-zero elements. Label the R_o extents corresponding to these column positions as *observable* and use the vector **o** to identify their positions in x. These extents are observable because one can compute a unique value based on the available information (measurements, extent-based measurement matrix, and initial conditions).

(c) Label the R_a extents that are sensed but not observable as *ambiguous*. These extents are ambiguous because one cannot compute a unique value based on the available information. The vector **a** is used to identify their positions in x.

Based on this labeling, the vectors $x_\mathrm{s}(t)$, $x_\mathrm{n}(t)$, $x_\mathrm{o}(t)$, and $x_\mathrm{a}(t)$ are defined to represent (i) the R_s sensed extents, (ii) the $R_\mathrm{n} = R - R_\mathrm{s}$ non-sensed extents; (iii) the R_o observable extents; and (iv) the $R_\mathrm{a} = R_\mathrm{s} - R_\mathrm{o}$ ambiguous extents. This is illustrated in Figure 1.

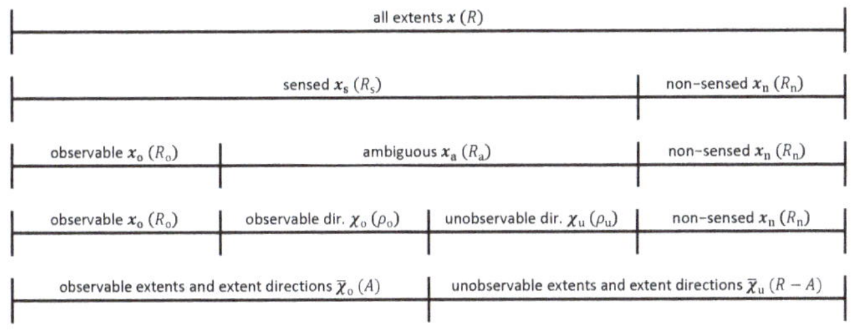

Figure 1. Labeling of the R extents as sensed, non-sensed, observable, and ambiguous. The ambiguous extents are spanned by ρ_o observable and ρ_u unobservable extent directions. This way, the extent space can be represented by $A = R_\mathrm{o} + \rho_\mathrm{o}$ observable extents and extent directions, and $R - A = R_\mathrm{n} + \rho_\mathrm{u}$ unobservable extents and extent directions.

With the above definitions, (5) can be reformulated as:

$$\begin{aligned} n(t) &= n_0 + \mathbf{N}_{\mathrm{s},\bullet}{}^{\mathrm{T}}\, x_\mathrm{s}(t) + \mathbf{N}_{\mathrm{n},\bullet}{}^{\mathrm{T}}\, x_\mathrm{n}(t) \\ &= n_0 + \mathbf{N}_{\mathrm{o},\bullet}{}^{\mathrm{T}}\, x_\mathrm{o}(t) + \mathbf{N}_{\mathrm{a},\bullet}{}^{\mathrm{T}}\, x_\mathrm{a}(t) + \mathbf{N}_{\mathrm{n},\bullet}{}^{\mathrm{T}}\, x_\mathrm{n}(t). \end{aligned} \tag{12}$$

Furthermore, it follows from $\mathbf{G}_{\bullet,\mathrm{n}} = \mathbf{0}_{M \times R_\mathrm{n}}$ that

$$\begin{aligned}\tilde{y}_h &= y_0 + \frac{1}{V}\,\mathbf{G}_{\bullet,\mathrm{s}}\,x_{\mathrm{s}}(t_h) + \epsilon_h \\ &= y_0 + \frac{1}{V}\,\mathbf{G}_{\bullet,\mathrm{o}}\,x_{\mathrm{o}}(t_h) + \frac{1}{V}\,\mathbf{G}_{\bullet,\mathrm{a}}\,x_{\mathrm{a}}(t_h) + \epsilon_h. \end{aligned} \tag{13}$$

3.2.3. Practical Cases

Depending on the values of A, M, and R, one can distinguish the following cases:

I. Full-rank extent-based measurement matrix ($A = R$). This occurs when there are at least as many measurements as reactions ($M \geq R$) and the matrix $\mathbf{G}$ is full column rank. As a result, $R_{\mathrm{n}} = 0$, $R_{\mathrm{o}} = R$, and $R_{\mathrm{a}} = 0$. This is the most frequently studied case, e.g., in [5–7]. This case enables computing unique values for the R extents by means of a linear transformation [5], thus allowing the estimation of kinetic parameters for each reaction rate model individually.

II. Rank-deficient measurement matrix ($A < R$). If $A < R$, for example because there are fewer measurements than reactions ($M < R$), it is no longer possible to compute all R extents from M measurements without additional information such as an established kinetic model. One can distinguish two situations within this case:

   (a) No ambiguity ($R_{\mathrm{a}} = 0$). In this case, $R_{\mathrm{o}} = A$ and $R_{\mathrm{o}} + R_{\mathrm{n}} = R$. As shown in [8], it is possible in this case to implement efficient parameter estimation by identifying subsystems of the complete model that include a subset of the kinetic rate laws and their parameters.

   (b) Ambiguity present ($0 < R_{\mathrm{a}} \leq A$). This situation results in $R_{\mathrm{o}} < A$. For this case, no generally applicable method for incremental parameter estimation is available until now.

The method proposed in this work is developed with the aim of handling all the above cases in a single framework for extent-based kinetic parameter estimation.

3.3. *Observable and Unobservable Extent Directions*

The ambiguous extents are now investigated in more detail to determine observable *directions* among them.

3.3.1. Factorization of $\mathbf{G}_{\bullet,\mathrm{a}}$

Let ρ_{o} be the rank of the $(M \times R_{\mathrm{a}})$-dimensional measurement matrix $\mathbf{G}_{\bullet,\mathrm{a}}$. Then, $\mathbf{G}_{\bullet,\mathrm{a}}$ can be factorized into the $(M \times \rho_{\mathrm{o}})$-dimensional measurement matrix $\mathcal{G}_{\mathrm{o}}$ and the $(R_{\mathrm{a}} \times \rho_{\mathrm{o}})$-dimensional matrix $\mathbf{V}_{\mathrm{o}}$:

$$\mathbf{G}_{\bullet,\mathrm{a}} = \mathcal{G}_{\mathrm{o}}\,\mathbf{V}_{\mathrm{o}}^{\mathrm{T}}. \tag{14}$$

Using the reduced row echelon form $\mathbf{B}_{\bullet,\mathrm{a}}$, the matrices $\mathcal{G}_{\mathrm{o}}$ and $\mathbf{V}_{\mathrm{o}}$ are computed as follows:

- $\mathbf{V}_{\mathrm{o}}^{\mathrm{T}}$ is obtained by selecting the ρ_{o} non-zero rows in $\mathbf{B}_{\bullet,\mathrm{a}}$,
- $\mathcal{G}_{\mathrm{o}}$ is the matrix composed of the columns of $\mathbf{G}_{\bullet,\mathrm{a}}$ corresponding to the column positions of the first non-zero elements in the rows of $\mathbf{V}_{\mathrm{o}}^{\mathrm{T}}$.

3.3.2. Definition of Observable and Unobservable Extent Directions

The term $\mathbf{G}_{\bullet,\mathrm{a}}\,x_{\mathrm{a}}(t_h)$ in (13) can be expressed as:

$$\mathbf{G}_{\bullet,\mathrm{a}}\,x_{\mathrm{a}}(t) = \mathcal{G}_{\mathrm{o}}\,\mathbf{V}_{\mathrm{o}}^{\mathrm{T}}\,x_{\mathrm{a}}(t) = \mathcal{G}_{\mathrm{o}}\,\chi_{\mathrm{o}}(t), \tag{15}$$

with $\chi_o(t)$ the ρ_o-dimensional vector of *observable extent directions* among the ambiguous extents:

$$\chi_o(t) := \mathbf{V_o}^T \mathbf{x_a}(t) = \mathbf{V_o}^T \mathbf{I_{a,\bullet}}\, \mathbf{x}(t), \tag{16}$$

where the $(R_a \times R)$-dimensional matrix $\mathbf{I_{a,\bullet}}$ includes the rows $\mathbf{a}$ of the identity matrix $\mathbf{I}_R$ such that

$$\mathbf{x_a}(t) = \mathbf{I_{a,\bullet}}\, \mathbf{x}(t). \tag{17}$$

Equation (15) indicates that while the extents $\mathbf{x_a}(t)$ cannot be observed individually, their combined effects on the measurements can be observed as the linear combinations $\chi_o(t)$.

Remark 1. *The unobservable extent directions span the null space of $\mathbf{G_{\bullet,a}}$, which is also the null space of $\mathbf{V_o}^T$. Denoting this null space by the $(R_a \times \rho_u)$-dimensional matrix $\mathbf{V_u}$,*

$$\mathbf{V_u} = \text{null}\left(\mathbf{V_o}^T\right), \tag{18}$$

with $\rho_u = (R_a - \rho_o)$, one can define the ρ_u-dimensional vector of unobservable extent directions *$\chi_u(t)$ as:*

$$\chi_u(t) := \mathbf{V_u}^T \mathbf{x_a}(t) = \mathbf{V_u}^T \mathbf{I_{a,\bullet}}\, \mathbf{x}(t). \tag{19}$$

3.4. Observable Extents and Extent Directions

We further define the vector $\overline{\chi}_o(t)$ consisting of the $A = R_o + \rho_o$ observable extents and extent directions as follows:

$$\overline{\chi}_o(t) := \begin{bmatrix} \mathbf{x_o}(t) \\ \chi_o(t) \end{bmatrix} = \begin{bmatrix} \mathbf{x_o}(t) \\ \mathbf{V_o}^T \mathbf{x_a}(t) \end{bmatrix}. \tag{20}$$

With this definition, the measurement Equation (13) can be rewritten as:

$$\begin{aligned} \tilde{y}_h &= y_0 + \frac{1}{V}\, \mathbf{G_{\bullet,o}}\, \mathbf{x_o}(t_h) + \frac{1}{V}\, \mathcal{G}_o\, \chi_o(t) + \epsilon_h \\ &= y_0 + \frac{1}{V}\, \overline{\mathcal{G}}_o\, \overline{\chi}_o(t_h) + \epsilon_h, \end{aligned} \tag{21}$$

with $\overline{\mathcal{G}}_o$, an $(M \times A)$-dimensional matrix, constructed as

$$\overline{\mathcal{G}}_o := \begin{bmatrix} \mathbf{G_{\bullet,o}} & \mathcal{G}_o \end{bmatrix}. \tag{22}$$

Since the $(R_a \times A)$-dimensional matrix $\overline{\mathcal{G}}_o$ has full column rank, (21) can be used to compute the maximum-likelihood estimates $\tilde{\overline{\chi}}_{o,h}$ of the observable extents and extent directions directly from the measurements:

$$\tilde{\overline{\chi}}_{o,h} = \mathbf{P}\left(\tilde{y}_h - y_0\right), \tag{23}$$

with the $(A \times M)$-dimensional matrix $\mathbf{P}$ given by:

$$\mathbf{P} = V \left(\overline{\mathcal{G}}_o^{\,T}\, \mathbf{\Sigma}_\epsilon^{-1}\, \overline{\mathcal{G}}_o\right)^{-1} \overline{\mathcal{G}}_o^{\,T}\, \mathbf{\Sigma}_\epsilon^{-1}. \tag{24}$$

The associated expected variance-covariance matrix of the estimation errors becomes:

$$\mathbf{\Sigma}_{\overline{\chi}} = \mathbf{P}\, \mathbf{\Sigma}_\epsilon\, \mathbf{P}^T = \left(\overline{\mathcal{G}}_o^{\,T}\, \mathbf{\Sigma}_\epsilon^{-1}\, \overline{\mathcal{G}}_o\right)^{-1}. \tag{25}$$

3.5. Unobservable Extents and Extent Directions

We finally define the vector $\overline{\chi}_u(t)$ consisting of the $R - A = R_n + \rho_u$ unobservable extents and non-sensed extent directions as follows:

$$\overline{\chi}_u(t) := \begin{bmatrix} x_n(t) \\ \chi_u(t) \end{bmatrix} = \begin{bmatrix} x_n(t) \\ \mathbf{V}_u{}^T x_a(t) \end{bmatrix} = \begin{bmatrix} \mathbf{I}_{n,\bullet} \\ \mathbf{V}_u{}^T \mathbf{I}_{a,\bullet} \end{bmatrix} x(t). \tag{26}$$

With this definition, the expression for the number of moles (12) can be rewritten as:

$$\begin{aligned} n(t) &= n_0 + \mathbf{N}_{o,\bullet}{}^T x_o(t) + \mathbf{N}_{a,\bullet}{}^T \left((\mathbf{V}_o)^{+T} \chi_o(t) + (\mathbf{V}_u)^{+T} \chi_u(t) \right) + \mathbf{N}_{n,\bullet}{}^T x_n(t) \\ &= n_0 + \overline{\mathcal{N}}_o{}^T \overline{\chi}_o(t) + \overline{\mathcal{N}}_u{}^T \overline{\chi}_u(t) \end{aligned} \tag{27}$$

with:

$$\overline{\mathcal{N}}_o := \begin{bmatrix} \mathbf{N}_{o,\bullet} \\ (\mathbf{V}_o)^{+} \mathbf{N}_{a,\bullet} \end{bmatrix} \tag{28}$$

$$\overline{\mathcal{N}}_u := \begin{bmatrix} \mathbf{N}_{n,\bullet} \\ (\mathbf{V}_u)^{+} \mathbf{N}_{a,\bullet} \end{bmatrix} \tag{29}$$

3.6. System Partitioning

Incremental parameter estimation is based on the possibility of separating the parameter estimation problem into smaller problems. Ideally, if all extents of reaction could be computed from measurements, each reaction could be identified individually, that is, independently of the other reactions. This way, the parameter estimation problem reduces to the solution of R smaller problems. As discussed in the previous section, some extents may not be observable. For these situations, it would still be nice to be able to separate the parameter estimation problem into J smaller problems ($J \leq R$). The objective is therefore to partition the reaction system effectively—with as many small groups of reactions as possible—to simplify the parameter estimation task. The first step to achieve this consists of system partitioning.

An algorithm is developed to group the kinetic parameters into J parameter subsets ($j = 1, \ldots, J$), each represented as a $T^{(j)}$-dimensional vector $\boldsymbol{\theta}^{(j)}$ satisfying the following properties:

- The size $T^{(j)}$ of each parameter subset should be as small as possible.
- The estimates $\hat{\boldsymbol{\theta}}^{(j)}$ in the jth parameter subset can be computed without consideration of any other parameter subset $\boldsymbol{\theta}^{(i)}$, $i \neq j$.
- Each parameter in $\boldsymbol{\theta}$ appears in at most one of the parameter subsets $\boldsymbol{\theta}^{(j)}$.

This objective is achieved by means of model reformulation and graph-based system partitioning. The graph-based procedure can be interpreted as a symbolic manipulation of the process model. It does not require symbolic differentiation, however.

3.6.1. Step 1—Model Reformulation

An extended model is first defined to describe the dynamics of all extents and all observable directions among the ambiguous extents. To this end, the following procedure is applied:

(a) *Express $\dot{x}(t)$ as a function of $\overline{\chi}_o(t)$ and $x(t)$*

The dynamic model (6) is modified by replacing $n(t)$ with the right-hand side of (27):

$$\dot{x}(t) = V\, f\left(\overline{\chi}_o(t), \overline{\chi}_u(t), \boldsymbol{\theta}\right), \qquad x(0) = \mathbf{0}. \tag{30}$$

The vector $\overline{\chi}_{\mathrm{u}}(t)$ is now replaced with the right-hand side of (26). As a result, the above system becomes:

$$\dot{x}(t) = V\, f\left(\overline{\chi}_{\mathrm{o}}(t), x(t), \boldsymbol{\theta}\right), \qquad x(0) = \mathbf{0}. \tag{31}$$

(b) *State augmentation*

Define the ρ_{aug}-dimensional vector $\chi_{\mathrm{aug}}(t) := \begin{bmatrix} x(t) \\ \chi_{\mathrm{o}}(t) \end{bmatrix} = \begin{bmatrix} \mathbf{I}_R \\ \mathbf{V}_{\mathrm{o}}{}^{\mathrm{T}}\, \mathbf{I}_{\mathbf{a},\bullet} \end{bmatrix} x(t)$ that includes all extents and extent directions, with $\rho_{\mathrm{aug}} = R + \rho_{\mathrm{o}}$. The dynamic behavior of $\chi_{\mathrm{aug}}(t)$ can be described by a differential-algebraic system including the R differential Equation (31) and the ρ_{o} algebraic expressions (16):

$$\dot{x}(t) = V\, f\left(\overline{\chi}_{\mathrm{o}}(t), x(t), \boldsymbol{\theta}\right), \qquad x(0) = \mathbf{0} \tag{32}$$

$$\chi_{\mathrm{o}}(t) = \mathbf{V}_{\mathrm{o}}{}^{\mathrm{T}}\, \mathbf{I}_{\mathbf{a},\bullet}\, x(t)\,. \tag{33}$$

(c) *Interpolation of the observable extents and extent directions*

To increase the efficiency of system partitioning, it is useful to account for the fact that the observable extents and extent directions can be expressed in terms of measurements. However, since the observable extents and extent directions are only known at H discrete measurement points, their values always need to be obtained via interpolation. In this work, we apply piece-wise linear interpolation as follows:

$$\tilde{\overline{\chi}}_{\mathrm{o,i}}(t) := \tilde{\overline{\chi}}_{\mathrm{o},h} + \frac{t - t_h}{t_{h+1} - t_h}\left(\tilde{\overline{\chi}}_{\mathrm{o},h+1} - \tilde{\overline{\chi}}_{\mathrm{o},h}\right), \qquad t_h \le t < t_{h+1}, \qquad h = 1, \ldots, H, \tag{34}$$

with which the system (32) and (33) becomes:

$$\dot{x}(t) = V\, f\left(\tilde{\overline{\chi}}_{\mathrm{o,i}}(t), x(t), \boldsymbol{\theta}\right), \qquad x(0) = \mathbf{0} \tag{35}$$

$$\chi_{\mathrm{o}}(t) = \mathbf{V}_{\mathrm{o}}{}^{\mathrm{T}}\, \mathbf{I}_{\mathbf{a},\bullet}\, x(t)\,. \tag{36}$$

3.6.2. Step 2—Graph-Based System Partitioning

The equation system (35) and (36) is now analyzed by means of a graph partitioning procedure to determine the smallest groups of kinetic parameters that can be estimated separately. To this end, the following steps are performed:

(a) *Create a graph*
One creates a directed graph $\mathcal{F}$ with a vertex for every state variable in $\chi_{\mathrm{aug}}(t)$ and every parameter in $\boldsymbol{\theta}$. Hence, this graph has $R + \rho_{\mathrm{o}} + T$ vertices. A directed arc is added from vertex v to vertex w if the vth element of $\begin{bmatrix} \chi_{\mathrm{aug}}(t) \\ \boldsymbol{\theta} \end{bmatrix}$ appears in the right-hand side of the wth equation in (35) and (36) ($v = 1, \ldots, R + \rho_{\mathrm{o}} + T$, $w = 1, \ldots, R + \rho_{\mathrm{o}}$). This graph represents the information flow for simulating (35) and (36). Additional arcs and vertices may be added to describe the influence of known inputs and the links between extents and measured variables. For system partitioning, this is however unnecessary and omitted for clarity.

(b) *Extents predicted from measurements or simulation*
The simultaneous approach uses a complete model of the reaction system to predict the extents (or concentrations) via simulation. If one wants to partition the reaction system into small groups of reactions, only the extents belonging to a given group can be generated via the simulation of that group. The other extents that enter the rate laws must be provided by the user as quantities known from measurements.

That information can be included in the graph $\mathcal{F}$ by annotating the various arcs. The arcs that originate at a vertex corresponding to an observable extent or an observable extent direction are labeled *observation arcs*, considering that observable extent or an observable extent direction can be replaced with their measured values (34). They are visualized as dashed-line arrows. The remaining arcs are labeled *simulation arcs* and visualized as solid-line arrows. The observation arcs represent the idea that the elements of $\tilde{\bar{\chi}}_{\mathrm{o,i}}(t)$ can be regarded as known inputs for simulating (35) and (36).

(c) *Subgraph selection*
Identify the J subgraphs $\mathcal{F}^{(j)}$ consisting of arcs and vertices in $\mathcal{F}$ on directed paths that (i) lead to a vertex representing an observable extent or an observable extent direction; and (ii) consist of simulation arcs only. The selected vertices represent an $R^{(j)}$-dimensional vector of extents $\boldsymbol{x}^{(j)}(t)$, a $\rho_{\mathrm{o}}^{(j)}$-dimensional vector of directions $\chi_{\mathrm{o}}^{(j)}(t)$, and a $T^{(j)}$-dimensional vector of parameters $\boldsymbol{\theta}^{(j)}$. The positions of $\boldsymbol{x}^{(j)}(t)$ in $\boldsymbol{x}(t)$ are given by the vector $\mathbf{j}$ so that:

$$\boldsymbol{x}^{(j)}(t) := \mathbf{I}_{\mathbf{j},\bullet}\, \boldsymbol{x}(t) \tag{37}$$

$$\boldsymbol{f}^{(j)}(\bullet) := \mathbf{I}_{\mathbf{j},\bullet}\, \boldsymbol{f}(\bullet), \tag{38}$$

and the selection matrices $\boldsymbol{\Lambda}_{\mathrm{o}}^{(j)}$ and $\boldsymbol{\Lambda}_{\theta}^{(j)}$ are defined so that:

$$\chi_{\mathrm{o}}^{(j)}(t) := \boldsymbol{\Lambda}_{\mathrm{o}}^{(j)}\, \chi_{\mathrm{o}}(t) \tag{39}$$

$$\boldsymbol{\theta}^{(j)} := \boldsymbol{\Lambda}_{\theta}^{(j)}\, \boldsymbol{\theta}. \tag{40}$$

This means that each subgraph $\mathcal{F}^{(j)}$ represents a subset of Equations (35) and (36) that describes the dynamics of $\boldsymbol{x}^{(j)}(t)$ and $\chi_{\mathrm{o}}^{(j)}(t)$ without reference to any other state variable:

$$\dot{\boldsymbol{x}}^{(j)}(t) = V\, \boldsymbol{f}^{(j)}\left(\tilde{\bar{\chi}}_{\mathrm{o,i}}(t), \boldsymbol{x}^{(j)}(t), \boldsymbol{\theta}^{(j)}\right), \qquad \boldsymbol{x}^{(j)}(0) = \boldsymbol{x}_{\mathbf{0}} \tag{41}$$

$$\chi_{\mathrm{o}}^{(j)}(t) = \mathbf{U}^{(j)}\, \boldsymbol{x}^{(j)}(t), \tag{42}$$

with $\mathbf{U}^{(j)} := \boldsymbol{\Lambda}_{\mathrm{o}}^{(j)}\, \mathbf{V}_{\mathrm{o}}^{\mathrm{T}}\, \mathbf{I}_{\mathbf{a},\bullet}\, \mathbf{I}_{\mathbf{j},\bullet}^{\mathrm{T}}$.

(d) *Add observation arcs and vertices*
For every graph $\mathcal{F}^{(j)}$, add (i) the observation arcs that have a target vertex belonging to $\mathcal{F}^{(j)}$ and (ii) the source vertices of the added observation arcs. These added source vertices represent the minimal subset of interpolants in $\tilde{\bar{\chi}}_{\mathrm{o,i}}(t)$ (34) that are required to simulate $\boldsymbol{x}^{(j)}(t)$ and $\chi_{\mathrm{o}}^{(j)}(t)$ and are referred to as $\tilde{\bar{\chi}}_{\mathrm{o,i}}^{(j)}(t)$. This means that the graph $\mathcal{F}^{(j)}$ now represents all information required to simulate the observable extents $\boldsymbol{x}^{(j)}(t)$ and the observable extent directions $\chi_{\mathrm{o}}^{(j)}(t)$. Accordingly, one can rewrite the jth equation subsystem as:

$$\dot{\boldsymbol{x}}^{(j)}(t) = V\, \boldsymbol{f}^{(j)}\left(\tilde{\bar{\chi}}_{\mathrm{o,i}}^{(j)}(t), \chi_{\mathrm{o}}^{(j)}(t), \boldsymbol{x}^{(j)}(t), \boldsymbol{\theta}^{(j)}\right), \qquad \boldsymbol{x}^{(j)}(0) = \boldsymbol{x}_{\mathbf{0}} \tag{43}$$

$$\chi_{\mathrm{o}}^{(j)}(t) = \mathbf{U}^{(j)}\, \boldsymbol{x}^{(j)}(t). \tag{44}$$

At the end of this procedure, the equation system (35) and (36) is approximated by J smaller equation subsystems, each including a subset of the kinetic parameters. As intended, every kinetic parameter appears in at most one of the J subsystems. In addition, the identified subsystems do not share any of the observable extents or extent directions as state variables, that is, each observable extent and extent direction is simulated in only one of the identified subsystems.

3.7. Parameter Estimation Methods

In this work, we solve the parameter estimation problem in two distinct ways. The first way consists in a simultaneous estimation of all parameters in the maximum-likelihood sense. The second way consists in an extent-based incremental parameter estimation. The next paragraphs describe how incremental parameter estimation approximates the simultaneous estimation procedure to minimize the number of parameters that are estimated together.

3.7.1. Simultaneous Parameter Estimation

Maximum-likelihood estimation of the kinetic parameters can be obtained by solving minimization of the weighted mean squared error:

Problem P_0

$$\hat{\boldsymbol{\theta}} = \arg\min_{\boldsymbol{\theta}} Q_0 := \frac{1}{H \cdot M} \sum_{h}^{H} \left(\tilde{\boldsymbol{y}}_h - \boldsymbol{y}(t_h)\right)^{\mathrm{T}} \boldsymbol{\Sigma}_{\epsilon}^{-1} \left(\tilde{\boldsymbol{y}}_h - \boldsymbol{y}(t_h)\right), \tag{45}$$

subject to (1)–(3). This estimation problem can be equivalently formulated in terms of the computed extents as follows:

Problem P_0^*

$$\hat{\boldsymbol{\theta}} = \arg\min_{\boldsymbol{\theta}} Q_0^* := \sum_{h}^{H} \left(\tilde{\overline{\boldsymbol{\chi}}}_{\mathbf{o},h} - \overline{\boldsymbol{\chi}}_{\mathbf{o}}(t_h)\right)^{\mathrm{T}} \mathbf{W} \left(\tilde{\overline{\boldsymbol{\chi}}}_{\mathbf{o},h} - \overline{\boldsymbol{\chi}}_{\mathbf{o}}(t_h)\right), \tag{46}$$

subject to (6)–(8) and (20) and with $\mathbf{W} := \boldsymbol{\Sigma}_{\overline{\chi}}^{-1}$.

3.7.2. Incremental Parameter Estimation

The incremental parameter estimation procedure is obtained by applying two modifications, A and B, to problem P_0^*.

(a) *Modification A: Removing correlation terms.* Let $A^{(j)} = R_{\mathbf{o}}^{(j)} + \rho_{\mathbf{o}}^{(j)}$ and define the $A^{(j)}$-dimensional vector of observable extents and extent directions $\overline{\boldsymbol{\chi}}_{\mathbf{o}}^{(j)}(t)$. This vector includes all observable extents $\boldsymbol{x}_{\mathbf{o}}^{(j)}(t)$, whose positions in $\boldsymbol{x}^{(j)}(t)$ are given by the vector $\mathbf{o}^{(j)}$, and all observable extent directions $\boldsymbol{\chi}_{\mathbf{o}}^{(j)}(t)$ in Subsystem j. We further define the matrix $\overline{\mathbf{U}}^{(j)}$ so that:

$$\boldsymbol{x}_{\mathbf{o}}^{(j)}(t) := \mathbf{I}_{\mathbf{o}^{(j)},\bullet}\, \boldsymbol{x}^{(j)}(t) \tag{47}$$

$$\overline{\boldsymbol{\chi}}_{\mathbf{o}}^{(j)}(t) = \begin{bmatrix} \boldsymbol{x}_{\mathbf{o}}^{(j)}(t) \\ \boldsymbol{\chi}_{\mathbf{o}}^{(j)}(t) \end{bmatrix} = \overline{\mathbf{U}}^{(j)}\, \boldsymbol{x}^{(j)}(t) \tag{48}$$

$$\overline{\mathbf{U}}^{(j)} := \begin{bmatrix} \mathbf{I}_{\mathbf{o}^{(j)},\bullet} \\ \mathbf{U}^{(j)} \end{bmatrix}. \tag{49}$$

The $A^{(j)}$-dimensional vector $\tilde{\overline{\boldsymbol{\chi}}}_{\mathbf{o},h}^{(j)}$ is then defined by selecting the elements of $\tilde{\overline{\boldsymbol{\chi}}}_{\mathbf{o},h}$ as:

$$\tilde{\overline{\boldsymbol{\chi}}}_{\mathbf{o},h}^{(j)} := \mathbf{I}_{\mathbf{e}^{(j)},\bullet}\, \tilde{\overline{\boldsymbol{\chi}}}_{\mathbf{o},h}, \tag{50}$$

where the vector $\mathbf{e}^{(j)}$ gives the positions of $\tilde{\overline{\chi}}^{(j)}_{\mathrm{o},h}$ in $\tilde{\overline{\chi}}_{\mathrm{o},h}$. Now define the residuals $\boldsymbol{d}_h^{(j)}$ associated with subsystem j:

$$\boldsymbol{d}_h^{(j)} := \tilde{\overline{\chi}}^{(j)}_{\mathrm{o},h} - \overline{\chi}_{\mathrm{o}}^{(j)}(t_h) \tag{51}$$

This way, the objective function Q_0^* defined above can be reformulated as:

$$\begin{aligned} Q_0^* &= \sum_h^H \sum_i^J \sum_j^J \boldsymbol{d}_h^{(i)\,\mathrm{T}} \mathbf{I}_{\mathbf{e}^{(i)},\bullet} \mathbf{W} \mathbf{I}_{\mathbf{e}^{(j)},\bullet}{}^{\mathrm{T}} \boldsymbol{d}_h^{(j)} \\ &= \sum_h^H \left(\sum_j^J \boldsymbol{d}_h^{(j)\,\mathrm{T}} \mathbf{W}^{(j)} \boldsymbol{d}_h^{(j)} + \sum_{i,j:\, i \neq j} \boldsymbol{d}_h^{(i)\,\mathrm{T}} \mathbf{I}_{\mathbf{e}^{(i)},\bullet} \mathbf{W} \mathbf{I}_{\mathbf{e}^{(j)},\bullet}{}^{\mathrm{T}} \boldsymbol{d}_h^{(j)} \right) \end{aligned} \tag{52}$$

and can subsequently approximated with Q_1:

$$Q_0^* \approx Q_1 := \sum_j^J \sum_h^H \boldsymbol{d}_h^{(j)\,\mathrm{T}} \mathbf{W}^{(j)}\,, \boldsymbol{d}_h^{(j)} \tag{53}$$

with $\mathbf{W}^{(j)} := \mathbf{I}_{\mathbf{e}^{(j)},\bullet} \mathbf{W} \mathbf{I}_{\mathbf{e}^{(j)},\bullet}{}^{\mathrm{T}}$. This first modification results in:

Problem P_1

$$\hat{\boldsymbol{\theta}} = \arg\min_{\boldsymbol{\theta}} Q_1 := \sum_j^J \sum_h^H \boldsymbol{d}_h^{(j)\,\mathrm{T}} \mathbf{W}^{(j)} \boldsymbol{d}_h^{(j)}, \tag{54}$$

subject to (6)–(8) and (20). This error due to this approximation is referred to as *Type A* approximation error. This error is zero if all matrices $\mathbf{I}_{\mathbf{e}^{(i)},\bullet} \mathbf{W} \mathbf{I}_{\mathbf{e}^{(j)},\bullet}{}^{\mathrm{T}}, i \neq j$, are zero matrices. This is true when the correlation between the estimation errors of any element in $\tilde{\overline{\chi}}^{(i)}_{\mathrm{o},h}$ and any element of $\tilde{\overline{\chi}}^{(j)}_{\mathrm{o},h}$ $(i \neq j)$ is zero.

(b) *Modification B: Separation of problem P_1 into J smaller problems $P_1^{(j)}$.* An approximation to problem P_1 is now obtained by optimizing each of the J terms in (54) separately and simulating the values of $\overline{\chi}_{\mathrm{o}}^{(j)}(t_h)$ with (43) and (44). We refer to each problem as $P_1^{(j)}$:

Problem $P_1^{(j)}$

$$\hat{\boldsymbol{\theta}}^{(j)} = \arg\min_{\boldsymbol{\theta}} Q_1^{(j)} := \sum_h^H \boldsymbol{d}_h^{(j)\,\mathrm{T}} \mathbf{W}^{(j)} \boldsymbol{d}_h^{(j)} \tag{55}$$

subject to

$$\dot{\boldsymbol{x}}^{(j)}(t) = V\, \boldsymbol{f}^{(j)} \left(\tilde{\overline{\chi}}^{(j)}_{\mathrm{o,i}}(t), \chi_{\mathrm{o}}^{(j)}(t), \boldsymbol{x}^{(j)}(t), \boldsymbol{\theta}^{(j)} \right), \qquad \boldsymbol{x}^{(j)}(0) = \mathbf{0} \tag{56}$$

$$\chi_{\mathrm{o}}^{(j)}(t) = \mathbf{U}^{(j)} \boldsymbol{x}^{(j)}(t) \tag{57}$$

$$\boldsymbol{d}_h^{(j)} = \tilde{\overline{\chi}}^{(j)}_{\mathrm{o},h} - \overline{\chi}_{\mathrm{o}}^{(j)}(t_h) = \tilde{\overline{\chi}}^{(j)}_{\mathrm{o},h} - \overline{\mathbf{U}}^{(j)} \boldsymbol{x}^{(j)}(t_h)\,. \tag{58}$$

The most important feature of problems $P_1^{(j)}$, $j = 1, \ldots, J$, is that each of them involves only a small number of parameters $\boldsymbol{\theta}^{(j)}$. Please note that the approximation of problem P_1 by this set of problems $P_1^{(j)}$ is perfect in the special case where the right-hand sides (56) do not involve interpolated extents, that is, if the vectors $\tilde{\tilde{\boldsymbol{\chi}}}_{o,i}^{(j)}(t)$ $(j = 1, \ldots, J)$ are empty. Another error, named *Type B* approximation error, is introduced when this is not the case.

Thanks to two modifications of problem P_0, one obtains J smaller problems $P_1^{(j)}$, each of which includes only a fraction of the original set of parameters. The price to pay for such a simplification is a deviation from maximum likelihood due to the introduction of approximation errors (type A & B). These errors are often marginal as demonstrated below.

As in previous work [5–7], the parameter estimates obtained by solving problem $P_1^{(j)}$ $(j = 1, \ldots, J)$ can serve as reliable initial guesses to initiate the solution to problem P_0. The problem P_0 when solved with these initial guesses is named problem P_{1+0}.

3.8. Implementation

All results can be reproduced with the open-source Efficient Model Identification (EMI) MATLAB package for efficient model identification [6] that includes all methods and simulations used in this study. This package is added in the Supplementary Information (Section A).

4. Results

We first explain the results obtained within an extensive simulation study to demonstrate the method. After that, results obtained with an experimental data set are used to demonstrate real-world applicability.

4.1. Simulation Study

The developed methods are demonstrated via a batch reaction system and investigating 5 different measurement scenarios (A–E). The reaction system is described first. Then, scenario A and its results are discussed in detail. The results for scenarios B to E are only summarized, with the details described in the Supplementary Information (Section B).

4.1.1. Reaction System

This reaction system has $R = 5$ reactions involving $S = 6$ species (A to F) with the following reaction scheme:

$$\begin{aligned}
R_1 &: \quad \mathrm{A} + \mathrm{B} \longrightarrow \mathrm{C} \\
R_2 &: \quad 2\,\mathrm{A} \longrightarrow \mathrm{D} \\
R_3 &: \quad 2\,\mathrm{C} \longrightarrow \mathrm{B} + \mathrm{D} \\
R_4 &: \quad \mathrm{D} \longrightarrow \mathrm{E} \\
R_5 &: \quad 2\,\mathrm{D} \longrightarrow \mathrm{E} + \mathrm{F}
\end{aligned}$$

with

$$\mathbf{N} := \begin{bmatrix} -1 & -1 & +1 & 0 & 0 & 0 \\ -2 & 0 & 0 & +1 & 0 & 0 \\ 0 & +1 & -2 & +1 & 0 & 0 \\ 0 & 0 & 0 & -1 & +1 & 0 \\ 0 & 0 & 0 & -2 & +1 & +1 \end{bmatrix}. \tag{59}$$

4.1.2. Dynamic Model in Terms of Numbers of Moles

The simulated kinetic rate expressions are:

$$f\left(\boldsymbol{n}(t),\boldsymbol{\theta}\right) := \begin{bmatrix} k_1\left(\frac{n_1(t)}{V}\frac{n_2(t)}{V} - K_1\frac{n_3(t)}{V}\right) \\ k_2\left(\frac{n_1(t)}{V}\right)^2 \\ k_3\frac{n_3(t)}{V} \\ k_4\frac{n_4(t)}{V} \\ k_5\left(\frac{n_4(t)}{V}\right)^2 \end{bmatrix} \tag{60}$$

with $\boldsymbol{\theta} := \begin{bmatrix} k_1 & k_2 & k_3 & k_4 & k_5 & K_1 \end{bmatrix}^{\mathrm{T}}$. The ground truth values of the kinetic parameters are given in Table 2.

Table 2. Scenario A—Parameter estimates. Model simulation and parameter estimation results: kinetic parameters and model fit. All parameter estimates are reported with their standard deviation based on the Laplacian approximation of the likelihood function (i.e., $exp\,(-Q_0)$, $exp\left(-Q_1^{(j)}\right)$).

	Name	Unit	True Value	P_0	$P_1^{(j)}$	P_{1+0}
θ_1	k_1	$\mathrm{L\,mol^{-1}\,h^{-1}}$	2.0	1.9490 ($\pm$0.0003)	2.0010 ($\pm$3.3625)	1.9857 ($\pm$0.0001)
θ_2	k_2	$\mathrm{L\,mol^{-1}\,h^{-1}}$	0.5	0.4929 ($\pm$0.0002)	0.5006 ($\pm$2.4090)	0.4771 ($\pm$0.0001)
θ_3	k_3	$\mathrm{h^{-1}}$	1.0	1.0056 ($\pm$0.0827)	0.8955 ($\pm$0.0002)	0.9974 ($\pm$0.0805)
θ_4	k_4	$\mathrm{h^{-1}}$	0.4	0.4342 ($\pm$0.4275)	0.4003 ($\pm$2.6400)	0.4606 ($\pm$0.1309)
θ_5	k_5	$\mathrm{L\,mol^{-1}\,h^{-1}}$	1.6	1.2285 ($\pm$2.5160)	1.6009 ($\pm$2.9587)	1.2111 ($\pm$0.8018)
θ_6	K_1	$\mathrm{mol\,L^{-1}}$	1.4	1.3435 ($\pm$0.1016)	1.4001 ($\pm$2.9081)	1.3986 ($\pm$0.1135)
$\sqrt{\frac{Q_0}{H \cdot M}}$	WRMSR	–		0.44731	0.47934	0.44771

The initial conditions are $\boldsymbol{n}_0 := \begin{bmatrix} 0.73 & 0.42 & 0 & 0 & 0 & 0 \end{bmatrix}^{\mathrm{T}}$ mol, and the reactor volume is $V = 1L$. In all investigated scenarios, measurements are taken at intervals of 5 min during 10 h ($H = 121$, $t_h = 0, 1/12, 2/12, \ldots, 10$).

4.1.3. Scenario A

The first scenario considers the case where the concentrations of B and C and the sum of the concentrations of E and F are measured, that is,

$$\mathbf{M} := \begin{bmatrix} 0 & 1 & 0 & 0 & 0 & 0 \\ 0 & 0 & 1 & 0 & 0 & 0 \\ 0 & 0 & 0 & 0 & 1 & 1 \end{bmatrix}. \tag{61}$$

The measurement error-covariance matrix is $\boldsymbol{\Sigma}_\epsilon = \mathrm{diag}\left(\begin{bmatrix} 1 & 1 & 2 \end{bmatrix}^{\mathrm{T}}\right) 10^{-4}\,\mathrm{mol^2 L^{-2}}$. Figure 2 shows the simulated noise-free and noisy measurements.

Extent labeling

To specify the measurement model in terms of extents (7), one needs to specify $\boldsymbol{y}_0$ and $\mathbf{G}$:

$$\boldsymbol{y}_0 = \mathbf{M}\,\boldsymbol{n}_0 = \begin{bmatrix} 0.42 & 0 & 0 \end{bmatrix}^{\mathrm{T}} \tag{62}$$

$$\mathbf{G} = \mathbf{M}\,\mathbf{N}^{\mathrm{T}} = \begin{bmatrix} -1 & 0 & +1 & 0 & 0 \\ +1 & 0 & -2 & 0 & 0 \\ 0 & 0 & 0 & +1 & +2 \end{bmatrix}. \tag{63}$$

The reduced row echelon form of $\mathbf{G}$ is

$$\mathbf{B} = \begin{bmatrix} +1 & 0 & 0 & 0 & 0 \\ 0 & 0 & +1 & 0 & 0 \\ 0 & 0 & 0 & +1 & +2 \end{bmatrix}, \tag{64}$$

which indicates that there are:

- $R_\mathrm{n} = 1$ non-sensed extent, $\boldsymbol{x}_\mathbf{n}(t) = x_2(t)$,
- $R_\mathrm{o} = 2$ observable extents, $\boldsymbol{x}_\mathbf{o}(t) = \begin{bmatrix} x_1(t) & x_3(t) \end{bmatrix}^\mathrm{T}$,
- $R_\mathrm{a} = 2$ ambiguous extents, $\boldsymbol{x}_\mathbf{a}(t) = \begin{bmatrix} x_4(t) & x_5(t) \end{bmatrix}^\mathrm{T}$,

which gives:

$$\mathbf{G}_{\bullet,\mathbf{a}} = \begin{bmatrix} 0 & 0 \\ 0 & 0 \\ +1 & +2 \end{bmatrix}. \tag{65}$$

Observable extents and extent directions

The factorization of $\mathbf{G}_{\bullet,\mathbf{a}}$ gives:

$$\mathcal{G}_\mathrm{o} = \begin{bmatrix} 0 \\ 0 \\ +1 \end{bmatrix}, \qquad \mathbf{V}_\mathrm{o}^\mathrm{T} = \begin{bmatrix} +1 & +2 \end{bmatrix}, \tag{66}$$

so that

$$\chi_\mathrm{o}(t) = \mathbf{V}_\mathrm{o}^\mathrm{T}\,\mathbf{I}_{\mathbf{a},\bullet}\,\boldsymbol{x}(t) = \begin{bmatrix} 0 & 0 & 0 & 1 & 2 \end{bmatrix} \boldsymbol{x}(t) = x_4(t) + 2\,x_5(t) \tag{67}$$

$$\overline{\chi}_\mathrm{o}(t) = \begin{bmatrix} x_1(t) \\ x_3(t) \\ \chi_\mathrm{o}(t) \end{bmatrix} = \begin{bmatrix} x_1(t) \\ x_3(t) \\ x_4(t) + 2\,x_5(t) \end{bmatrix}. \tag{68}$$

The values of the observable extents and extent directions $\tilde{\overline{\chi}}_{\mathrm{o},h} := \begin{bmatrix} \tilde{x}_{1,h}\; \tilde{x}_{3,h}\; \tilde{\chi}_{\mathrm{o},h} \end{bmatrix}^\mathrm{T}$ can be computed from (23) with

$$\mathbf{P} = \begin{bmatrix} -2 & -1 & 0 \\ -1 & -1 & 0 \\ 0 & 0 & +1 \end{bmatrix}. \tag{69}$$

The expected estimation error variance-covariance matrix $\boldsymbol{\Sigma}_{\overline{\chi}}$ is:

$$\boldsymbol{\Sigma}_{\overline{\chi}} = \begin{bmatrix} +5 & +3 & 0 \\ +3 & +2 & 0 \\ 0 & 0 & +2 \end{bmatrix} 10^{-4}. \tag{70}$$

This demonstrates that the estimation errors in the first and second observable extents are uncorrelated with the estimation error in the computed observable extent direction. Please note that the correlation between the estimation errors of the first and second observable extents is fairly high ($\frac{3}{\sqrt{5\cdot 2}} = 0.95$). Figure 3 shows the computed values of the observable extents and extent direction.

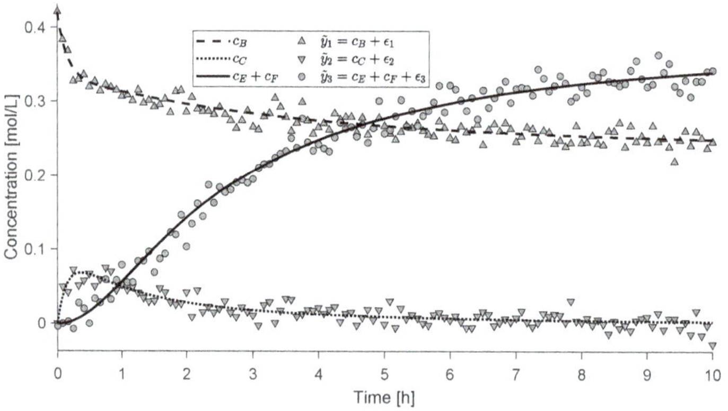

Figure 2. Scenario A—Simulation. Noise-free and noisy measurements as a function of time.

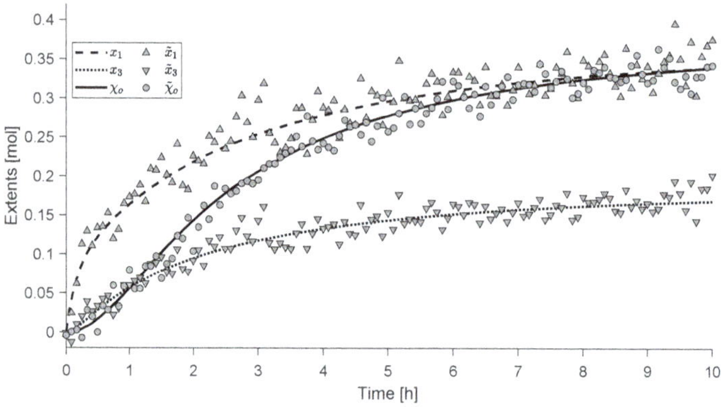

Figure 3. Scenario A—Extent computation. Observable extents (x_1, x_3) and observable extent direction (χ_o) and their computed equivalents ($\tilde{x}_1$, $\tilde{x}_3$, $\tilde{\chi}_o$).

Expression (27) is then fully specified with the following vectors and matrices:

$$\boldsymbol{n}_0 = \begin{bmatrix} 1 \\ 1 \\ 0 \\ 0 \\ 0 \\ 0 \end{bmatrix} \tag{71}$$

$$\overline{\boldsymbol{\mathcal{N}}}_{\mathrm{o}} := \begin{bmatrix} -1 & -1 & +1 & 0 & 0 & 0 \\ 0 & +1 & -2 & +1 & 0 & 0 \\ 0 & 0 & 0 & -1 & \frac{3}{5} & \frac{2}{5} \end{bmatrix}. \tag{72}$$

$$\overline{\boldsymbol{\mathcal{N}}}_{\mathrm{u}} := \begin{bmatrix} 0 & 0 & 0 & 0 & 0 & 0 \\ -2 & 0 & 0 & +1 & 0 & 0 \\ 0 & 0 & 0 & 0 & 0 & 0 \\ 0 & 0 & 0 & 0 & +\frac{2}{5} & -\frac{2}{5} \\ 0 & 0 & 0 & 0 & -\frac{1}{5} & +\frac{1}{5} \end{bmatrix} \tag{73}$$

Since the 2nd and 3rd column of $\boldsymbol{\mathcal{N}}_{\mathrm{u}}$ are null vectors, it follows that values of $\tilde{n}_2(t)$ and $\tilde{n}_3(t)$ can be computed based on the interpolated estimates $\tilde{\boldsymbol{\chi}}_{\mathrm{o,i}}(t)$ without the need to simulate $\boldsymbol{x}(t)$. A further exploration of this idea remains out of the scope of this paper, however.

System partitioning

For simplicity of notation, we omit the time dependence in what follows. For example, the interpolants $\tilde{\boldsymbol{\chi}}_{\mathrm{o,i}}(t)$ are given as $\tilde{\boldsymbol{\chi}}_{\mathrm{o,i}} := \begin{bmatrix} \tilde{x}_1 & \tilde{x}_3 & \tilde{\chi}_{\mathrm{o}} \end{bmatrix}^{\mathrm{T}}$. Following Steps 1(a)–(b) in Section 3.6.1, the augmented equation system becomes:

$$\dot{\boldsymbol{x}} = \begin{bmatrix} \frac{k_1}{V^2}\left[(n_{0,1} - \tilde{x}_1 - 2x_2)\,(n_{0,2} - \tilde{x}_1 + \tilde{x}_3) - K_1\,(n_{0,3} + \tilde{x}_1 - 2\tilde{x}_3)\right] \\ \frac{k_2}{V^2}\,(n_{0,1} - \tilde{x}_1 - 2x_2)^2 \\ \frac{k_3}{V}\,(n_{0,3} + \tilde{x}_1 - 2\tilde{x}_3) \\ \frac{k_4}{V}\,(n_{0,4} + \tilde{x}_3 - \tilde{\chi}_{\mathrm{o}} + x_2) \\ \frac{k_5}{V^2}\,(n_{0,4} + \tilde{x}_3 - \tilde{\chi}_{\mathrm{o}} + x_2)^2 \end{bmatrix}, \qquad \boldsymbol{x}(0) = \boldsymbol{0} \tag{74}$$

$$\chi_{\mathrm{o}} = \mathbf{V}_{\mathrm{o}}^{\mathrm{T}}\,\mathbf{I}_{\mathbf{a},\bullet}\,\boldsymbol{x} = \begin{bmatrix} 0 & 0 & 0 & 1 & 2 \end{bmatrix} \boldsymbol{x} = x_4 + 2\,x_5\,. \tag{75}$$

Figure 4 shows the graph corresponding to the above equation system. The vertices corresponding to the observable extents and extent direction are shaded, while the other vertices are white. The simulation arcs are shown with full-line arrows, while the observation arcs are shown as dashed-line arrows. To identify possible subsystems, one removes all the observation arcs, which results in two subgraphs. The first subgraph includes the parameters k_1, k_2, k_4, k_5, and K_1, which affect the observable quantities x_1 and χ_{o} via a network that also involves the unobservable extents x_2, x_4, and x_5. The second subgraph is much smaller and includes the parameter k_3 that influences the observable extent x_3.

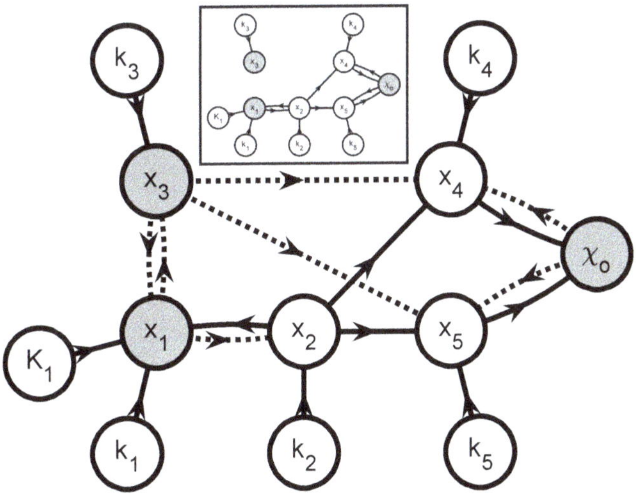

Figure 4. Scenario A—Graph $\mathcal{F}$. There are three shaded vertices corresponding the observable extents (x_1, x_3) and extent direction (χ_{o}). The remaining vertices represent the unobservable extents (x_2, x_4, x_5) and the parameters (k_1, k_2, k_3, k_4, k_5, K_1). The simulation and observation arcs are shown as solid-line and dashed-line arrows, respectively. Removing the observation arcs and graph partitioning results in 2 subgraphs for the parameters as shown in the inset: one graph in which k_1, k_2, k_4, k_5, and K_1 are connected to the observable quantities x_1 and χ_{o}, and another graph in which k_3 is connected to the observable extent x_3.

Accordingly, the $J = 2$ subsystems are:

$$\dot{x}^{(1)} = \begin{bmatrix} \dot{x}_1 \\ \\ \dot{x}_2 \\ \dot{x}_4 \\ \dot{x}_5 \end{bmatrix} = \begin{bmatrix} \frac{k_1}{V^2} \left[(n_{0,1} - x_1 - 2x_2)\, (n_{0,2} - x_1 + \tilde{x}_3) \right. \\ \left. - K_1\, (n_{0,3} + x_1 - 2\tilde{x}_3) \right] \\ \frac{k_2}{V^2}\, (n_{0,1} - x_1 - 2x_2)^2 \\ \frac{k_4}{V}\, (n_{0,4} + \tilde{x}_3 - \chi_o + x_2) \\ \frac{k_5}{V^2}\, (n_{0,4} + \tilde{x}_3 - \chi_o + x_2)^2 \end{bmatrix}, \quad x^{(1)}(0) = \begin{bmatrix} x_{0,1} \\ \\ x_{0,2} \\ x_{0,4} \\ x_{0,5} \end{bmatrix} = \mathbf{0} \tag{76}$$

$$\chi_o^{(1)} = \chi_o = x_4 + 2\,x_5, \tag{77}$$

and

$$\dot{x}^{(2)} = \dot{x}_3 = \frac{k_3}{V}\, (n_{0,3} + \tilde{x}_1 - 2x_3)\,, \qquad x^{(2)}(0) = x_{0,3} = 0\,. \tag{78}$$

The information flows in these subsystems are shown in the inset of Figure 4.

Incremental parameter estimation

The resulting partitioning means that estimates of the first set of parameters $\theta^{(1)} := \left[\, k_1, k_2, k_4, k_5, K_1 \,\right]$ can be obtained by minimizing a weighted least-squares deviation between the predicted $x_1(t_h)$ and $\chi_o(t_h)$ and their measured counterparts $\tilde{x}_{1,h}$ and $\tilde{\chi}_{o,h}$, that is, by solving problem $P_1^{(1)}$:

$$\hat{\theta}^{(1)} = \arg\min_{\theta^{(1)}} \sum_h^H d_h^{(1)\,\mathrm{T}}\, \mathbf{W}^{(1)}\, d_h^{(1)} \tag{79}$$

subject to

$$\begin{bmatrix} \dot{x}_1(t) \\ \\ \dot{x}_2(t) \\ \dot{x}_4(t) \\ \dot{x}_5(t) \end{bmatrix} = \begin{bmatrix} \frac{k_1}{V^2} \left[(n_{0,1} - x_1(t) - 2x_2(t))\, (n_{0,2} - x_1(t) + \tilde{x}_3(t)) \right. \\ \left. - K_1\, (n_{0,3} + x_1(t) - 2\tilde{x}_3(t)) \right] \\ \frac{k_2}{V^2}\, (n_{0,1} - x_1(t) - 2x_2(t))^2 \\ \frac{k_4}{V}\, (n_{0,4} + \tilde{x}_3(t) - \chi_o(t) + x_2(t)) \\ \frac{k_5}{V^2}\, (n_{0,4} + \tilde{x}_3(t) - \chi_o(t) + x_2(t))^2 \end{bmatrix}, \quad \begin{bmatrix} x_1(0) \\ \\ x_2(0) \\ x_4(0) \\ x_5(0) \end{bmatrix} = \mathbf{0} \tag{80}$$

$$\chi_o(t) = x_4(t) + 2\,x_5(t) \tag{81}$$

$$d_h^{(1)} = \begin{bmatrix} \tilde{x}_{1,h} \\ \tilde{\chi}_{o,h} \end{bmatrix} - \begin{bmatrix} x_1(t_h) \\ \chi_o(t_h) \end{bmatrix}, \tag{82}$$

where

$$\mathbf{W}^{(1)} := \mathbf{L}_e^{(1)}\, \boldsymbol{\Sigma}_{\tilde{x}}^{-1}\, \mathbf{L}_e^{(1)\,\mathrm{T}} = \begin{bmatrix} +1 & 0 & 0 \\ 0 & 0 & +1 \end{bmatrix} \left(\begin{bmatrix} 2 & -3 & 0 \\ -3 & 5 & 0 \\ 0 & 0 & 1/2 \end{bmatrix} 10^4 \right) \begin{bmatrix} 1 & 0 & 0 \\ 0 & 0 & 1 \end{bmatrix}^{\mathrm{T}} \tag{83}$$

$$= \begin{bmatrix} 2 & 0 \\ 0 & 1/2 \end{bmatrix} 10^4. \tag{84}$$

The second set of parameters consists of k_3 only, $\theta^{(2)} := k_3$. Its value can be obtained by minimizing the least-squares deviation between $x_3(t_h)$ and its measured counterparts $\tilde{x}_{3,h}$, that is, by solving problem $P_1^{(2)}$:

$$\hat{\theta}^{(2)} = \arg\min_{\theta^{(2)}} \sum_{h}^{H} \left(d_h^{(2)}\right)^2 \tag{85}$$

subject to

$$\dot{x}_3(t) = \frac{k_3}{V}\left(n_{0,3} + \tilde{x}_1(t) - 2x_3(t)\right), \qquad x_3(0) = 0 \tag{86}$$

$$d_h^{(2)} = \tilde{x}_{3,h} - x_3(t_h). \tag{87}$$

Figure 5 shows the measured concentrations and the simulated profiles obtained with (i) the true parameters; (ii) simultaneous parameter estimation (P_0); and (iii) incremental parameter estimation ($P_1^{(1)}$ and $P_1^{(2)}$). All simulated profiles describe the experiment well and are practically indistinguishable. The parameter values obtained by solving P_0, $P_1^{(1)}$, $P_1^{(2)}$, and P_{1+0} are shown in Table 2. All parameter estimates are in close agreement to each other. It is worth noting that the parameter k_2 can be estimated even if the corresponding extent x_2 is labeled non-sensed, i.e., our results suggest that this parameter is both structurally and practically identifiable.

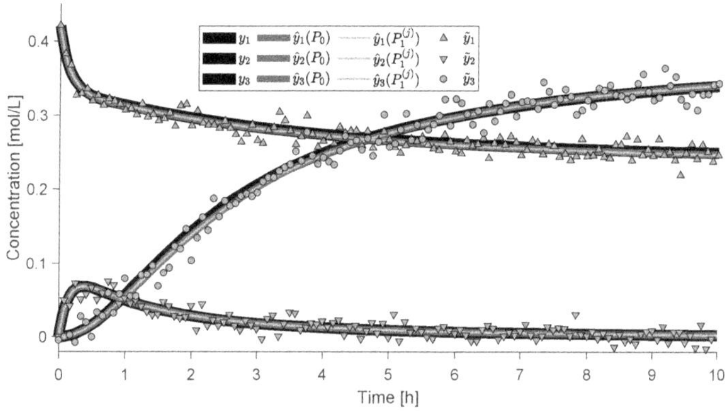

Figure 5. Scenario A—Parameter estimation. Measured concentrations and simulated profiles obtained with (i) true parameters; (ii) simultaneous parameter estimation (P_0); and (iii) incremental parameter estimation ($P_1^{(1)}$ and $P_1^{(2)}$). The produced simulation results exhibit strong overlap, meaning that the identified models closely approximate the ground truth model.

4.1.4. Scenario B

Scenario B assumes concentration measurements for the species B and C only:

$$\mathbf{M} := \begin{bmatrix} 0 & 1 & 0 & 0 & 0 & 0 \\ 0 & 0 & 1 & 0 & 0 & 0 \end{bmatrix}. \tag{88}$$

The measurement error-covariance matrix is $\mathbf{\Sigma}_\epsilon = \mathrm{diag}\left(\begin{bmatrix} 1 & 1 \end{bmatrix}^{\mathrm{T}}\right) 10^{-4}\,\mathrm{mol}^2\mathrm{L}^{-2}$. The reduced row echelon form of $\mathbf{G}$ is:

$$\mathbf{B} = \begin{bmatrix} +1 & 0 & 0 & 0 & 0 \\ 0 & 0 & +1 & 0 & 0 \end{bmatrix}, \tag{89}$$

which leads to the following extent labeling:

- $R_n = 3$ non-sensed extents, $\boldsymbol{x}_\mathbf{n}(t) = \begin{bmatrix} x_2(t) & x_4(t) & x_5(t) \end{bmatrix}^T$,
- $R_o = 2$ observable extents, $\boldsymbol{x}_\mathbf{o}(t) = \begin{bmatrix} x_1(t) & x_3(t) \end{bmatrix}^T$,
- $R_a = 0$ ambiguous extents.

Please note that the extent-based method proposed in [8] applies in this case since $R_a = 0$.

Figure 6 shows the corresponding graph $\mathcal{F}$, which can be partitioned into $J = 2$ subsystems and a leftover part. The first subgraph includes the parameters k_1, k_2, and K_1 since they all have some effect on the observable extent x_1 via a network that also involves the unobservable extent x_2. The second subgraph includes the parameter k_3 that influences the observable extent x_3. The information flows in these subsystems are shown in the inset of Figure 6. The leftover part includes the parameters k_4 and k_5 and the extents x_4 and x_5. These parameters and variables are not part of any of the identified subsystems because there are no directed paths composed of simulation arcs from the corresponding vertices to any of the observable extents. Hence, it follows that these parameters are unidentifiable. This is consistent with the method proposed in [8] (not shown). Clearly, the graph-based partitioning procedure shows potential for parameter identifiability analysis, which is discussed as an opportunity below.

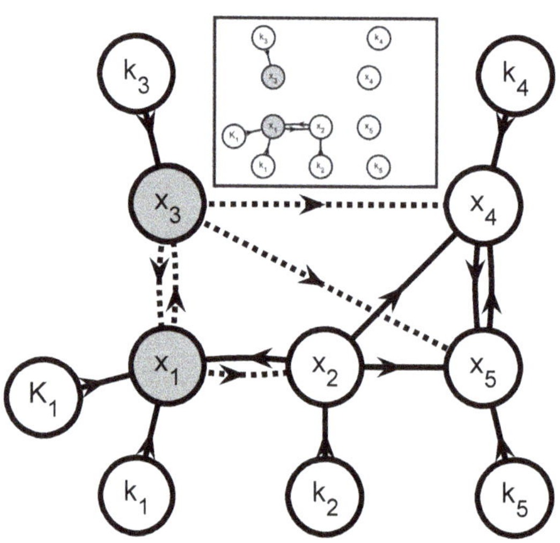

Figure 6. Scenario B—Graph $\mathcal{F}$. There are two shaded vertices corresponding the observable extents (x_1, x_3). Removing the observation arcs and graph partitioning results in 2 subgraphs for the parameters as shown in the inset: one graph in which k_1, k_2, and K_1 affect the observable extent x_1, a second graph in which k_3 affects the observable extent x_3. The vertices k_4 and k_5 have no effect on the observable extents x_1 and x_3.

4.1.5. Scenario C

Scenario C assumes concentration measurements for the species B, C, E, and F:

$$\mathbf{M} := \begin{bmatrix} 0 & 1 & 0 & 0 & 0 & 0 \\ 0 & 0 & 1 & 0 & 0 & 0 \\ 0 & 0 & 0 & 0 & 1 & 0 \\ 0 & 0 & 0 & 0 & 0 & 1 \end{bmatrix}. \quad (90)$$

The measurement error-covariance matrix is $\mathbf{\Sigma}_\epsilon = \text{diag}\left(\begin{bmatrix} 1 & 1 & 1 & 1 \end{bmatrix}^\text{T}\right) 10^{-4}\,\text{mol}^2\text{L}^{-2}$. The reduced row echelon form of $\mathbf{G}$ is:

$$\mathbf{B} = \begin{bmatrix} +1 & 0 & 0 & 0 & 0 \\ 0 & 0 & +1 & 0 & 0 \\ 0 & 0 & 0 & +1 & 0 \\ 0 & 0 & 0 & 0 & +1 \end{bmatrix}, \tag{91}$$

which leads to the following extent labeling:

- $R_\text{n} = 1$ non-sensed extents, $\mathbf{x_n}(t) = x_2(t)$.
- $R_\text{o} = 4$ observable extents, $\mathbf{x_o}(t) = \begin{bmatrix} x_1(t) & x_3(t) & x_4(t) & x_6(t) \end{bmatrix}^\text{T}$.
- $R_\text{a} = 0$ ambiguous extents.

Figure 7 shows the corresponding graph $\mathcal{F}$, which can be partitioned into 2 subsystems. The first subgraph includes the parameters k_1, k_2, k_4, k_5, and K_1 since they all have some effect on the observable extents x_1, x_4, and x_5 via a network that also involves the unobservable extents x_2. The second subgraph includes the parameter k_3 that influences the observable extent x_3. The information flows in these subsystems are shown in the inset of Figure 7. Although the assignment of the parameters is the same as in scenario A, the parameter estimation problems are different since the fit of subsystem 1 is determined with respect to $\tilde{x}_1$, $\tilde{x}_4$, and $\tilde{x}_5$ in scenario C, compared to only $\tilde{x}_1$ and $\tilde{\chi}_\text{o}$ in scenario A.

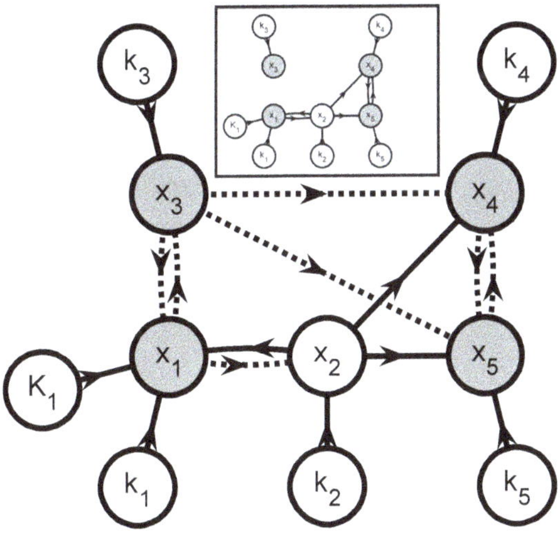

Figure 7. Scenario C—Graph $\mathcal{F}$. There are 4 shaded vertices corresponding to the observable extents x_1, x_3, x_4, and x_5. Removing the observation arcs results in 2 subgraphs for the parameters as shown in the inset: one graph in which k_1, k_2, k_4, k_5, and K_1 affect the observable extents x_1, x_4, and x_5, and another graph in which k_3 affect the observable extent x_3.

4.1.6. Scenario D

Scenario D assumes concentration measurements for the species A, C, and E:

$$\mathbf{M} := \begin{bmatrix} 1 & 0 & 0 & 0 & 0 & 0 \\ 0 & 0 & 1 & 0 & 0 & 0 \\ 0 & 0 & 0 & 0 & 1 & 0 \end{bmatrix}. \tag{92}$$

The measurement error-covariance matrix is $\mathbf{\Sigma}_\epsilon = \mathsf{diag}\left(\left[\begin{array}{ccc} 1 & 1 & 1 \end{array}\right]^{\mathrm{T}}\right) 10^{-4}\,\mathrm{mol}^2\mathrm{L}^{-2}$. The reduced row echelon form of $\mathbf{G}$ is:

$$\mathbf{B} = \begin{bmatrix} +1 & 0 & -2 & 0 & 0 \\ 0 & +1 & +1 & 0 & 0 \\ 0 & 0 & 0 & +1 & +1 \\ 0 & 0 & 0 & 0 & 0 \end{bmatrix}, \tag{93}$$

which leads to the following extent labeling:

- $R_{\mathsf{n}} = 0$ non-sensed extents,
- $R_{\mathsf{o}} = 0$ observable extents,
- $R_{\mathsf{a}} = 5$ ambiguous extents, $\mathbf{x}_{\mathbf{a}}(t) = \left[\begin{array}{ccccc} x_1(t) & x_2(t) & x_3(t) & x_4(t) & x_6(t) \end{array}\right]^{\mathrm{T}}$.

From $A = \mathsf{rank}\,(\mathbf{B}) = 3$ and $R_{\mathsf{o}} = 0$, one sees that there are $\rho_{\mathsf{o}} = A - R_{\mathsf{o}} = 3$ observable extent directions among the 5 ambiguous extents. The first two are linear combinations of x_1, x_2, and x_3, while the third one is a linear combination of x_4 and x_5. This particular appearance of extents in subsets of the observable directions stems from the subspace clustering property of the reduced row echelon form and will be discussed in some more detail below.

Figure 8 shows the corresponding graph $\mathcal{F}$, which can be partitioned into 2 subsystems. The first subgraph includes the parameters k_1, k_2, k_3, and K_1 since they all have some effect on the observable extent directions $\chi_{\mathsf{o},1}$ and $\chi_{\mathsf{o},2}$ via a network that also involves the unobservable extents x_1, x_2, and x_3. The second subgraph includes the parameters k_4 and k_5 that influence the observable extent direction $\chi_{\mathsf{o},3}$ via a network that also involves the unobservable extents x_4 and x_5. Hence, this leads to two parameter estimation problems involving 4 and 2 parameters, respectively. The information flows to simulate each of the corresponding subsystems are shown in the inset of Figure 8.

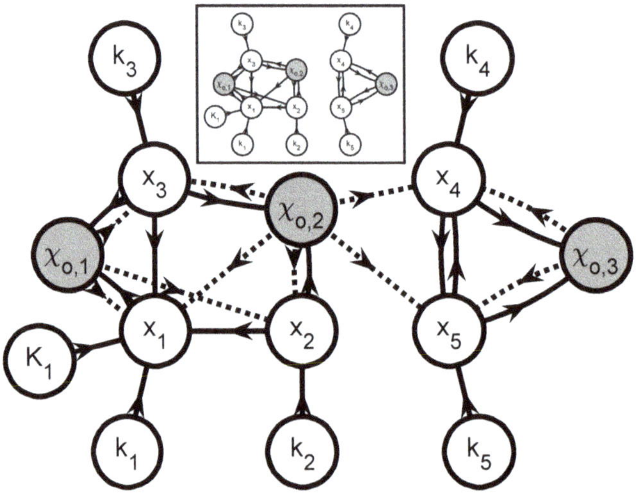

Figure 8. Scenario D—Graph $\mathcal{F}$. There are 3 shaded vertices corresponding the observable extent directions $\chi_{\mathsf{o},1}$, $\chi_{\mathsf{o},2}$, and $\chi_{\mathsf{o},3}$. Removing the observation arcs results in 2 subgraphs for the parameters as shown in the inset: one graph in which k_1, k_2, k_3, and K_1 affect the observable extent directions $\chi_{\mathsf{o},1}$ and $\chi_{\mathsf{o},2}$, and another graph in which k_4 and k_5 affect the observable extent direction $\chi_{\mathsf{o},3}$.

4.1.7. Scenario E

Scenario E assumes concentration measurements of all species:

$$\mathbf{M} := \mathbf{I}_6 . \tag{94}$$

The measurement error-covariance matrix is $\mathbf{\Sigma}_\epsilon = \text{diag}\left(\begin{bmatrix} 1 & 1 & 1 & 1 & 1 & 1 \end{bmatrix}^T\right) 10^{-4}\,\text{mol}^2\text{L}^{-2}$. The reduced row echelon form of $\mathbf{G}$ is:

$$\mathbf{B} = \begin{bmatrix} +1 & 0 & 0 & 0 & 0 \\ 0 & +1 & 0 & 0 & 0 \\ 0 & 0 & +1 & 0 & 0 \\ 0 & 0 & 0 & +1 & 0 \\ 0 & 0 & 0 & 0 & +1 \\ 0 & 0 & 0 & 0 & 0 \end{bmatrix}, \tag{95}$$

which leads to the following extent labeling:

- $R_n = 0$ non-sensed extents,
- $R_o = 5$ observable extents, $\mathbf{x_o}(t) = \begin{bmatrix} x_1(t) & x_2(t) & x_3(t) & x_4(t) & x_6(t) \end{bmatrix}^T$,
- $R_a = 0$ ambiguous extents.

This means that the original framework for extent computation, which assumes $A = R$ (Section 3.2.3), applies.

Figure 9 shows the graph $\mathcal{F}$, which can be partitioned into 5 subsystems that include the parameters as follows: (i) k_1 and K_1; (ii) k_2; (iii) k_3; (iv) k_4; and (v) k_5. Hence, one obtains 5 entirely decoupled parameter estimation problems. This set of optimization problems is the same as those obtained with the original extent-based model identification method as is also clear from the inset of Figure 9.

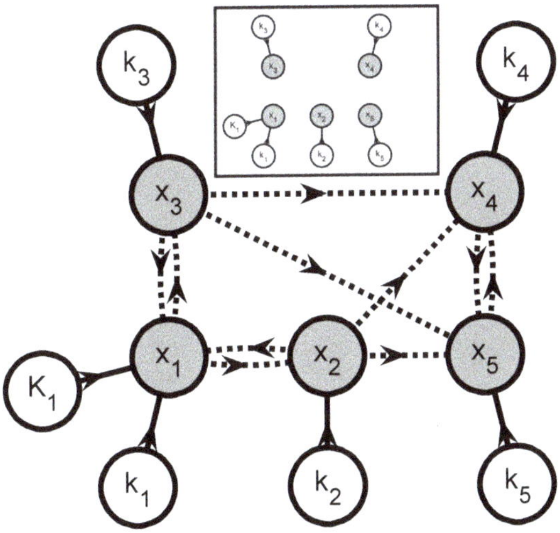

Figure 9. Scenario E—Graph $\mathcal{F}$. All extent vertices are observable and thus shaded. Removing the observation arcs results in 5 subgraphs for the parameters as shown in the inset: k_1 and K_1 are estimated from x_1, k_2 from x_2, k_3 from x_3, k_4 from x_4, and k_5 from x_5.

4.2. Experimental Study

To explore the applicability of the novel incremental parameter estimation method with realistic laboratory data, we execute parameter estimation for the α-pinene batch experiment first reported in [11] and later studied in [12–16]. The reaction network consists of 5 species. These species are labeled A, B, C, D, and E and correspond to α-pinene, dipentene, allo-ocimene, pyronene, and dimer. The reaction network is:

$$\begin{aligned} R_1 &: \; \mathrm{A} \longrightarrow \mathrm{B} \\ R_2 &: \; \mathrm{A} \longrightarrow \mathrm{C} \\ R_3 &: \; \mathrm{C} \longrightarrow \mathrm{D} \\ R_4 &: \; \mathrm{C} \longrightarrow \mathrm{E} \\ R_5 &: \; \mathrm{E} \longrightarrow \mathrm{C} \end{aligned}$$

As in [12], the model is defined with the following stoichiometric matrix and process rates:

$$\mathbf{N} := \begin{bmatrix} -1 & +1 & 0 & 0 & 0 \\ -1 & 0 & +1 & 0 & 0 \\ 0 & 0 & -1 & +1 & 0 \\ 0 & 0 & -1 & 0 & +1 \\ 0 & 0 & +1 & 0 & -1 \end{bmatrix} \tag{96}$$

$$\boldsymbol{r} := \begin{bmatrix} k_1 \frac{n_1(t)}{V} \\ k_2 \frac{n_1(t)}{V} \\ k_3 \frac{n_3(t)}{V} \\ k_4 \frac{n_3(t)}{V} \\ k_5 \frac{n_5(t)}{V} \end{bmatrix} . \tag{97}$$

At the start of the batch experiment, only A is present. As explained in [12], the concentrations of all species but D are measured. The concentrations of D are obtained by assuming that 3% of the transformed A is present as D at all times. Finally, all concentrations are normalized to 1 (100%). Figure 10 shows the experimental data reported in [12]. As the initial numbers of moles and the volume are unknown, we express both the concentrations and the extents as dimensionless fractions of the numbers of moles relative to the initial numbers of moles of A. Hence, we can write $\boldsymbol{n}_0 := \begin{bmatrix} 1 & 0 & 0 & 0 & 0 \end{bmatrix}^{\mathrm{T}}$.

<u>Extent labeling</u>

For extent labeling, we assume that all species are measured with independent zero-mean Gaussian measurement errors, similar to the least-squares approach of [12]:

$$\mathbf{M} := \mathbf{I}_5 . \tag{98}$$

It follows that

$$\mathbf{G} = \mathbf{M}\,\mathbf{N}^{\mathrm{T}} = \begin{bmatrix} -1 & -1 & 0 & 0 & 0 \\ 1 & 0 & 0 & 0 & 0 \\ 0 & 1 & -1 & -1 & 1 \\ 0 & 0 & 1 & 0 & 0 \\ 0 & 0 & 0 & 1 & -1 \end{bmatrix} . \tag{99}$$

The reduced row echelon form of $\mathbf{G}$ is

$$\mathbf{B} = \begin{bmatrix} +1 & 0 & 0 & 0 & 0 \\ 0 & +1 & 0 & 0 & 0 \\ 0 & 0 & +1 & 0 & 0 \\ 0 & 0 & 0 & +1 & -1 \end{bmatrix}, \tag{100}$$

which indicates that there are:

- $R_\mathrm{n} = 0$ non-sensed extents,
- $R_\mathrm{o} = 3$ observable extents, $\boldsymbol{x}_\mathbf{o}(t) = \begin{bmatrix} x_1(t) & x_2(t) & x_3(t) \end{bmatrix}^\mathrm{T}$,
- $R_\mathrm{a} = 2$ ambiguous extents, $\boldsymbol{x}_\mathbf{a}(t) = \begin{bmatrix} x_4(t) & x_5(t) \end{bmatrix}^\mathrm{T}$,

which delivers:

$$\mathbf{G}_{\bullet,\mathbf{a}} = \begin{bmatrix} 0 & 0 \\ 0 & 0 \\ -1 & +1 \\ 0 & 0 \\ +1 & -1 \end{bmatrix}. \tag{101}$$

<u>Observable extents and extent directions</u>

The factorization of $\mathbf{G}_{\bullet,\mathbf{a}}$ gives:

$$\mathcal{G}_\mathrm{o} = \begin{bmatrix} 0 \\ 0 \\ 0 \\ +1 \end{bmatrix}, \qquad \mathbf{V}_\mathrm{o}^\mathrm{T} = \begin{bmatrix} +1 & -1 \end{bmatrix}, \tag{102}$$

so that

$$\chi_\mathrm{o}(t) = \mathbf{V}_\mathrm{o}^\mathrm{T}\,\mathbf{I}_{\mathbf{a},\bullet}\,\boldsymbol{x}(t) = \begin{bmatrix} 0 & 0 & 0 & 1 & -1 \end{bmatrix} \boldsymbol{x}(t) = x_4(t) - x_5(t) \tag{103}$$

$$\overline{\chi}_\mathrm{o}(t) = \begin{bmatrix} x_1(t) \\ x_2(t) \\ x_3(t) \\ \chi_\mathrm{o}(t) \end{bmatrix} = \begin{bmatrix} x_1(t) \\ x_2(t) \\ x_3(t) \\ x_4(t) - x_5(t) \end{bmatrix}. \tag{104}$$

Figure 11 shows the computed values of the observable extents and extent direction assuming these definitions. The corresponding error variance-covariance matrix is:

$$\mathbf{\Sigma}_{\overline{\chi}} = \begin{bmatrix} +0.8 & -0.6 & -0.2 & -0.2 \\ -0.6 & +1.2 & +0.4 & +0.4 \\ -0.2 & +0.4 & +0.8 & -0.2 \\ -0.2 & +0.4 & -0.2 & +0.8 \end{bmatrix} \sigma \tag{105}$$

with σ the (unknown) measurement error variance.

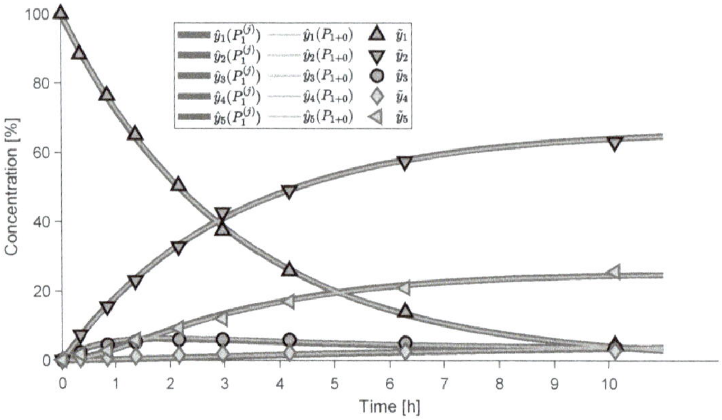

Figure 10. α-pinene—Parameter estimation. Measured concentrations and simulated profiles obtained with (i) incremental parameter estimation ($P_1^{(j)}, j = 1, \ldots, 4$) followed by (ii) simultaneous parameter estimation (P_{1+0}). The produced simulation results exhibit strong overlap, meaning that both models produce very similar results.

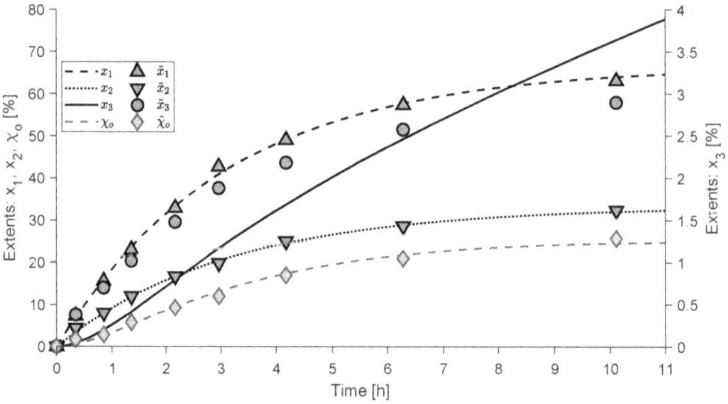

Figure 11. α-pinene—Extents. Computed extents and simulated profiles obtained following incremental parameter estimation ($P_1^{(j)}, j = 1, \ldots, 4$).

System partitioning

As above, we omit the time dependence in what follows. Following Steps 1(a)–(b) in Section 3.6.1, the augmented equation system becomes:

$$\dot{x} = \begin{bmatrix} k_1 \left(n_{0,1} - \tilde{x}_1 - 2\tilde{x}_2\right) \\ k_2 \left(n_{0,1} - \tilde{x}_1 - 2\tilde{x}_2\right) \\ k_3 \left(n_{0,3} + \tilde{x}_2 - \tilde{x}_3 - \tilde{\chi}_\text{o}\right) \\ k_4 \left(n_{0,4} + \tilde{x}_3\right) \\ k_5 \left(n_{0,5} + \tilde{\chi}_\text{o}\right) \end{bmatrix}, \qquad x(0) = \mathbf{0}, \tag{106}$$

$$\chi_\text{o} = x_4 - x_5 \,. \tag{107}$$

Figure 12 shows the graph corresponding to the above equation system. Removing all the observation arcs generates 4 subgraphs. The first three subgraphs represent the first three reactions and include: k_1 and x_1 $(j = 1)$; k_2 and x_2 $(j = 2)$; and k_3 and x_3 $(j = 3)$. The fourth graph includes the parameters k_4 and k_5 via a network that also involves the observable extent direction χ_o and the unobservable extents x_4 and x_5 ($j = 4$).

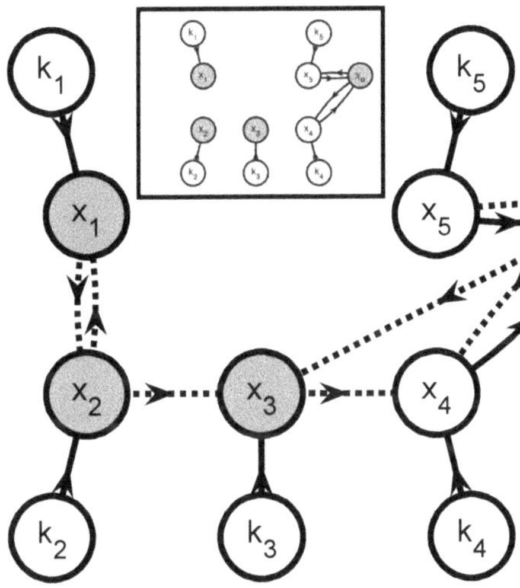

Figure 12. α-pinene—Graph $\mathcal{F}$. There are four shaded vertices corresponding the observable extents (x_1, x_2, x_3) and extent direction (χ_o). The remaining vertices represent the unobservable extents (x_4, x_5) and the parameters (k_1, k_2, k_3, k_4, k_5). The simulation and observation arcs are shown as solid-line and dashed-line arrows, respectively. Removing the observation arcs and graph partitioning results in 4 subgraphs for the parameters as shown in the inset: (i) one with vertices for k_1 and x_1; (ii) one with vertices for k_2 and x_2; (iii) one with vertices for k_3 and x_3; and (iv) one with vertices k_4, k_5, x_4, x_5, and χ_o.

Accordingly, the $J = 4$ subsystems are:

$$\dot{x}^{(1)} = \dot{x}_1 = k_1\,(n_{0,1} - x_1 - 2\tilde{x}_2)\,, \qquad x^{(1)}(0) = x_{0,1} = 0\,, \tag{108}$$

$$\dot{x}^{(2)} = \dot{x}_2 = k_2\,(n_{0,1} - \tilde{x}_1 - 2x_2)\,, \qquad x^{(2)}(0) = x_{0,2} = 0\,, \tag{109}$$

$$\dot{x}^{(3)} = \dot{x}_3 = k_3\,(n_{0,3} + \tilde{x}_2 - x_3 - \tilde{\chi}_o)\,, \qquad x^{(3)}(0) = x_{0,3} = 0\,, \tag{110}$$

$$\dot{x}^{(4)} = \begin{bmatrix} \dot{x}_4 \\ \dot{x}_5 \end{bmatrix} = \begin{bmatrix} k_4\,(n_{0,3} + \tilde{x}_2 - \tilde{x}_3 - \tilde{\chi}_o) \\ k_5\,(n_{0,5} + \tilde{\chi}_o) \end{bmatrix}, \qquad x^{(4)}(0) = \begin{bmatrix} x_{0,4} \\ x_{0,5} \end{bmatrix} = \mathbf{0}\,, \tag{111}$$

$$\chi_o^{(4)} = \chi_o = x_4 - x_5, \tag{112}$$

The information flows in these subsystems are shown in the inset of Figure 12.

Incremental parameter estimation

The resulting partitioning means that estimates of the k_1, k_2, and k_3 can be obtained by solving the following three single-parameter estimation problems ($P_1^{(1)}$, $P_1^{(2)}$, $P_1^{(3)}$):

$$P_1^{(1)}: \quad \hat{k}_1 = \arg\min_{k_1} \sum_{h}^{H} (\tilde{x}_{1,h} - x_1(t_h))^{\mathrm{T}}\, W^{(1)}\, (\tilde{x}_{1,h} - x_1(t_h)) \tag{113}$$

subject to

$$\dot{x}^{(1)} = \dot{x}_1 = k_1\,(n_{0,1} - x_1 - 2\tilde{x}_2)\,, \qquad x^{(1)}(0) = x_{0,1} = 0 \tag{114}$$

where

$$W^{(1)} := 5/4 \tag{115}$$

$$P_1^{(2)}: \quad \hat{k}_2 = \arg\min_{k_2} \sum_h^H (\tilde{x}_{2,h} - x_2(t_h))^{\mathrm{T}} W^{(2)} \; (\tilde{x}_{2,h} - x_2(t_h)) \tag{116}$$

subject to

$$\dot{x}^{(2)} = \dot{x}_2 = k_2 \; (n_{0,1} - \tilde{x}_1 - 2x_2) \,, \qquad x^{(2)}(0) = x_{0,2} = 0 \tag{117}$$

where

$$W^{(2)} := 5/3 \tag{118}$$

$$P_1^{(3)}: \quad \hat{k}_3 = \arg\min_{k_3} \sum_h^H (\tilde{x}_{3,h} - x_3(t_h))^{\mathrm{T}} W^{(3)} \; (\tilde{x}_{3,h} - x_3(t_h)) \tag{119}$$

subject to

$$\dot{x}^{(3)} = \dot{x}_3 = k_3 \; (n_{0,3} + \tilde{x}_2 - x_3 - \tilde{\chi}_{\mathrm{o}}) \,, \qquad x^{(3)}(0) = x_{0,3} = 0 \tag{120}$$

where

$$W^{(3)} := 5/4 \tag{121}$$

The fourth set of parameters, k_4 and k_5, are estimated by solving the following problem:

$$P_1^{(4)}: \quad \hat{k}_3, \hat{k}_4 = \arg\min_{k_3,k_4} \sum_h^H (\tilde{\chi}_{\mathrm{o},h} - \chi_{\mathrm{o}}(t_h))^{\mathrm{T}} W^{(4)} \; (\tilde{\chi}_{\mathrm{o},h} - \chi_{\mathrm{o}}(t_h)) \tag{122}$$

subject to

$$\dot{\boldsymbol{x}}^{(4)} = \begin{bmatrix} \dot{x}_4 \\ \dot{x}_5 \end{bmatrix} = \begin{bmatrix} k_4 \; (n_{0,3} + \tilde{x}_2 - \tilde{x}_3 - \tilde{\chi}_{\mathrm{o}}) \\ k_5 \; (n_{0,5} + \tilde{\chi}_{\mathrm{o}}) \end{bmatrix}, \qquad \boldsymbol{x}^{(4)}(0) = \begin{bmatrix} x_{0,4} \\ x_{0,5} \end{bmatrix} = \mathbf{0} \tag{123}$$

$$\chi_{\mathrm{o}}^{(4)} = \chi_{\mathrm{o}} = x_4 - x_5, \tag{124}$$

where

$$W^{(4)} := 5/4 \tag{125}$$

Please note that the (unknown) measurement error variance σ can arbitrarily be set equal to one during parameter estimation without affecting the parameter estimates. This, however, means that parameter uncertainties cannot be quantified in this case without estimating σ as well. This estimation and subsequent uncertainty analysis is considered out of scope for this study. Figure 11 shows the simulated extent profiles obtained after solving the optimization problems ($P_1^{(j)}, j = 1, \ldots, 4$). There is a reasonable fit in all cases, except for the third extent of reaction (x_3). This, in the mind of the authors, demonstrates a tangible benefit of the extent-based model framework. Indeed, one can identify which parts of the model could be improved, e.g., by reconsidering the reaction rate expression corresponding to the computed extents that are approximated poorly. However, this is not explored further in this work.

Figure 10 shows the measured concentrations and the simulated profiles obtained with (i) incremental parameter estimation ($P_1^{(j)}, j = 1, \ldots, 4$), followed by (ii) simultaneous parameter estimation with estimates obtained through incremental parameter estimation (P_{1+0}). The parameter values obtained by solving P_0, $P_1^{(1)}$, $P_1^{(2)}$, and P_{1+0} are shown in Table 3. Note the final values obtained with both P_0 and P_{1+0} are equal within numerical precision.

Table 3. α-pinene—Parameter estimates. The parameters obtained with simultaneous estimation (P_0) and incremental estimation (P_{1+0}) are practically the same.

	Name	Unit	P_0	$P_1^{(j)}$	P_{1+0}
θ_1	k_1	$\% h^{-1}$	0.213	0.214	0.213
θ_2	k_2	$\% h^{-1}$	0.107	0.106	0.107
θ_3	k_3	$\% h^{-1}$	0.074	0.074	0.074
θ_4	k_4	$\% h^{-1}$	0.989	1.037	0.989
θ_5	k_5	$\% h^{-1}$	0.144	0.148	0.144
$\sqrt{\frac{Q_0}{H \cdot M}}$	WRMSR	–	0.66	0.67	0.66

5. Discussion

A novel procedure for model parameter estimation is proposed. The procedure combines two important features: (i) a new framework for extent computation based on the computation of observable directions among the set of sensed but ambiguous extents; and (ii) a graph-based system partitioning procedure that identifies the groups of equations and kinetic parameters that can be simulated independently from the remaining part of the dynamic model. The latter is possible by approximating the original equation system via interpolation of observed extents and extent directions. The proposed procedure enables splitting the parameter estimation problem into smaller problems in cases that could so far only be handled with simultaneous model identification methods (e.g., scenario A, C, and D in Section 4.1). It also subsumes preexisting methods under special conditions such as complete observability (e.g., scenario E in Section 4.1, [5–7]) or absence of ambiguity (e.g., scenario B in Section 4.1, [8]). Thus, this procedure is applicable to parameter estimation for any multivariate differential equation model under a large range of structural observability conditions.

This work removes an important bottleneck for the implementation of incremental model identification to biological systems. Indeed, by providing a method that can be applied to any known stoichiometric matrix **N** and measurement matrix **M**, incremental model identification is now applicable to biological processes where the number of measurements M is lower than the number of modeled reactions R. Computationally speaking, the extent computation and graph-based partitioning are expected to scale well with the number of modeled reactions. The effects on the computational efficiency of the parameter estimation step will however depend primarily on the lengths of the state vectors in the identified subsystems. In turn, these lengths depend strongly on model structure and the available measured variables and less so on the number of states in the complete system.

A remaining obstacle is the fact that both **M** and **N** are assumed to be known. While several techniques exist to handle this, including methods based on extents [17], some coefficients in these matrices may be unidentifiable based on atomic and stoichiometric balances alone. For this reason, dealing with unknown stoichiometry (**M**) or ill-defined measurements (**N**) deserves continued attention.

Another aspect open for exploration is that no global optimization method has been tested so far to solve the smaller parameter estimation problems $P_1^{(j)}$ that result from system partitioning. Inspiration can be drawn from existing parameter estimation methods even if these methods (i) might be complex yet without publicly available implementation [18]; or (ii) remain limited to cases where all extents are observable and therefore modeled as univariate processes for parameter estimation [19,20].

5.1. Generalized Framework for Extent Computation

The addition of new concepts, such as ambiguous extents and observable extent directions expands the framework for extent computation beyond its original range of applicability. In this study, the focus has been given to the problem of parameter estimation. However, the same generalized framework is likely to be applicable and useful for challenges that have been handled with earlier extent-based methods such as data reconciliation [7], model structure identification [6], and process control [21]. Moreover, there are no obvious limitations that hinder applications involving multi-phase or distributed system models [22,23] or models including algebraic constraints [6,22].

5.2. Optimal System Partitioning

According to our understanding, the proposed procedure delivers the most efficient system partitioning given the available measurements and model information $\mathbf{M}$, $f(\cdot)$, $\mathbf{N}$, and $\boldsymbol{n}_0$. In other words, the procedure generates the subsystems with the smallest number of parameters that can be estimated independently from parameters in other subsystems. However, this is stated here without formal proof. The factorization of $\mathbf{G}_{\bullet,\mathbf{a}}$ based on the reduced row echelon form of $\mathbf{G}$, as described in Section 3.3.1, plays a crucial role. Indeed, it follows from work on subspace clustering methods for noise-free data [24,25] that this factorization minimizes the presence of simulation arcs from the ambiguous extents to the observable directions, thereby leading to the best possible system partitioning.

Please note that the obtained generalized extent framework is optimal only for the purpose of parameter estimation. For instance, one could consider a factorization of $\mathbf{G}_{\bullet,\mathbf{a}}$ which identifies directions that exhibit uncorrelated estimation errors, thereby improving the statistical quality of the computed observable directions and possibly avoiding Type A approximation error. This could be achieved via singular value decomposition of $\mathbf{G}_{\bullet,\mathbf{a}}$. However, singular value decomposition cannot guarantee optimal system partitioning [26]. Consequently, it follows that the optimal factorization of $\mathbf{G}_{\bullet,\mathbf{a}}$ depends on the objective of this factorization. This stands in contrast to the existing body of work concerning extents, which do not involve any such purpose-dependent factorization steps.

Note also that the proposed procedure is designed for the definition of optimality used here. For instance, if one relaxes the constraint that parameters can only appear in one subsystem (Section 3.6), then the proposed graph partitioning procedure is not optimal. For example, in scenario A, one may be able to estimate the parameters in 3 subsets rather than 2, namely, (i) k_1, k_2, and K_1 by simulation of $x_1(t)$ and $x_2(t)$; (ii) k_2, k_3, k_4, and k_5 by simulation of $x_2(t)$, $x_4(t)$, $x_5(t)$, and $\chi_{o,1}(t)$; and (iii) k_3 by simulation of $x_3(t)$. This leads to 3 parameter estimation problems with 3, 4, and 1 parameters, which may be easier to solve than the 2 parameter estimation problems with 5 and 1 parameters as obtained with the proposed procedure. It is however non-trivial to identify a procedure, including both factorization of $\mathbf{G}_{\bullet,\mathbf{a}}$ and system partitioning, that guarantees the best partitioning when applying this relaxed definition of optimality, nor is it clear whether such a procedure exists.

5.3. New Opportunities

This work also hints at several new applications of the generalized framework for extent computation:

(a) **Identifiability analysis.** While not a core objective of this work, it has been shown that the graph partitioning method can help identify unidentifiable parameters. Given that this labeling is based on a model reformulation and does not depend on the temporal resolution or quality of the collected measurements, it follows that the method identifies structurally unidentifiable parameters. Unlike other methods [27], the proposed approach does not require symbolic differentiation. It remains to be explored whether this can also be used to positively identify structurally identifiable parameters. For a discussion on the evaluation and use of indicators of structural and practical parameter identifiability we refer to [28].

(b) **Soft-sensing.** The appearance of observable directions among the ambiguous extents is closely related to the concepts of observability and detectability in the context of state estimation [29]. Note however that the observability labels in this work are based on the stoichiometric balances and measurement equations alone, thus excluding the dynamic model. At the same time, it is suspected that the extents corresponding to vertices that are not on directed paths to vertices representing observable extents or extent directions can be labeled as structurally unobservable, again observing that the exact timing and quality of the measurements does not play any role in this labeling. Similarly to the identifiability analysis discussed above, such an approach would not rely on symbolic differentiation. Whether this can be used to unambiguously determine observability and detectability for all model states remains to be studied.

(c) **Experimental design.** The labeling of extents and directions as observable or unobservable suggests that experimental design may be used to optimize the selection of measured variables. A method to do so has been applied in [30] to enumerate all Pareto-optimal flow sensor layouts in wastewater treatment plants. In [31], symbolic computation enabled the identification of optimal experimental designs. Similar approaches could be applied as a measurement selection method for metabolic flux analysis and the monitoring of complex systems.

6. Conclusions

In this work, an incremental parameter identification procedure has been developed and tested based on a generalized extent-based framework. This generalized framework enables incremental parameter estimation in cases where the previous methods based on the computation of extents did not permit this. Importantly, through study of simulated and experimental medium-sized examples, the generality of the developed procedure has been demonstrated and new opportunities offered by this framework have been identified.

Supplementary Materials: The following are available online at http://www.mdpi.com/2227-9717/7/2/75/s1: The current version of the Efficient Model Identification toolbox (EMI), including all code to reproduce our results in MATLAB R2017b, Figure S1: Scenario B—Simulation. Noise-free and noisy measurements as a function of time, Figure S2: Scenario B—Parameter estimation. Observable extents (x_1, x_3) and their computed equivalents ($\tilde{x}_1$, $\tilde{x}_3$), Figure S3: Scenario B—Parameter estimation. Measured concentrations and simulated profiles obtained with (i) true parameters; (ii) simultaneous parameter estimation P_0; and (iii) incremental parameter estimation $P_1^{(1)}$ and $P_1^{(2)}$, Figure S4: Scenario C—Simulation. Noise-free and noisy measurements as a function of time, Figure S5: Scenario C—Extent computation. Observable extents (x_1, x_3, x_4, x_5) and their computed equivalents ($\tilde{x}_1$, $\tilde{x}_3$, $\tilde{x}_4$, $\tilde{x}_5$), Figure S6: Scenario C—Parameter estimation. Measured concentrations and simulated profiles obtained with (i) true parameters; (ii) simultaneous parameter estimation P_0; and (iii) incremental parameter estimation $P_1^{(1)}$ and $P_1^{(2)}$, Figure S7: Scenario D—Simulation. Noise-free and noisy measurements as a function of time, Figure S8: Scenario D—Extent computation. Observable extent directions ($\chi_{o,1}$, $\chi_{o,2}$, $\chi_{o,3}$) and their computed equivalents ($\tilde{\chi}_{o,1}$, $\tilde{\chi}_{o,2}$, $\tilde{\chi}_{o,3}$), Figure S9: Scenario D—Parameter estimation. Measured concentrations and simulated profiles obtained with (i) true parameters; (ii) simultaneous parameter estimation P_0; and (iii) incremental parameter estimation $P_1^{(1)}$ and $P_1^{(2)}$, Figure S10: Scenario E—Simulation. Noise-free and noisy measurements as a function of time, Figure S11: Scenario E—Extent computation. Observable extents (x_1, x_3) and observable extent direction (χ_o) and their computed equivalents ($\tilde{x}_1$, $\tilde{x}_3$, $\tilde{\chi}_o$), Figure S12: Scenario E—Parameter estimation. Measured concentrations and simulated profiles obtained with (i) true parameters; (ii) simultaneous parameter estimation P_0; and (iii) incremental parameter estimation $P_1^{(1)}$ and $P_1^{(2)}$

Author Contributions: Conceptualization, K.V., J.B. and D.B.; Methodology, K.V., J.B. and D.B.; Results and software implementation, K.V.; Validation, K.V., J.B. and D.B.; Writing—Original Draft Preparation, K.V.; Writing—Review & Editing, K.V., J.B. and D.B.; Funding Acquisition, K.V.

Funding: This study was made possible by Eawag Discretionary Funds (grant No.: 5221.00492.009.03, project: DF2015/EMISSUN).

Acknowledgments: The authors thank Alma Mašić for inputs during initiation of this work and Ivan Miletic for suggesting the study of the α-pinene case.

Conflicts of Interest: The authors declare no conflict of interest.

References

1. Rieger, L.; Gillot, S.; Langergraber, G.; Ohtsuki, T.; Shaw, A.; Takács, I.; Winkler, S. *Guidelines for Using Activated Sludge Models. IWA Task Group on Good Modelling Practice. IWA Scientific and Technical Report*; IWA Publishing: London, UK, 2012.
2. Jakeman, A.J.; Letcher, R.A.; Norton, J.P. Ten iterative steps in development and evaluation of environmental models. *Environ. Model. Softw.* **2006**, *21*, 602–614.
3. Bhatt, N.; Amrhein, M.; Bonvin, D. Incremental identification of reaction and mass-transfer kinetics using the concept of extents. *Ind. Eng. Chem. Res.* **2011**, *50*, 12960–12974.
4. Bhatt, N.; Kerimoglu, N.; Amrhein, M.; Marquardt, W.; Bonvin, D. Incremental identification of reaction systems—A comparison between rate-based and extent-based approaches. *Chem. Eng. Sci.* **2012**, *83*, 24–38.
5. Rodrigues, D.; Srinivasan, S.; Billeter, J.; Bonvin, D. Variant and invariant states for chemical reaction systems. *Comput. Chem. Eng.* **2015**, *73*, 23–33.
6. Mašić, A.; Srinivasan, S.; Billeter, J.; Bonvin, D.; Villez, K. Identification of biokinetic models using the concept of extents. *Environ. Sci. Technol.* **2017**, *51*, 7520–7531.
7. Srinivasan, S.; Billeter, J.; Narasimhan, S.; Bonvin, D. Data reconciliation for chemical reaction systems using vessel extents and shape constraints. *Comput. Chem. Eng.* **2017**, *101*, 44–58.
8. Mašić, A.; Billeter, J.; Bonvin, D.; Villez, K. Extent computation under rank-deficient conditions. *IFAC-PapersOnLine* **2017**, *50*, 3929–3934.
9. Kretsovalis, A.; Mah, R.S.H. Observability and redundancy classification in multicomponent process networks. *AIChE J.* **1987**, *33*, 70–82.
10. Crowe, C.M. Observability and redundancy of process data for steady state reconciliation. *Chem. Eng. Sci.* **1989**, *44*, 2909–2917.
11. Fuguitt, R.E.; Hawkins, J.E. Rate of the thermal isomerization of α-Pinene in the liquid phase1. *J. Am. Chem. Soc.* **1947**, *69*, 319–322.
12. Box, G.E.P.; Hunter, W.G.; MacGregor, J.F.; Erjavec, J. Some problems associated with the analysis of multiresponse data. *Technometrics* **1973**, *15*, 33–51.
13. Tjoa, I.B.; Biegler, L.T. Simultaneous solution and optimization strategies for parameter estimation of differential-algebraic equation systems. *Ind. Eng. Chem. Res.* **1991**, *30*, 376–385.
14. Rodriguez-Fernandez, M.; Egea, J.A.; Banga, J.R. Novel metaheuristic for parameter estimation in nonlinear dynamic biological systems. *BMC Bioinform.* **2006**, *2006*, 483.
15. Brunel, N.J.; Clairon, Q. A tracking approach to parameter estimation in linear ordinary differential equations. *Electr. J. Stat.* **2015**, *9*, 2903–2949.
16. Dattner, I.; Gugushvili, S. Application of one-step method to parameter estimation in ODE models. *Stat. Neerl.* **2018**, *72*, 126–156.
17. Bonvin, D.; Rippin, D.W.T. Target factor analysis for the identification of stoichiometric models. *Chem. Eng. Sci.* **1990**, *45*, 3417–3426.
18. Sahlodin, A.M.; Chachuat, B. Convex/concave relaxations of parametric ODEs using Taylor models. *Comput. Chem. Eng.* **2011**, *35*, 844–857.
19. Mašić, A.; Udert, K.; Villez, K. Global parameter optimization for biokinetic modeling of simple batch experiments. *Environ. Model. Softw.* **2016**, *85*, 356–373.
20. Rodrigues, D.; Billeter, J.; Bonvin, D. Maximum-likelihood estimation of kinetic parameters via the extent-based incremental approach. *Comput. Chem. Eng.* **2018**, doi:10.1016/j.compchemeng.2018.05.024.
21. Billeter, J.; Rodrigues, D.; Srinivasan, S.; Amrhein, M.; Bonvin, D. On decoupling rate processes in chemical reaction systems—Methods and applications. *Comput. Chem. Eng.* **2017**, *114*, 296–305.
22. Srinivasan, S.; Billeter, J.; Bonvin, D. Identification of multiphase reaction systems with instantaneous equilibria. *Ind. Eng. Chem. Res.* **2016**, *29*, 8034–8045.
23. Rodrigues, D.; Billeter, J.; Bonvin, D. Generalization of the concept of extents to distributed reaction systems. *Chem. Eng. Sci.* **2017**, *171*, 558–575.
24. Aldroubi, A.; Sekmen, A. Reduced row echelon form and non-linear approximation for subspace segmentation and high-dimensional data clustering. *Appl. Comput. Harmon. Anal.* **2014**, *37*, 271–287.

25. Vidal, R. Subspace clustering. *IEEE Signal Process. Mag.* **2011**, *28*, 52–68.
26. Billeter, J.; Bonvin, D.; Villez, K. *Extent-Based Model Identication under Incomplete Observability Conditions*; Technical Report No. 6, v3.0; Eawag: Dübendorf, Switzerland, 2018.
27. Petersen, B.; Gernaey, K.; Devisscher, M.; Dochain, D.; Vanrolleghem, P.A. A simplified method to assess structurally identifiable parameters in Monod-based activated sludge models. *Water Res.* **2003**, *37*, 2893–2904.
28. Bonvin, D.; Georgakis, C.; Pantelides, C.C.; Barolo, M.; Grover, M.A.; Rodrigues, D.; Schneider, R.; Dochain, D. Linking models and experiments. *Ind. Eng. Chem. Res.* **2016**, *55*, 6891–6903.
29. Sontag, E.D. *Mathematical Control Theory: Deterministic Finite Dimensional Systems*; Springer Science & Business Media: Berlin, Germany, 2013; Volume 6.
30. Villez, K.; Vanrolleghem, P.A.; Corominas, L. Optimal flow sensor placement on wastewater treatment plants. *Water Res.* **2016**, *101*, 75–83.
31. Billeter, J.; Neuhold, Y.M.; Hungerbuehler, K. Systematic prediction of linear dependencies in the concentration profiles and implications on the kinetic hard-modelling of spectroscopic data. *Chemom. Intell. Lab. Syst.* **2009**, *95*, 170–187.

processes

MDPI

Article

Sequential Parameter Estimation for Mammalian Cell Model Based on In Silico Design of Experiments

Zhenyu Wang, Hana Sheikh, Kyongbum Lee and Christos Georgakis *

Department of Chemical and Biological Engineering and Systems Research Institute for Chemical and Biological Processes Tufts University, Medford, MA 02155, USA; zwang12@dow.com (Z.W.); Hana.Sheikh@gmail.com (H.S.); Kyongbum.Lee@tufts.edu (K.L.)

* Correspondence: christos.georgakis@tufts.edu; Tel.: +1-617-627-2573

Received: 18 May 2018; Accepted: 20 July 2018; Published: 24 July 2018

Abstract: Due to the complicated metabolism of mammalian cells, the corresponding dynamic mathematical models usually consist of large sets of differential and algebraic equations with a large number of parameters to be estimated. On the other hand, the measured data for estimating the model parameters are limited. Consequently, the parameter estimates may converge to a local minimum far from the optimal ones, especially when the initial guesses of the parameter values are poor. The methodology presented in this paper provides a systematic way for estimating parameters sequentially that generates better initial guesses for parameter estimation and improves the accuracy of the obtained metabolic model. The model parameters are first classified into four subsets of decreasing importance, based on the sensitivity of the model's predictions on the parameters' assumed values. The parameters in the most sensitive subset, typically a small fraction of the total, are estimated first. When estimating the remaining parameters with next most sensitive subset, the subsets of parameters with higher sensitivities are estimated again using their previously obtained optimal values as the initial guesses. The power of this sequential estimation approach is illustrated through a case study on the estimation of parameters in a dynamic model of CHO cell metabolism in fed-batch culture. We show that the sequential parameter estimation approach improves model accuracy and that using limited data to estimate low-sensitivity parameters can worsen model performance.

Keywords: Pharmaceutical Processes; Mammalian Cell Culture; sensitivity analysis; parameter estimation; Design of Experiments

---

1. Introduction

The use of biologics, including antibiotics and antibodies, has increased across different therapeutic areas, and is poised to fuel pharmaceutical revenues and stimulate growth in the biopharmaceutical market. In 2012, the global sales of biologics reached 124.9 billion in US dollars, a 10.4% increase over 2011 [1]. More than half of the therapeutic recombinant proteins are produced in immortalized mammalian cell lines, including Chinese hamster ovary (CHO), baby hamster kidney (BHK), and mouse myeloma cells (NS0). Dynamic models of cellular metabolism have been developed to provide insight into the mechanism behind a process, and further enable prediction and optimization for the productivity of cell cultures [2–5]. These metabolic models in some cases contain hundreds of rate constants to describe the rate processes occurring in the cell and bioreactor. Very often, the experimental data set is not large enough to allow for the estimation of all the parameters in the metabolic model [6]. Usually, only a subset of the parameters might be estimable [7]. Moreover, parameter estimation presents a difficult challenge due to the complicated, nonlinear structure of the metabolic models. This problem is exacerbated when some of the parameters have little impact on the model's outputs. A method to systematically select the subset of parameters with the largest impact on the model outputs would greatly benefit not only parameter estimation, but also model validation.

Sensitivity analysis methodologies, including local and global sensitivity analysis, are widely applied to select the subset of parameters with the largest impact on the outputs of a metabolic model to be estimated with available data [8]. The local sensitivity analysis (LSA) approaches calculate the sensitivity coefficients via partial derivatives of output variables with respect to each model parameter with given values of input variables and nominal values of other parameters. Once the sensitivity matrix with sensitivity coefficients as elements has been constructed, a variety of methods, including orthogonalization algorithm [9,10], the Mean Squared Error-based method [11,12], and the Principal Component Analysis-based method [13], can be applied to rank the importance of the parameters. The recent progresses in LSA has been well-reviewed by [7]. As the sensitivity coefficients are calculated based on the impacts by individual parameters, the LSA approaches do not account for the interaction impact of multiple parameters, which may significantly affect the output variables of nonlinear and complicated models.

On the other hand, Global Sensitivity Analysis (GSA), also known as Sobol's method [14], calculates the sensitivity coefficients by simultaneously varying all the parameters in the range of interest. The obtained sensitivity coefficients account for the interaction impact of all the parameters, as well as the impact of individual parameters. GSA is frequently employed to identify the most important and sensitive parameters of metabolic models [15–17]. The major limitation of such a method based on Monte Carlo simulations is that it incurs high computational cost. To reduce the computational effort, [18] proposed using a meta-model, a Response Surface Methodology (RSM) model [19], to approximate the original dynamic model comprising a large set of differential and algebraic equations. Then, the sensitivity analysis is conducted on the simplified meta-model via Monte Carlo simulations similar to GSA. As the meta-model is an algebraic model, the computational cost for simulating such a model is drastically less than a dynamic model of cellular metabolism. Consequently, the sensitivity analysis is completed much faster.

In this paper, we present a new sequential parameter estimation methodology, prioritizing the estimation of parameters with descending sensitivity indices. We first improved the sensitivity analysis approach proposed by [18] by eliminating the step of running Monte Carlo simulations on the meta-model. Instead, we calculate the sensitivity index analytically using the estimated RSM model. Moreover, we propose a systematic approach to discriminate the parameters into four categories based on the sensitivity of the model outputs. This allows the modeler to prioritize the estimation of the model parameters by initially focusing on those with the highest importance or largest sensitivity indices. We demonstrate the power of the proposed method by successfully identifying the important parameters in a well-received model of CHO cell metabolism [4] using experimental data. We show that our sequential parameter estimation method results in a more accurate model compared to when all of the parameters are estimated simultaneously.

2. Global Sensitivity Analysis

Sobol's method, or Global Sensitivity Analysis (GSA), varies the parameters of interest simultaneously over their entire domain to examine the interaction effects among parameters. The dynamic model of interest is decomposed into sums of orthogonal functions (also known as summand), as given below.

$$y = f(\boldsymbol{\theta}) = g_0 + \sum_{s=1}^{n} \sum_{i_1 < i_2 < \cdots < i_s}^{n} g_{i_1 \cdots i_s}(\theta_{i_1}, \cdots, \theta_{i_s}) \tag{1}$$

where y is the output variable, while $\boldsymbol{\theta} = \begin{bmatrix} \theta_1 & \theta_2 & \cdots & \theta_n \end{bmatrix}^T$ is a column vector with n parameters as elements. Each orthogonal function $g(\cdot)$ represents the effect of corresponding parameters on the output. If the output, y, is affected by two parameters, θ_1 and θ_2, the expansion of Equation (1) is given by

$$y = g_0 + g_1(\theta_1) + g_2(\theta_2) + g_{12}(\theta_1, \theta_2) \tag{2}$$

The magnitude of the effect on the output by each variable is related to the variance of the corresponding orthogonal function, $g(\cdot)$, which is calculated as follows:

$$\begin{aligned} D &= \int_{a}^{b} f(\theta)^2 d\theta - g_0^2 \\ D_i &= \int_{a_i}^{b_i} g_i^2(\theta_i) d\theta_i \\ D_{i_1\cdots i_s} &= \int_{a}^{b} g_{i_1\cdots i_s}^2(\theta_{i_1\cdots i_s}) d\theta_{i_1\cdots i_s} \end{aligned} \quad (3)$$

where $D = \sum_{s=1}^{n} \sum_{i_1<\cdots<i_s}^{n} D_{i_1\cdots i_s}$ represents the total variance in outputs due to all parameters in the domain defined by $[\boldsymbol{a}, \boldsymbol{b}]$. The $n \times 1$ vectors $\boldsymbol{a}$ and $\boldsymbol{b}$ are the lower and upper bounds for the n parameters, while D_i represents the variance in outputs due to parameter θ_i in the corresponding range of $[a_i,\ b_i]$. The sensitivity of an output to a parameter is quantified by the Total Sensitivity Index (TSI), as given below

$$TSI_i = \frac{D_i + \sum_{j=1}^{n} D_{ij} + \sum_{j=1}^{n} \sum_{k=1}^{n} D_{ijk} + \cdots}{D} \quad (4)$$

The larger the TSI_i, the more strongly the corresponding parameter affects the output. Therefore, a parameter is considered more important if its TSI is larger. For a complicated model comprising a system of many nonlinear differential equations, the explicit solution for the summands, $g(\cdot)$, cannot be obtained. The corresponding variance, and therefore the sensitivity indices, are instead estimated through Monte Carlo simulations, e.g., using Satelli's algorithm [20].

3. Sensitivity Analysis Based on In Silico Design of Experiments

To reduce the computational cost, the method proposed by [18] first estimates a Response Surface Methodology (RSM) model, and then determines the sensitivity index by running Monte Carlo Simulations on the estimated RSM model. To estimate the RSM model, the parameters, $\boldsymbol{\theta}$, are first coded in the range of $[-1,\ +1]$. For each model parameter θ_i, the corresponding coded parameter x_i is given by Equation (5).

$$x_i = (\theta_i - \theta_{0,i})/\Delta\theta_i \quad (5)$$

Here, $\theta_{0,i}$ is the reference value of parameter θ_i and $\Delta\theta_i$ is the half interval in which we expect the parameter's optimal value will lie. Both θ_i and $\Delta\theta_i$ are selected by the modeler based on the modeler's understanding of the process. If the estimated value of the parameter is at an endpoint of this interval, the initial choice of the interval might need to be corrected.

To minimize the number of detailed simulations, a D-optimal design [19] of in silico experiments is performed, where the model parameters are systematically varied in the range of $\theta_0 \pm \Delta\theta$ around their nominal values θ_0. Customarily, these inputs are transformed into their dimensionless coded form as defined above. The resulting time-resolved output values, $y(t|\boldsymbol{x})$, for the defined combinations of the coded inputs, $\boldsymbol{x}$, are collected through the simulation of the metabolic model and are used to estimate the RSM model of $y(t|\boldsymbol{x})$ to $\boldsymbol{x}$. An example of a quadratic RSM model with n coded inputs or factors is given by

$$y(t|\boldsymbol{x}) = f(\boldsymbol{x}) = \beta_0 + \sum_{i=1}^{n} \beta_i x_i + \sum_{i=1}^{n} \sum_{j>i}^{n} \beta_{ij} x_i x_j + \sum_{i=1}^{n} \beta_{ii} x_i^2 \quad (6)$$

The β_i, β_{ij}, and β_{ii} are coefficients of the RSM model and are estimated by stepwise regression [21] to avoid overfitting of the model. Instead of conducting a GSA on the complicated system of differential equations, the GSA is applied to the relatively simple algebraic RSM model as given above for each of the outputs of interest. Specifically, sensitivity indices are calculated using the outputs obtained by varying the coded inputs simultaneously. As shown in [18], the Design of Experiment-based method significantly reduces the required computational time.

4. Sequential Parameter Estimation

Here we present a new sequential parameter estimation method consisting of an improved method for the aforementioned sensitivity analysis approach and a systematic way to prioritize the estimation of subsets of parameters. We first improve the above sensitivity analysis method by eliminating the step of running Monte Carlo simulations on the estimated RSM model. In the improved method, the sensitivity index of each parameter is estimated analytically following Sobol's method, as detailed below. We here illustrate the proposed approach using a quadratic RSM model, given in Equation (6), as an example. If a higher-order RSM model is at hand, the sensitivity index of each parameter will be determined in a similar manner.

Using the definition given by Sobol [14], the orthogonal summands can be solved as follows:

$$\begin{aligned} g_0 &= \tfrac{1}{L^n}\int_{-1}^{+1} f(\mathbf{x})d\mathbf{x} = \beta_0 + \tfrac{1}{3}\sum_{i=1}^{n}\beta_{ii} \\ g_i(x_i) &= \tfrac{1}{L^{n-1}}\int_{-1}^{+1} f(\mathbf{x})dx_{k\neq i} - g_0 = \beta_i x_i + \beta_{ii}(x_i^2 - \tfrac{1}{3}) \\ g_{i,j}(x_i, x_j) &= \tfrac{1}{L^{n-2}}\int_{-1}^{+1} f(\mathbf{x})dx_{k\neq i,j} - g_0 - g_i - g_j = \beta_{ij}x_i x_j \end{aligned} \quad (7)$$

Here L is the range of the input variables in the RSM model. The input variables here are the parameters of the metabolic model. These are coded into the range of $[-1, +1]$. We use $L = 2$ to derive the orthogonal functions as shown in Equation (7). By substituting the orthogonal functions into Equation (3), we express the variance functions as follows:

$$\begin{aligned} D_i &= \tfrac{1}{L}\int_{-1}^{+1} g_i^2(x_i)dx_i = \tfrac{1}{3}\beta_i^2 + \tfrac{4}{45}\beta_{ii}^2 \\ D_{i,j} &= \tfrac{1}{L^2}\int_{-1}^{+1} g_{i,j}^2(x_i, x_j)dx_i dx_j = \tfrac{1}{9}\beta_{ij}^2 \end{aligned} \quad (8)$$

Then the total variance is calculated as $D = \sum\limits_{s=1}^{n}\sum\limits_{i_1<\cdots<i_s}^{n} D_{i_1\cdots i_s}$. The total sensitivity indices are calculated by substituting the variances calculated above into Equation (4). As the quadratic RSM model accounts for the interaction effect of up to two inputs, the sensitivity index is calculated as follows.

$$TSI_i = \frac{D_i + \sum_{j=1}^{n} D_{ij}}{D} \quad (9)$$

The results in Equations (8) and (9) can be easily generalized to estimate higher order (>2) sensitivity indices. However, this will require the estimation of RSM of higher order.

Each parameter is ranked from most to least important, according to its sensitivity index. Next, we classify the parameters into three subsets, most important (subset A), important (subset B) and least important (subset C), based on the percentage of explained output variance. We would like subset A to explain at least α% of the total output variance, subset B to explain an additional β%, and subset C to explain another γ% of the variance. The restriction is that $(\alpha + \beta + \gamma) < 100\%$. This could leave out a small percentage of the overall variance, i.e., $100\% - (\alpha + \beta + \gamma)$. This small percentage typically corresponds to the noise in the data and, thus, is of low importance in terms of parameter estimation. The set of parameters that are not selected in the subsets A, B and C are grouped into subset D. In the present study, we set the values of α, β, and γ to 50%, 30% and 10%, respectively. The remaining unexplained variance, $100\% - (\alpha + \beta + \gamma)$, is then 10% of the overall and is related to the parameters in subset D, i.e., parameters which have the smallest effect on the model's predictions and whose values it might not be worth adjusting further beyond their initial estimates. More generally, the specific values of the parameters α, β, and γ are left to the discretion of the modeler, as well as the $100\% - (\alpha + \beta + \gamma)$ fraction of the variance that could have been addressed by minute adjustments to a possibly large number of parameters, each one of which has a negligible impact on the model predictions.

Sobol's sensitivity index represents the ratio of the output variances caused by a certain variable change to the output variances caused by all variable changes. Therefore, we can quantify the output variance by a given subset of parameters by directly summing the corresponding sensitivity indices. Given a set of N total parameters and a subset of S parameters of interest, the percentage of variance is calculated using the following equation.

$$\eta = \frac{\sum_{i\in S} TSI_i}{\sum_{j=1}^{N} TSI_j} \times 100\% \tag{10}$$

Using Equation (10) and the selected values for the thresholds α, β, and γ, we divide the parameters into four subsets, A, B, C, and D. We then sequentially estimate the values of the parameter's starting with those in subset A, then those in subsets A and B and finally those in subsets A, B and C. When we estimate the most important subset of parameters, subset A, we hold the remaining parameters in subset B, C and D at their nominal values. These nominal values can be obtained from the literature or approximately estimated based on available knowledge about the process. They will most likely be equal to the reference values, θ_0, defined above. This reduces the dimensionality of the optimization problem that has to be solved in each parameter estimation task, substantially alleviating the challenge caused by local minima, especially prevalent when the number of decision variables is very large. The parameter estimation problem can be defined mathematically as follows:

$$\theta_S^* = \underset{\theta_S}{\operatorname{argmin}} \sum_{m=1}^{M} \sum_{k=1}^{K} \left(\frac{\hat{y}_{m,k}(\theta_S, \theta_{i\notin S} = \theta_0) - y_{m,k}}{y_{m,k}} \right)^2 \tag{11}$$

where θ_S are the parameters to be estimated and $\theta_{i\notin S}$ are the parameters to be held at their fixed nominal values, θ_0. Also $\hat{y}_{m,\,k}$ and $y_{m,k}$ are the values predicted by the model and the corresponding measured values of species m at time instant k. The above parameter estimation problem is solved in Matlab's Optimization toolbox [22] using an interior-point algorithm [23] and *fmincon* function. Once the most important parameters in subset A have been estimated, we can estimate the values of parameters in subset B. In this round of estimation, we will take the optimal values for the parameters in subset A, $\theta^*_{S\in A}$, and the nominal values for parameters in subset B, $\theta_{S\in B}$, as the initial guess to solve the optimization problem given in Equation (10). Note that the newly estimated parameters in subset A may be slightly different from the originally estimated nominal values. Next, the parameters in subset C are estimated together with subset A and B parameters using a similar approach. The initial guesses for subset C parameters are their nominal values while the initial guesses for subset A and B parameters are their optimal values obtained in the previous round of parameter estimation. By sequentially estimating the subsets of parameters of descending importance, we obtain a dynamic model of improved accuracy compared to the model where all parameters are estimated simultaneously.

5. Results and Discussion

In this section, we apply the proposed method to estimate the model parameters of a complicated dynamic model of CHO cell metabolism in a fed-batch reactor [4]. The CHO cell model consists of 34 reactions and 51 parameters, and covers major pathways of central carbon metabolism. The model explicitly accounts for redox- and temperature-dependent changes to pathway activities, and directly calculates the measured variables, i.e., metabolite concentration time profiles in the reactor, by defining rate expressions based on extracellular metabolites. The reactions and metabolites involved in the model are visualized in Figure 1.

There are two types of parameters of the CHO cell model: four parameters related to the process operation and 47 kinetic parameters. The process parameters are the shift temperature, shift day, seed density and harvest day. These parameters relate to the operation of the reactor and are selected based on the choice of the typical operational ranges. In this work, we fix the process parameter ranges to their default values [4] and estimate only the kinetic parameters. However, to identify which process parameters significantly affect the metabolites, we conduct sensitivity analysis for the process parameters as well. The intervals over which the process parameters (shift temperature, shift day, seed density and harvest day) will be varied are 31 ± 3 , 3 ± 1 day, $(3.6 \pm 1.8) \times 10^6$ cell/mL, and 9 $\pm$ 1 day, respectively. We label the process parameters #1 to #4.

In Table 1, we define parameters #5 to #51, which refer to the kinetic parameters that have to be estimated from the experimental data. This table also identifies which model parameters are assigned to subsets A, B, C, and D. The 47 kinetic parameters include the maximal reaction velocities (v_{max}), the half saturation constants (K_m), the inhibition constants (K_i) and the temperature dependency constants (TC). The nominal values for the kinetic parameters are set to the values defined in the original model [4]. For the present analysis, these parameters are scaled to be in the same order of magnitude by multiplying each of the parameters with a corresponding scaling factor as given below.

$$\theta_{0,i} = c\theta_i' \tag{12}$$

where θ_i' is the ith parameter in the original model, while $\theta_{0,i}$ is the corresponding scaled parameter and c is the scaling factor. The values of c and $\theta_{0,i}$ are given in columns 2 and 3 of Table 1, respectively. In each simulation-based sensitivity analysis experiment, the kinetic parameters are varied by ±20% of the nominal values. To estimate the linear and nonlinear sensitivities of the 51 parameters, we design a set of 1398 experiments using the D-Optimal design. Of these, 1378 runs are used to estimate the parameters in a quadratic RSM model, 10 center point runs to represent the expected fed-batch process variability, and 10 additional runs to estimate the Lack-of-Fit (LoF) statistics. A 4% normally distributed error is added to all simulation results to reflect the expected normal variability of the process. This is also the accuracy we expect of the model.

Table 1. Parameter values of CHO Cell model obtained via simultaneous and sequential parameter estimation.

Parameter	Scaling Factor	Scaled Parameter	Subset A	Subset A + B	Subset A + B + C	Simultaneous
V_{max1} (#05)	0.001	3.8		3.75	3.77	3.46
K_{i1} (#06)	0.1	2				1.78
K_{m1} (#07)	0.1	1		1.02	1.01	0.87
Exp_{1a} (#08)	1	3		3.02	3.02	2.67
Exp_{1b} (#09)	1	1				1.11
TB_{1b} (#10)	1	5				4.39
V_{max2} (#11)	0.001	2.2				2.12
K_{m2} (#12)	1	6			6.00	5.73
V_{max3f} (#13)	0.01	3.5	4.10	4.06	4.07	3.97
V_{max3r} (#14)	0.01	1.5			1.51	1.36
K_{m3a} (#15)	1	4		3.97	3.99	3.49
K_{m3b} (#16)	1	2.5				2.16
K_{m3c} (#17)	1	5			5.04	4.55
TC_3 (#18)	1	2				2.18
V_{max8f} (#19)	0.001	2.2		2.24	2.25	2.21
V_{max8r} (#20)	0.01	2	2.31	2.26	2.26	1.97
K_{m8a} (#21)	1	2.5		2.51	2.49	2.27
K_{m8b} (#22)	1	1			0.99	0.95
K_{m8c} (#23)	1	1		1.00	1.01	1.11
TC_{8b} (#24)	1	5				5.33
V_{max9f} (#25)	1	1				1.07
V_{max9r} (#26)	1	1				0.87
K_{m9} (#27)	10	7		7.06	7.06	6.06
V_{max10f} (#28)	0.01	4.75		4.79	4.82	4.45
V_{max10r} (#29)	0.1	2			2.01	2.21
K_{m10z} (#30)	10	3		3.01	3.02	3.06
K_{m10b} (#31)	1	1				0.87
K_{m10c} (#32)	1	2			2.01	1.73
TC_{10b} (#33)	1	1.5		1.49	1.49	1.30
V_{max11} (#34)	10	5.5		5.55	5.58	5.69

Table 1. *Cont.*

Parameter	Scaling Factor	Scaled Parameter	Subset A	Subset A + B	Subset A + B + C	Simultaneous
V_{max12f} (#35)	10	0.9		0.90	0.89	1.00
V_{max12r} (#36)	0.1	2.5				2.35
K_{m12a} (#37)	1	1		0.99	0.99	1.11
K_{m12b} (#38)	1	3				2.60
V_{max13} (#39)	0.1	3				3.33
K_{m13} (#40)	1	1			1.01	1.07
V_{max16} (#41)	0.001	2.5	2.93	2.91	2.91	2.84
K_{m16a} (#42)	10	4		4.00	4.03	4.44
K_{m16b} (#43)	10	3		3.04	3.06	3.20
K_{m16c} (#44)	0.1	2		2.02	2.01	2.22
TC_{16b} (#45)	1	3				3.09
V_{max17} (#46)	0.01	5.25	5.63	5.67	5.69	5.82
K_{i17} (#47)	0.1	3				2.80
Exp_{17a} (#48)	10	5				5.64
Exp_{17b} (#49)	1	1				1.06
V_{max33a} (#50)	10	2				2.22
V_{max33b} (#51)	10	2		2.03	2.01	1.96

The model calculates values for 48 outputs: 34 reaction or exchange fluxes, 14 external metabolite concentrations, including biomass and antibody titer. Since the reaction and exchange fluxes depend on the metabolite concentrations, and the total cell density is directly proportional to biomass, there are only 14 independent outputs. Therefore, 14 quadratic RSMs are developed for the 14 independent metabolite concentrations. The inputs, or factors, in these RSM models are the 51 parameters (4 process and 47 kinetic constants) of the metabolic model whose sensitivity we are trying to assess. The same fractional error is added to the 10 replicated center point runs, through which the Analysis of Variance (ANOVA) [19] estimates the normal variability of the process.

Using the 14 RSMs, the sensitivity indices are calculated by applying Equations (8) and (9). For the case of a single output model, one can simply rank the importance of the parameters according to their sensitivity indices. For the case of multiple outputs, there could be different rankings for each parameter with respect to different outputs. Thus, we need to consider the importance of a parameter to multiple outputs and determine its overall importance. We separate the outputs into two classes: product class and product-relevant class. The product class comprises only the product output itself. Based on the sensitivity of each parameter with respect to the product output, we rank the parameters into four subsets as described previously. The parameters ranked in this fashion are assigned into subsets A1, B1, C1 and D1, in the order of decreasing importance. We then perform the parameter rankings again, this time based on the *average* sensitivities of the parameters with respect to the product-relevant outputs, and assign the parameters into subsets A2, B2, C2 and D2. We combine subsets A1 and A2 to form subset A, which contains the most important parameters overall. Subsets B and C are obtained similarly by combining the corresponding subsets (B1 and B2 and C1 and C2) subject to the condition that the parameters already assigned to a more important subset are excluded. For example, the subset B is the union of subsets B1 and B2 but excludes parameters that are in subset A.

In the CHO cell model case study, the antibody is the desired product. Therefore, we assign this model output to the product class. The remaining model outputs, i.e., metabolites such as glucose and lactate, are either substrates utilized by the cell or highly correlated with the antibody. These model outputs are assigned to the product-relevant class. We calculate the average of sensitivity indices of the kinetic parameters with respect to these outputs and rank the parameters. We apply the aforementioned described threshold criteria to identify the two most important subsets of parameters, A1 and A2 for two classes of outputs that contribute to the desired α fraction (50%) of the variance. The two subsets are combined to define subset A. Subsets B, C, and D are obtained in a similar manner.

The sensitivity indices of the parameters on the antibody concentration and the averaged sensitivity indices on the remaining model outputs are plotted in Figure 2a,b, respectively. In general, the outputs are strongly sensitive to all four process parameters. In comparison, the sensitivities of the outputs to the kinetic parameters are more varied.

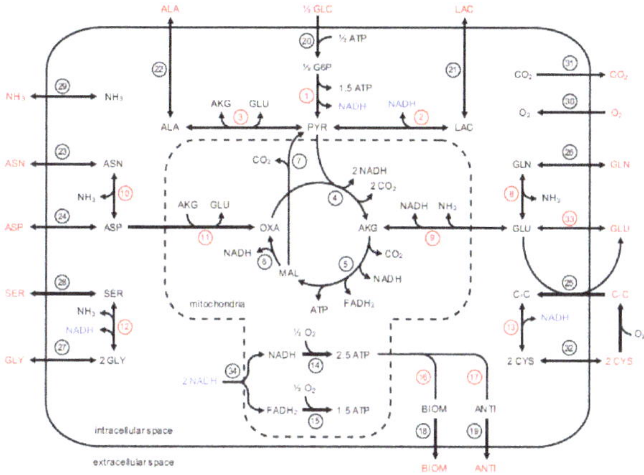

Figure 1. Metabolic pathways of CHO cell model. The 34 reactions are indexed. 14 extracellular metabolites, except for CO_2 and O_2, are modeled.

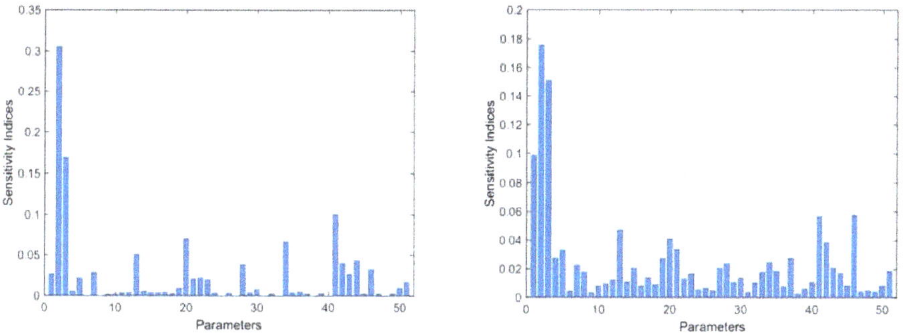

(**a**) (**b**)

Figure 2. Sensitivity indices of (**a**) Antibody Concentration (**b**) Other Measured Metabolites Concentration to the parameters in the CHO cell model. In both Figure 2a,b, Parameters 1 to 4 correspond to the four process variables and parameter 5 to 51 correspond to the 47 kinetic parameters.

We plot the percentage variance explained of antibody and metabolite concentrations as a function of number of parameters in Figure 3a,b, respectively. The four most important parameters are two process variables, (#2 and #3), and two kinetic parameters (#20 and #41). Together, these account for 50% of the variance of antibody concentration, as shown in Figure 3a. We assign these parameters to subset A1. Subset A2 consists of seven parameters, including three process variables, #1-3, and four kinetic parameters, #13, #20, #41, and #46. Together, these 7 parameters account for 50% of the variance of the metabolite outputs, as shown in Figure 3b. We obtain the most important subset of parameters, subset A, by combining the kinetic parameters in A1 and A2. We arrive at subset A consisting of four parameters, #13, #20, #41, and #46. The process parameters, #1-3, are not included in this subset, because they are externally defined by the operating conditions and thus are not estimated from the experimental data. In a similar manner, 18 additional important parameters are identified and assigned to subset B. Together with the 4 most important parameters of subset A, the subset B parameters explain 80% of the variance of the antibody and metabolite concentrations. Seven additional parameters are identified in subset C. The remaining 18 parameters explain the last 10% of the overall variance and are part of subset D.

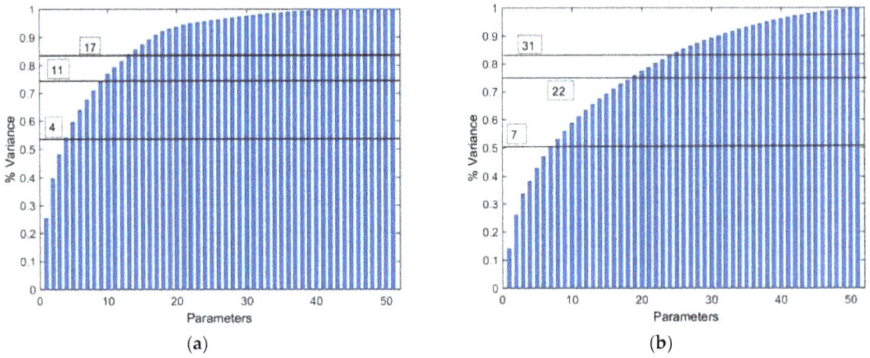

Figure 3. Accumulated sum of sensitivity indices: (**a**) Antibody Concentration (**b**) Other Measured Metabolites Concentration to 51 parameters in the CHO Model.

We first estimate the values of subset A parameters, while holding the other parameters fixed at their nominal values. The estimated values of these parameters in the first cycle of estimation are listed in Table 1, column 4. The parameters in the other subsets (B, C, and D) are at their nominal values as given in column 3. The Sum of Squared Error (SSE) between the predicted and measured outputs is 6.85 as listed in column 2 of Table 2. For comparison, we estimate all 47 parameters simultaneously using the same data. The SSE associated with the obtained model is 8.36, as given in column 5 of Table 2. This value is 22% larger than the SSE resulting from the estimation of only the 4 most important parameters in subset A. Moreover, the computational time for estimating all 47 parameters simultaneously is 5.48 h, whereas estimating the four parameters in subset A required only 0.48 h. All nonlinear parameter estimation tasks were performed in MATLAB using the *fmincon* function on a personal computer with 4 GB RAM memory and Intel Core i5-2500 (3.3 GHz) CPU. These results show that as the dimensionality of the parameter estimation problem is reduced, the accuracy of the model is improved, while the computational time is reduced. The SSE of the original model [4] is 11.68, which was obtained by estimating all of the model parameters simultaneously using simulated annealing. The difference between the SSEs of the models with the simultaneously estimated parameters (this study vs. [4]) suggests that when the dimensionality of the parameter estimation problem is large relative to the available data, the optimization may converge to different local minima depending on the algorithm.

Table 2. Comparison Sum of Squared Error of obtained models by sequentially and simultaneously re-estimating model parameters.

	Subset A	**Subset A + B**	**Subset A + B + C**	**Simultaneous**
Sum of Squared Error	6.85	6.63	6.60	8.36
Difference in SSE (%) [1]	0	−3.2	−3.7	22.0
Computational Time (h)	0.48	1.25	3.16	5.48

[1] The percentage difference uses SSE for subset A as the reference. A positive value means the corresponding SSE is larger than the SSE of subset A while negative value indicates a smaller SSE than the one of subset A.

By using the estimated values of the parameters in subset A in the first round as the initial guess, we estimate the parameters in subsets A and B in the next round. The initial values of the subset B parameters are their scaled values in column 3 of Table 1. The parameter values estimated in the second round are given in column 5 of Table 1. The corresponding SSE is 6.63 and it is given in column

3 of Table 2. With 18 additional parameters in subset B estimated, the SSE is only slightly (3.3%) smaller than the SSE obtained after the first round, while the computational time increases to 1.25 h from 0.48 h. This confirms that the parameters in subset B have a smaller impact on the model accuracy compared to subset A. With the best parameter values for subsets A and B as the initial guesses for the respective parameters, we estimate in the third round the 7 parameters in subset C, as well as the parameters in subsets A and B. After the third round, the SSE is further reduced by 0.5%. This very modest improvement in SSE again underscores that the most important parameters identified in subset A have the largest impact on model accuracy. Indeed, estimating just the 22 parameters in subsets A and B, while keeping the remaining 25 parameters fixed at their nominal values would have yielded a model that is just as accurate as the model obtained by estimating all parameters in subsets A, B, and C.

Taken together, the above results suggested that the sequential parameter estimation strategy could yield a more accurate model, while also reducing the computational time required for parameter estimation. To visually assess the accuracy of the sequentially estimated model, we plotted the model outputs obtained with the different parameter estimation strategies (Figure 4). For comparison, the experimental data used to estimate the original model [4] are also plotted in the same figure. The concentrations predicted by the model with simultaneously estimated parameters (Model 1) are shown in dashed lines, while the concentrations predicted by the model in which only the four most important parameters (subset A) are estimated (Model 2) are shown in solid line. Model 2 has a more accurate prediction in the concentrations of antibody (ANTI), the output of the greatest interest. In addition, Model 2 more accurately predicted the BIOM, GLC, ASP and SER concentration profiles.

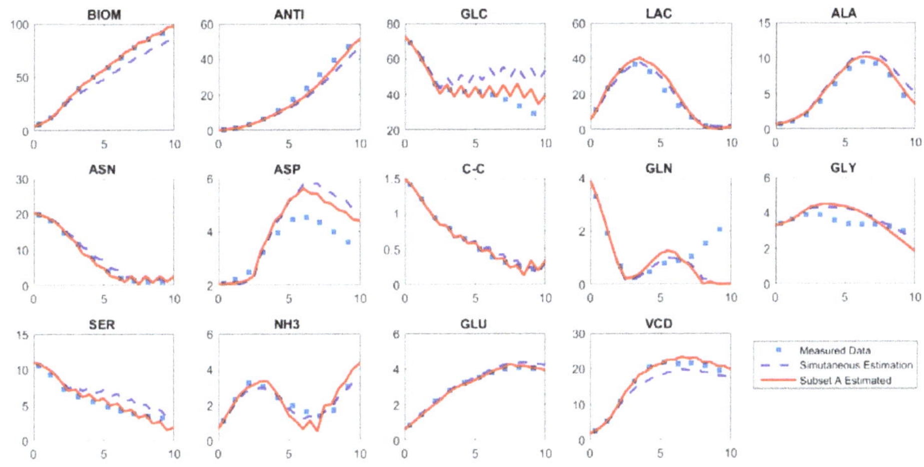

Figure 4. Comparison between predictions by the new model with 4 most sensitive parameters estimated (—) with predictions by the model with all parameters estimated simultaneously (–). The experimental data from [4] are shown, as well (■). The x-axis is the culture time in days and y-axis is the concentration in mM, except for the viable cell density (VCD) which is 10^6 cells/mL.

6. Conclusions

In this paper, we propose a sequential parameter estimation approach to improve the accuracy of the obtained model. The parameters to be estimated are first assigned to four subsets, A, B, C, and D, based on how sensitive the model predications are on the parameter values, quantified by their sensitivity indices. The parameters in subsets A, B and C correspond to the most important, important, and less important parameters, respectively. The least important parameters are grouped in subset D, and contribute only 10% to the model's outputs. The sensitivity indices are calculated using a refined approach originating from the global sensitivity analysis via RSM model proposed by [18]. Instead of running Monte Carlo simulations on the estimated RSM models, we analytically calculate the sensitivity indices of each parameter. This further reduces the computational cost.

In the proposed sequential estimation of model parameters, one initially estimates the parameters in subset A, then in sets A and B. If enough data is available, the parameters in subsets A, B and C are then estimated. As shown in this paper, fitting the parameters in subset D will have negligible impact on the model's accuracy. When we estimated all the parameters simultaneously, including those in subset D, we obtained a statistically less accurate model, which confirms the efficacy of the proposed method. Avoidance of local minima in the related optimization task is conjectured to be the main reason for the superior performance of the sequential estimation of the model's parameters.

We demonstrate the benefits of the proposed method using a case study on a dynamic model of CHO cell metabolism in fed-batch culture. The model parameters are separated into four subsets: A, B, C, and D of deceasing importance. The first subset accounts for 50% of the outputs' variance, while subsets B and C account for an additional 30% and 10% variance, respectively. The corresponding SSE indicates that by estimating the very small subset of most important parameter (subset A), we can obtain an accurate starting model. If we then follow up with the estimation of the parameters in subsets A and B, an even more accurate model can be obtained. If additional parameters of less importance are estimated, the further improvement on model accuracy is minimal. If we estimate all of the parameters simultaneously, a less accurate model is achieved compared to the sequentially estimated model. We speculate that this might be due to the existence of several local minima, although additional work is warranted to more thoroughly explain this result. At least for the CHO model investigated in this paper, the observation that the simultaneously estimated model affords lower accuracy and highlights the potential benefit of the sequential estimation approach proposed here.

Author Contributions: Conceptualization, C.G. and K.L.; Methodology, C.G., Z.W.; Investigation, Z.W., C.G., H.S. and K.L.; Formal analysis, Z.W., C.G., H.S. and K.L.; Validation, K.L.; Software, Z.W., H.S.; Writing-Original Draft, Z.W., C.G. and K.L.; Writing-Review, H.S.

Funding: This research received no external funding.

Conflicts of Interest: The authors declare no conflicts of interest.

References

1. Zhou, W.; Kantardjieff, A. Mammalian Cell Cultures for Biologics Manufacturing. In *Mammalian Cell Cultures for Biologics Manufacturing*; Zhou, W., Kantardjieff, A., Eds.; Springer: Berlin/Heidelberger, Germany, 2014.
2. Nolan, R.P.; Lee, K. Dynamic model for CHO cell engineering. *J. Biotechnol.* **2012**, *158*, 24–33. [CrossRef] [PubMed]
3. Sanderson, C.S.; Barford, J.P.; Barton, G.W. A structured, dynamic model for animal cell culture systems. *Biochem. Eng. J.* **1999**, *3*, 203–211. [CrossRef]
4. Nolan, R.P.; Lee, K. Dynamic model of CHO cell metabolism. *Metab. Eng.* **2011**, *13*, 108–124. [CrossRef] [PubMed]
5. Mulukutla, B.C.; Gramer, M.; Hu, W.-S. On metabolic shift to lactate consumption in fed-batch culture of mammalian cells. *Metab. Eng.* **2012**, *14*, 138–149. [CrossRef] [PubMed]
6. Raue, A.; Kreutz, C.; Maiwald, T.; Bachmann, J.; Schilling, M.; Klingmüller, U.; Timmer, J. Structural and practical identifiability analysis of partially observed dynamical models by exploiting the profile likelihood. *Bioinformatics* **2009**, *25*, 1923–1929. [CrossRef] [PubMed]
7. Kravaris, C.; Hahn, J.; Chu, Y. Advances and selected recent developments in state and parameter estimation. *Comput. Chem. Eng.* **2013**, *51*, 111–123. [CrossRef]
8. Saltelli, A.; Annoni, P. How to avoid a perfunctory sensitivity analysis. *Environ. Model. Softw.* **2010**, *25*, 1508–1517. [CrossRef]
9. Yao, K.Z.; Shaw, B.M.; Kou, B.; McAuley, K.B.; Bacon, D.W. Modeling Ethylene/Butene Copolymerization with Multi-site Catalysts: Parameter Estimability and Experimental Design. *Polym. React. Eng.* **2003**, *11*, 563–588. [CrossRef]

10. Lee, D.; Ding, Y.; Jayaraman, A.; Kwon, J. Mathematical Modeling and Parameter Estimation of Intracellular Signaling Pathway: Application to LPS-induced NFκB Activation and TNFα Production in Macrophages. *Processes* **2018**, *6*, 21. [CrossRef]
11. McLean, K.A.P.; Wu, S.; McAuley, K.B. Mean-Squared-Error Methods for Selecting Optimal Parameter Subsets for Estimation. *Ind. Eng. Chem. Res.* **2012**, *51*, 6105–6115. [CrossRef]
12. Eghtesadi, Z.; McAuley, K.B. Mean-squared-error-based method for parameter ranking and selection with noninvertible fisher information matrix. *AIChE J.* **2016**, *62*, 1112–1125. [CrossRef]
13. Degenring, D.; Froemel, C.; Dikta, G.; Takors, R. Sensitivity analysis for the reduction of complex metabolism models. *J. Process Control* **2004**, *14*, 729–745. [CrossRef]
14. Sobol, I.M. Global sensitivity indices for nonlinear mathematical models and their Monte Carlo estimates. *Math. Comput. Simul.* **2001**, *55*, 271–280. [CrossRef]
15. Ho, Y.; Varley, J.; Mantalaris, A. Development and analysis of a mathematical model for antibody-producing GS-NS0 cells under normal and hyperosmotic culture conditions. *Biotechnol. Prog.* **2006**, *22*, 1560–1569. [CrossRef] [PubMed]
16. Zheng, Y.; Rundell, A. Comparative study of parameter sensitivity analyses of the TCR-activated Erk-MAPK signalling pathway. *IEE Proc. Syst. Biol.* **2006**, *153*, 201–211. [CrossRef]
17. Mailier, J.; Delmotte, A.; Cloutier, M.; Jolicoeur, M.; Wouwer, A.V. Parametric Sensitivity Analysis and Reduction of a Detailed Nutritional Model of Plant Cell Cultures. *Biotechnol. Bioeng.* **2011**, *108*, 1108–1118. [CrossRef] [PubMed]
18. Kiparissides, A.; Georgakis, C.; Mantalaris, A.; Pistikopoulos, E.N. Design of In Silico Experiments as a Tool for Nonlinear Sensitivity Analysis of Knowledge-Driven Models. *Ind. Eng. Chem. Res.* **2014**, *53*, 7517–7525. [CrossRef]
19. Montgomery, D.C. *Design and Analysis of Experiments*, 8th ed.; Wiley: New York, NY, USA, 2013.
20. Saltelli, A. Making best use of model evaluations to compute sensitivity indices. *Comput. Phys. Commun.* **2002**, *145*, 280–297. [CrossRef]
21. Draper, N.R.; Smith, H. *Applied Regression Analysis*; Wiley: New York, NY, USA, 1998.
22. MathWorks. *Optimization Toolbox™ User's Guide (2015b)*; MathWorks Inc.: Natick, MA, USA, 2015.
23. Byrd, R.H.; Hribar, M.E.; Nocedal, J. An Interior Point Algorithm for Large-Scale Nonlinear Programming. *SIAM J. Optim.* **1999**, *9*, 877–900. [CrossRef]

processes

MDPI

Article

Toward a Comprehensive and Efficient Robust Optimization Framework for (Bio)chemical Processes

Xiangzhong Xie [1,2,3], René Schenkendorf [1,2,*] and Ulrike Krewer [1,2]

1 Institute of Energy and Process Systems Engineering, Technische Universität Braunschweig, Franz-Liszt-Straße 35, 38106 Braunschweig, Germany; x.xie@tu-braunschweig.de (X.X.); u.krewer@tu-braunschweig.de (U.K.)
2 Center of Pharmaceutical Engineering (PVZ), Technische Universität Braunschweig, Franz-Liszt-Straße 35a, 38106 Braunschweig, Germany
3 International Max Planck Research School (IMPRS) for Advanced Methods in Process and Systems Engineering, Sandtorstraße 1, 39106 Magdeburg, Germany
* Correspondence: r.schenkendorf@tu-braunschweig.de; Tel.: +49-531-391-65601

Received: 6 September 2018; Accepted: 26 September 2018; Published: 3 October 2018

Abstract: Model-based design principles have received considerable attention in biotechnology and the chemical industry over the last two decades. However, parameter uncertainties of first-principle models are critical in model-based design and have led to the development of robustification concepts. Various strategies have been introduced to solve the robust optimization problem. Most approaches suffer from either unreasonable computational expense or low approximation accuracy. Moreover, they are not rigorous and do not consider robust optimization problems where parameter correlation and equality constraints exist. In this work, we propose a highly efficient framework for solving robust optimization problems with the so-called point estimation method (PEM). The PEM has a fair trade-off between computational expense and approximation accuracy and can be easily extended to problems of parameter correlations. From a statistical point of view, moment-based methods are used to approximate robust inequality and equality constraints for a robust process design. We also apply a global sensitivity analysis to further simplify robust optimization problems with a large number of uncertain parameters. We demonstrate the performance of the proposed framework with two case studies: (1) designing a heating/cooling profile for the essential part of a continuous production process; and (2) optimizing the feeding profile for a fed-batch reactor of the penicillin fermentation process. According to the derived results, the proposed framework of robust process design addresses uncertainties adequately and scales well with the number of uncertain parameters. Thus, the described robustification concept should be an ideal candidate for more complex (bio)chemical problems in model-based design.

Keywords: robust optimization; uncertainty; point estimation method; equality constraints; parameter correlation

1. Introduction

Intensive competition in the (bio)chemical industry increases the requirements for better process performance. Thus, model-based tools are frequently applied to design (bio)chemical processes optimally, i.e., to optimize their performance while satisfying relevant system constraints [1,2]. However, external disturbances and process uncertainties might affect the performance of the plants, which then would deviate from the expected and simulated process characteristics or even result in operation failures [3]. The reliability of the designed processes under various conditions and disturbances is called robustness. Optimization problems that account for process performance and robustness must be tackled to provide solutions for real plants of industrial relevance.

The concept of robust optimization (RO) was first proposed by [4] and has been extensively applied to design upstream synthesis units [5,6] and downstream separation units [3,7] for bio(chemical) processes. RO concepts can be categorized into three groups: worst-case [7,8], probability-based [5,6,9] and possibility-based [10]. The worst-case and possibility-based approaches are a good choice for crude uncertainty expressions, but might lead to conservative results [11]. Probability-based concepts, which include detailed parameter uncertainty information regarding probability density functions (PDFs), are very relevant and have attracted considerable attention in the last decade [5,6,11]. However, the probability-based RO requires methods for uncertainty propagation and quantification (UQ), which pose obvious challenges in computational efficiency and approximation accuracy. Thus, the credibility and flexibility of the RO approach are determined by the underlying numerical UQ methods [4,12].

Various UQ methods for RO can be found in the literature. For instance, [13] used traditional sampling-based methods, i.e., (quasi) Monte Carlo (MC) simulation. Spectral methods, e.g., polynomial chaos expansion [14,15], have also been extensively used for RO [16–18], because of their fast convergence. Moreover, the desired statistical information can be calculated analytically. Gaussian quadrature (GQ), which was developed for solving numerical integration problems [19], is also a common approach for RO. These methods all have specific merits, but fall short in an essential aspect: they all suffer from the deficiency of computational expense. In this work, we propose the point estimate method (PEM) [2] for probability-based RO, because the PEM has superior efficiency compared to other UQ methods, as illustrated in Figure 1, and provides workable accuracy against various cubature methods, as concluded by [20,21]. Here, the computational demand (i.e., number of model evaluations) for different uncertainty quantification methods with the increasing number of uncertain parameters is illustrated to achieve similar approximation accuracy. The number of model evaluations for each method is determined based on the literature [15,20,22].

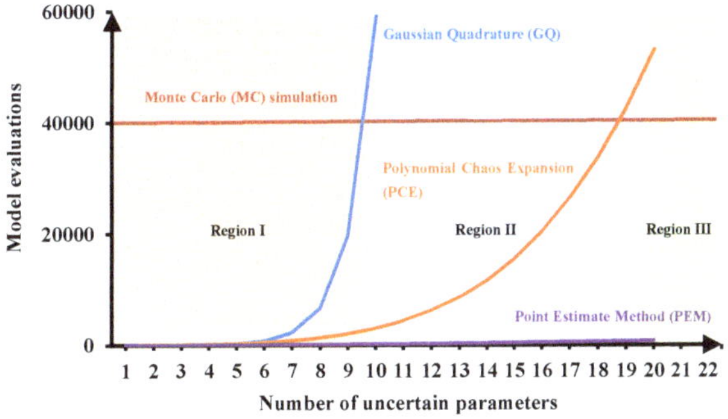

Figure 1. Computational demand (i.e., number of model evaluations) for different uncertainty quantification methods with increasing number of uncertain parameters and the same system complexity to achieve similar approximation accuracy.

The dependencies of parameter uncertainties, which is referred to as parameter dependencies in the following context, commonly exist in practical applications [23–25], but are generally not taken into account in RO studies. Recently, this issue has received more attention in the field of sensitivity analysis [25–27], where parameter correlation has a significant impact on parameter sensitivities and the resulting probability distributions of the model output [20,28]. Therefore, in this work, we adapted the PEM by implementing an isoprobabilistic transformation step [29] to include parameters dependencies properly. Thus, the effect of parameter dependencies on the RO result is investigated and critically compared with the reference case where parameter dependencies are neglected.

This paper also provides a holistic framework for probability-based RO with the PEM. The objective function is robustified by using its first and second statistical moments. The multi-objective optimization problem is transferred to a single-objective optimization problem by taking the weighted sum of these moments [5]. Moreover, we distinguish between *hard* and *soft* constraints where only the latter case needs to be robustified. *Soft* equality constraints might also be relevant in the design of (bio)chemical processes, but were rarely considered in previous RO studies [30,31]. In this work, we provide a robust formulation for *soft* inequality and equality constraints and investigate their effect on the objective function. With the statistical moments estimated by the PEM, the second and fourth moment methods introduced by [32] for structural reliability analysis are implemented to approximate the robustified *soft* constraints. The fourth moment method has a more rigorous structure than the second moment method, but requires knowledge about the third and fourth statistical moments, which might be challenging for the PEM, as the approximation accuracy degrades for higher order statistical moments. Therefore, we demonstrate and compare the performance of the two methods for approximating the robust *soft* inequality constraints. Additionally, the global sensitivity analysis technique [22] is utilized to obtain a better understanding of the process under study and provide information for simplifying and constructing the robust optimization problem systematically.

The paper is organized as follows. Section 2 refers to the basics of probability-based RO. The PEM and its extension to arbitrary and correlated parameters are described in Section 3. Section 4 provides details about robust inequality and equality constraints and approximation methods. The final structure of probability-based RO is given in Section 5. The basics of the global sensitivity analysis are given in Section 6. To demonstrate the performance of the proposed RO framework, two case studies are thoroughly discussed in Section 7: including a classic jacket tubular reactor and a fed-batch bioreactor for penicillin fermentation. Conclusions can be found in Section 8.

2. Background of Probability-Based Robust Optimization

This section starts with the problem formulation used throughout the paper and introduces the general structure of probability-based RO. First-principle models are used to describe physicochemical mechanisms of (bio)chemical processes mathematically. In the field of process system engineering, mathematical models typically consist of nonlinear different algebraic equations (DAEs) equal to:

$$\dot{\mathbf{x}}_{\mathbf{d}}(t) = \mathbf{g}_{\mathbf{d}}(\mathbf{x}(t), \mathbf{u}(t), \mathbf{p}), \qquad \mathbf{x}_{\mathbf{d}}(0) = \mathbf{x}_0, \tag{1}$$

$$\mathbf{0} = \mathbf{g}_{\mathbf{a}}(\mathbf{x}(t), \mathbf{u}(t), \mathbf{p}), \tag{2}$$

where $t \in [0, t_f]$ denotes the time, $\mathbf{u} \in \mathbb{R}^{n_u}$ the control input vector and $\mathbf{p} \in \mathbb{R}^{n_p}$ the time-invariant parameter vector. $\mathbf{x} = [\mathbf{x_d}, \mathbf{x_a}] \in \mathbb{R}^{n_x}$ is the state vector, while $\mathbf{x}_d \in \mathbb{R}^{n_{x_d}}$ and $\mathbf{x}_a \in \mathbb{R}^{n_{x_a}}$ are the differential and algebra states, respectively. $\mathbf{x}_0$ is the vector of the initial conditions for the differential states. Furthermore, two types of functions $\mathbf{g_d} : \mathbb{R}^{(n_{x_d}+n_{x_a})\times n_u \times n_p} \rightarrow \mathbb{R}^{n_{x_d}}$ and $\mathbf{g_a} : \mathbb{R}^{(n_{x_d}+n_{x_a})\times n_u \times n_p} \rightarrow \mathbb{R}^{n_{x_a}}$ are given, which denote the differential vector field and algebraic expressions of the process model.

Typically, the time-invariant parameters $\mathbf{p}$ and initial conditions $\mathbf{x}_0$ are not known exactly. Measurement and process noise give rise to uncertainties in model parameters, which are estimated through model fitting [2,23,33]. In addition, disturbances from the environment and the accuracy of the measurement devices result in uncertain initial conditions. As we intend to use random variables to describe the uncertainties in the parameters and the initial conditions, we define a probability space $(\Omega, \mathcal{F}, P)$ with the sample space Ω, σ-algebra $\mathcal{F}$, and the probability measure P. $\boldsymbol{\theta} = [\mathbf{p}(\omega), \mathbf{x}_0(\omega)]$ is the vector of random variables, which are functions of $\omega \in \Omega$ on the probability space and associated with continuous PDFs $\mathbf{f}(\boldsymbol{\theta}) = [f_1(\theta_1), \ldots, f_{n_\theta}(\theta_{n_\theta})]$ and correlation matrix Σ.

Parameter and initial condition uncertainties result in model-based prediction variations, i.e., the outcome of Equations (1) and (2) must be considered as random variables, as well. Therefore, nominal (i.e., ignoring given parameter variations) optimal control problems do not give reliable solutions for realistic processes as a single realization of the uncertain parameters is used [11]. To derive reliable solutions for almost all realizations of uncertain parameters, the following RO problem has to be solved.

Problem 1. *Probability-based robust optimization problem*

$$\min_{\mathbf{x}(\cdot),\mathbf{u}(\cdot)} \quad \mathbf{E}[M(\mathbf{x}_{t_f})] + \alpha \mathbf{Var}[M(\mathbf{x}_{t_f})], \tag{3}$$

subject to:

$$\dot{\mathbf{x}}_\mathbf{d}(t) = \mathbf{g}_\mathbf{d}(\mathbf{x}(t), \mathbf{u}(t), \mathbf{p}), \tag{4}$$

$$\mathbf{0} = \mathbf{g}_\mathbf{a}(\mathbf{x}(t), \mathbf{u}(t), \mathbf{p}), \tag{5}$$

$$\mathbf{x}_\mathbf{d}(0) = \mathbf{x}_0, \tag{6}$$

$$\mathbf{Pr}[\mathbf{h}_\mathbf{nq}(\mathbf{x}(t), \mathbf{u}(t), \mathbf{p}) \geq 0] \leq \varepsilon_{nq}, \tag{7}$$

$$\mathbf{Pr}[\mathbf{h}_\mathbf{eq}(\mathbf{x}(t), \mathbf{u}(t), \mathbf{p}) \neq 0] \leq \varepsilon_{eq}, \tag{8}$$

$$\mathbf{u}_{min} \leq \mathbf{u} \leq \mathbf{u}_{max}. \tag{9}$$

Here, $\mathbf{E}[\cdot]$ and $\mathbf{Var}[\cdot]$ denote the mean and the variance of the cost function $M(\mathbf{x}_{t_f})$, respectively, $\mathbf{Pr}[\cdot]$ denotes the probability measure, α denotes a scalar weight factor, ε_{nq} and ε_{eq} are tolerance factors, $[\mathbf{u}_{min}, \mathbf{u}_{max}]$ are the upper and lower boundaries for the control input vector and $\mathbf{x}_{t_f}$ is the state vector at the end of the time horizontal t_f. In detail, $M(\mathbf{x}_{t_f})$ denotes a Mayer objective term that is used for nominal optimal control problems. Please note that certain reformulations can be made to consider optimal Lagrange control problems, as well. The two functions $\mathbf{h}_\mathbf{nq} : \mathbb{R}^{(n_{x_d}+n_{x_a})\times n_u \times n_p} \rightarrow \mathbb{R}^{n_{nq}}$ and $\mathbf{h}_\mathbf{eq} : \mathbb{R}^{(n_{x_d}+n_{x_a})\times n_u \times n_p} \rightarrow \mathbb{R}^{n_{eq}}$ are used to represent the inequality and equality constraints, which come from process restrictions, such as temperature limitations. Equations (4) and (5) are the model equations that are considered as equality constraints as discussed in Section 4.

Problem 1 expresses the general formulation of the RO problem regarding probabilistic uncertainties. Equation (3) gives the robust form of the objective function $M(\mathbf{x}_{t_f})$, where $\mathbf{E}[M(\mathbf{x}_{t_f})]$ and $\mathbf{Var}[M(\mathbf{x}_{t_f})]$ represent the expected performance and the robustness of the objective function, respectively. The trade-off between the performance and the robustness is adjusted by the weight factor α. Equations (7) and (8) give the robust form of the inequality and equality constraints, respectively. They ensure that the probability of all constraint violations is less than or equal to a certain tolerance factor that can be adjusted according to given specifications and safety rules. However, to solve Problem 1 practically, we have to address the following two aspects. First, the estimation of the probabilities of both constraint violations cannot be solved in closed form, and standard numerical methods might be computationally demanding. Thus, highly efficient approximation routines have to be applied to ensure representative results. Second, the robust equality constraints in Equation (8) are infeasible and render RO insolvable. These two aspects are discussed and addressed in the following section.

3. Point Estimate Method

The point estimate method is a sample-based and an efficient cubature rule for approximating n-dimensional integrals [34–36]. It is analogous to the concept of the so-called unscented transformation presented by [37], which describes the parameter uncertainty with some deterministic sample points and approximate the statistics of outputs with the corresponding model evaluations, but has different deterministic sample points, associated weights and higher accuracy [34]. The PEM has been successfully applied in the field of sensitivity analysis [38] and optimal experimental design [39–41] to quantify the influence of measurement imperfections on system identification. A brief introduction to the PEM is given in Section 3.1. The concept of extending the PEM to problems with arbitrary and correlated parameter uncertainties is presented in Section 3.2.

3.1. Basics of the Point Estimate Method

The basic principle of the PEM is illustrated in Figure 2. Here, a nonlinear function $\mathbf{k}(\cdot)$ with a two-dimensional parameter $[\xi_1, \xi_2]$ and one model output y_1 is used for demonstration. We assume that the two parameters have a bivariate standard Gaussian distribution $\boldsymbol{\xi} \sim \mathcal{N}(\mathbf{0}, \mathbf{I})$. The probability distribution of the parameters does not have to be Gaussian and could follow a uniform, beta distribution or any other parametric distribution; if it is symmetric and independent [41]. First, nine deterministic sample points, i.e., the cross, circle and star points in Figure 2, are generated and used for function evaluations. Finally, the integral term is approximated by a weighted superposition of these function evaluations equal to:

$$\int_{\mathbb{I}_\xi} \mathbf{k}(\boldsymbol{\xi}) f(\boldsymbol{\xi}) d\boldsymbol{\xi} \approx \sum_{i=1}^{n_p} w_i \mathbf{k}(\boldsymbol{\xi}_i^s), \tag{10}$$

where $\boldsymbol{\xi}_i^s$ denotes the i-th sample point; n_ξ and n_p denote the number of random inputs and sample points, which are equal to two and nine in this example; w_i is a scalar weight factor; and $f(\boldsymbol{\xi})$ is the PDF of the uncertain parameters.

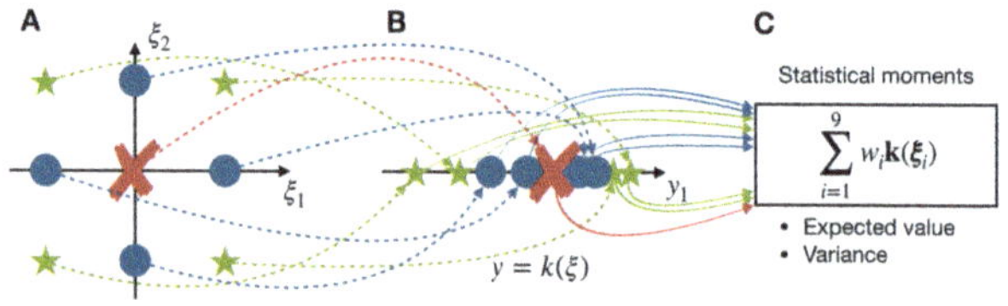

Figure 2. Illustration of the point estimation method (PEM) for a nonlinear function $\mathbf{y} = \mathbf{k}(\boldsymbol{\xi})$ that has (**A**) two random inputs; (**B**) one model output y_1; and (**C**) the resulting approximations of statistical moments of y_1.

The deterministic sample points used in this work are generated by the first three generator functions (GF[0], GF[$\pm\vartheta$], GF[$\pm\vartheta, \pm\vartheta$]) defined in [34], which leads to an overall number of $n_p = 2d^2 + 1$ sample points, where $d = n_\xi = n_\theta$. The specific weight factors are used for each generator function, which results in the final approximation scheme assuming standard Gaussian distributions:

$$\begin{aligned} &\int_{\mathbb{I}_\xi} \mathbf{k}(\boldsymbol{\xi}) f(\boldsymbol{\xi}) d\boldsymbol{\xi} \approx \\ &w_0 \mathbf{k}(GF[0]) + w_1 \sum \mathbf{k}(GF[\pm\vartheta]) + w_2 \sum \mathbf{k}(GF[\pm\vartheta, \pm\vartheta]), \end{aligned} \tag{11}$$

where $\vartheta = \sqrt{3}, w_0 = 1 + \frac{d^2-7d}{18}, w_1 = \frac{4-d}{18}, w_2 = \frac{1}{36}$ [38]. With these factors, Equation (11) provides suitable approximations for the integral of functions with moderate nonlinearities, i.e., system up to fifth-order [20,34]. Please note that the system with fifth-order means it can be accurately approximated with the sum of monomials up to order of five. In principle, we can also adapt the PEM to ensure lower or higher precision, but the proposed setting has the best trade-off between precision and computational costs [35].

3.2. Sampling Strategy for Independent/Correlated Random Variables of Arbitrary Distributions

As mentioned above, the proposed PEM is applicable only in the case of independent standard Gaussian distributions describing the parameter uncertainties. For most practical applications, however, we are confronted with problems of arbitrary and correlated probability distributions. Therefore, we extend the PEM by following Proposition 1.

Proposition 1. *For two random variables $(\boldsymbol{\theta}, \boldsymbol{\xi})$, where $\boldsymbol{\xi} \sim \mathcal{N}(\mathbf{0}, \mathbf{I})$ and $\boldsymbol{\theta}$ has an arbitrary distribution and the function $\Phi(\cdot) = F_{\theta}^{-1}(F_{\xi}(\cdot))$, the following relation for the integral terms of the nonlinear function $k(\boldsymbol{\theta})$ holds [20]:*

$$\int_{I_\theta} k(\boldsymbol{\theta}) f(\boldsymbol{\theta}) d\boldsymbol{\theta} = \int_{I_\xi} k(\Phi(\boldsymbol{\xi})) f(\boldsymbol{\xi}) d\boldsymbol{\xi}. \tag{12}$$

Based on Proposition 1, the integral expression with an arbitrary correlation function is approximated as:

$$\int_{\mathbb{I}_\theta} \mathbf{k}(\boldsymbol{\theta}) f(\boldsymbol{\theta}) d\boldsymbol{\theta} \approx w_0 \mathbf{k}(\Phi(\boldsymbol{\xi}^1)) + w_1 \sum_{i=2}^{2d+1} \mathbf{k}(\Phi(\boldsymbol{\xi}^i)) + w_2 \sum_{j=2d+2}^{2d^2+1} \mathbf{k}(\Phi(\boldsymbol{\xi}^j)), \tag{13}$$

where the samples from the original PEM for $\boldsymbol{\xi}$ are transformed via $\Phi(\cdot) = F_{\theta}^{-1}(F_{\xi}(\cdot))$ to the corresponding points in $\boldsymbol{\theta}$, which can be directly evaluated with function $k(\cdot)$. The joint cumulative density function (CDF) $F_{\boldsymbol{\theta}}(\boldsymbol{\theta})$ in $\Phi(\cdot)$ is typically unknown in practical applications and derived from marginal CDFs $[F_1(\theta_1), \ldots, F_d(\theta_d)]$ and the correlation matrix $\Sigma \in \mathbb{R}^{d \times d}$ for the uncertain parameter $\boldsymbol{\theta}$. Please note that it is actually infeasible to derive an analytical expression for $F_{\boldsymbol{\theta}}(\boldsymbol{\theta})$ and $\Phi(\cdot)$ [20]. Thus, we introduce Algorithm 1 to transform the samples from $\boldsymbol{\xi}$ to $\boldsymbol{\theta}$ numerically. The transformed sample points can be directly used for the approximation scheme:

$$\int_{\mathbb{I}_\theta} \mathbf{k}(\boldsymbol{\theta}) f(\boldsymbol{\theta}) d\boldsymbol{\theta} \approx w_0 \mathbf{k}(\boldsymbol{\theta}^1) + w_1 \sum_{i=2}^{2d+1} \mathbf{k}(\boldsymbol{\theta}^i) + w_2 \sum_{j=2d+2}^{2d^2+1} \mathbf{k}(\boldsymbol{\theta}^j). \tag{14}$$

Algorithm 1 is derived from the Nataf transformation procedure, which is based on Gaussian-copula [42]. By definition, the Gaussian-copula concept needs only the marginal distributions and the covariance matrix to approximate multivariate distributions. Technically, the Gaussian-copula is used for describing multivariate distributions with linear correlation, and thus might lose accuracy in describing multivariate distributions with non-linear correlations.

Algorithm 1 Sampling for correlated random variables

Initialization: Random variables $\boldsymbol{\xi} \sim \mathcal{N}(\mathbf{0}, \mathbf{I})$, $I \in \mathbb{R}^{d \times d}$; $\boldsymbol{\theta}$ have marginal cumulative density functions $[F_1(\theta_1), \ldots, F_d(\theta_d)]$ and correlation matrix $\Sigma \in \mathbb{R}^{d \times d}$;

1: Sample $\mathbf{U} = [\boldsymbol{\xi}^1, \cdots, \boldsymbol{\xi}^N]$ with size of $N = 2d^2 + 1$ from $\boldsymbol{\xi}$ and dimension d from Generator function $GF[\cdot]$;
2: Cholesky decomposition of $\Sigma = \mathbf{L}\mathbf{L}^T$, where L is a lower triangular matrix;
3: Correlate the sample, $\mathbf{V} = \mathbf{L}\mathbf{U}$;
4: Convert the sample to the corresponding cumulative density $\mathbf{W} = [F(V_1), \cdots, F(V_d)]^T$;
5: Transform into sample of $\boldsymbol{\theta}$, $[\boldsymbol{\theta}^1, \cdots, \boldsymbol{\theta}^N] = [F_1^{-1}(W_1), \cdots, F_d^{-1}(W_d)]^T$.

4. Moment Method for Approximating Robust Inequality and Equality Constraints

In this section, we discuss the details of inequality and equality constraints. In Section 4.1, we categorize the constraints into two special types, i.e., *hard* and *soft* constraints, and discuss the effects of parameter uncertainties on the constraints. In Sections 4.2 and 4.3, a robust formulation of *soft* inequality and *soft* equality constraints and methods for approximating the robustified expressions are presented.

4.1. Categorization of the Constraints

There are two types of robust inequality and equality constraints: *hard* and *soft* constraints [43]. *Hard* constraints must be satisfied regardless of uncertainties in the RO. *Hard* constraints ensure that optimized results satisfy physical laws. For instance, in Problem 1, equality constraints

Equations (4) and (5), i.e., the governing equations, are *hard* constraints as they describe the underlying (bio)chemical processes and have to be consistently satisfied when assuming deterministic simulation results. *Soft* constraints, in turn, do not have to be exactly satisfied under uncertainties. *Soft* constraints (e.g., Equations (7) and (8)) are typically imposed by the designer to restrict the design space and to satisfy additional process specifications. Therefore, *soft* constraints can be satisfied only in a probabilistic manner and might occasionally be violated, i.e., an acceptable violation probability has to be defined for RO. Please note that the performance of the objective function may decrease if a very low violation probability is required. *Soft* constraints are considerably affected by parameter uncertainties and are investigated in the following section.

4.2. Robust Formulation of Soft Inequality Constraints

Soft inequality constraints do not have to be strictly satisfied, but in a probabilistic manner. Inequality constraints $\mathbf{h_{nq}}(\mathbf{x}(t), \mathbf{u}(t), \mathbf{p}) \leq 0$ formulated on the probability space are also named chance constraints [44] and read as:

$$\mathbf{Pr}[\mathbf{h_{nq}}(\mathbf{x}(t), \mathbf{u}(t), \mathbf{p}) \leq 0] \geq 1 - \varepsilon_{nq}, \tag{15}$$

where the probability of constraint satisfaction must be higher or equal to $1 - \varepsilon_{nq}$. Please note that Equation (15) can also be equivalently transformed into Equation (16) when the probability of a constraint violation is used:

$$\mathbf{Pr}[\mathbf{h_{nq}}(\mathbf{x}(t), \mathbf{u}(t), \mathbf{p}) \geq 0] \leq \varepsilon_{nq}. \tag{16}$$

The probability of constraint violations is frequently estimated by MC simulations. A large number of samples are drawn from given parameter distributions, and the samples, where the constraints are violated, are counted. MC simulations are straightforward in implementation but require a considerable number of CPU-intensive model evaluations. The computational burden might be prohibitive, especially for the iterative nature of the RO. Moment-based approximation of failure probabilities has been widely applied in the field of reliability analysis [32], and thus, this method is used as an alternative concept to approximate the chance constraints in this work. In addition, it takes the advantage of the proposed PEM for estimating the needed statistical moments.

The basic idea of the moment-based approximation method is to transform the probability distribution of the constraint functions into some specific distributions, e.g., the standard normal distribution $\xi \sim \mathcal{N}(0,1)$ and to obtain the failure probability based on the probability. Here, the one-dimensional constraint function $-h_{nq}(\mathbf{x}(t), \mathbf{u}(t), \mathbf{p})$ with a negative sign is abbreviated as $\overline{h}_{nq}$ and used in the following. The isoprobabilistic transform given in Proposition 1 is applied to express the relation between the standard normal distribution and one random variable with given distribution as:

$$\xi = F_{\xi}^{-1}(F_{\overline{h}_{nq}}(\overline{h}_{nq})), \tag{17}$$

where F_{ξ}^{-1} indicates the inverse CDF of the standard normal distribution and $F_{\overline{h}_{nq}}$ indicates the CDF of h_{nq}. Based on this transformation, the failure probability of the constraint function $\overline{h}_{nq}$ is equivalent to the probability of $\xi \leq F_{\xi}^{-1}(F_{\overline{h}_{nq}}(0))$ as shown in Equation (18). As the CDF of ξ is known analytically, the failure probability of the constraint function can be determined if $F_{\xi}^{-1}(F_{\overline{h}_{nq}}(0))$ is given. However, the transformation function $F_{\xi}^{-1}(F_{\overline{h}_{nq}}(\cdot))$ is typically not available as the CDF of $\overline{h}_{nq}$ is unknown in practice. Thus, we aim at transformation rules that are based only on the statistical moments of $\overline{h}_{nq}$ [32]:

$$\begin{aligned} \mathbf{Pr}[h_{nq}(\mathbf{x}(t), \mathbf{u}(t), \mathbf{p}) \geq 0] &= \mathbf{Pr}[\overline{h}_{nq} \leq 0], \\ &= \mathbf{Pr}[\xi \leq F_{\xi}^{-1}(F_{\overline{h}_{nq}}(0))]. \end{aligned} \tag{18}$$

Two representative moment-based approximation methods [32], i.e., the second moment method and the fourth moment method, are used to estimate the failure probability with the first four statistical moments of the probability distribution of the constraint function $\overline{h}_{nq}$, which are the mean ($\mu_{\overline{h}_{nq}}$), variance ($\sigma^2_{\overline{h}_{nq}}$), skewness ($\alpha_{\overline{h}_{nq},3}$) and kurtosis ($\alpha_{\overline{h}_{nq},4}$). The second moment method approximates the transformation function with the first two moments as in Equation (19), while the fourth moment method utilizes all four moments and has a more complex structure; see Equation (20) [32]. The approximations are incorporated in Equation (18) to calculate the failure probability of the constraints:

$$F_{\xi}^{-1}(F_{\overline{h}_{nq}}(0)) = -\frac{\mu_{\overline{h}_{nq}}}{\sigma_{\overline{h}_{nq}}}, \tag{19}$$

$$F_{\xi}^{-1}(F_{\overline{h}_{nq}}(0)) = -\frac{3(\alpha_{\overline{h}_{nq},4}-1)(\frac{\mu_{\overline{h}_{nq}}}{\sigma_{\overline{h}_{nq}}}) + \alpha_{\overline{h}_{nq},3}((\frac{\mu_{\overline{h}_{nq}}}{\sigma_{\overline{h}_{nq}}})^2 - 1)}{\sqrt{(9\alpha_{\overline{h}_{nq},4} - 5\alpha^2_{\overline{h}_{nq},3} - 9)(\alpha_{\overline{h}_{nq},4} - 1)}}. \tag{20}$$

The accuracy of the moment-based approximation methods is determined by two factors. The first factor is the intrinsic approximation error, which results from the approximated transformation function (Equation (17)) using a limited number of statistical moments. By definition, the fourth moment method has a lower intrinsic approximation error because this method is more rigorously defined with higher order statistical moments. The second factor is the estimation error of the statistical moments, especially the higher order moments, e.g., skewness and kurtosis. The PEM introduced in Section 3 is used to calculate the needed statistical moments with considerably lower computational costs in comparison to MC simulations. However, the precision of the estimated statistical moments deteriorates with higher order statistical moments, because the PEM might fail for highly nonlinear problems of higher order terms. Thus, especially the fourth moment method may suffer from the estimation error. According to these two sources of approximation errors, it is difficult to determine which approximation method, i.e., the second or fourth moment method, is superior for robust process design. Therefore, we further analyze both concepts and investigate their benefits for efficient and credible robustification strategies in the following section.

4.3. Robust Formulation of soft Equality Constraints

Similar to the inequality constraints, *soft* equality constraints are considered in a probabilistic manner for the RO problem and are given as:

$$\mathbf{Pr}[\mathbf{h_{eq}}(\mathbf{x}(t), \mathbf{u}(t), \mathbf{p}) \neq 0] \leq \varepsilon_{eq} \tag{21}$$

However, Equation (21) is not directly solvable for most applications as the constraint function h_{eq} has a continuous probability distribution. In other words, the probability of a single point is equal to zero when the random space is continuous [45]. Thus, we can find that:

$$\mathbf{Pr}[\mathbf{h_{eq}}(\mathbf{x}(t), \mathbf{u}(t), \mathbf{p}) \neq 0] = 1, \tag{22}$$

which contradicts Equation (21) if $\varepsilon_{eq} \lnapprox 1$. Note that we aim to satisfy the equality constraint with high probability, and thus, $\varepsilon_{eq} \ll 1$. Figure 3a shows an example of the equality constraint in the random parameter space. Here, the samples are drawn from their distributions, and the curve shows the locations where the samples satisfy the constraints.

To solve the RO problem, the robust equality constraints must be relaxed as shown in Figure 3b. This idea is analogous to the relaxed margin used in support vector machines (SVMs), which have been applied extensively in machine learning [46]. We ease the restriction from the constraints by admitting that samples can lie within a certain range around the constraints. Based on the relaxation, the robust equality constraints in Equation (21) are substituted by:

$$\mathbf{Pr}[-\delta_{eq} \leq \mathbf{h_{eq}}(\mathbf{x}(t), \mathbf{u}(t), \mathbf{p}) \leq \delta_{eq}] \geq 1 - \varepsilon_{eq}, \tag{23}$$

where δ_{eq} indicates the relaxation factor and determines the range of relaxed equality constraints. As we can see in Figure 3b and Equation (23), we have a region rather than a single curve where the constraint is satisfied. Thus, we can have nonzero probability, and the RO problem becomes solvable. The robust equality constraints in Equation (23) have nearly the same structure as the robust inequality constraints in Equation (15). Therefore, the methods described in Section 4.2 can be used to solve Equation (23) in RO problems immediately.

As mentioned previously, there is a trade-off between the performance of the objective function and the satisfaction probability of *soft* inequality and *soft* equality constraints. The relevant factors, $\varepsilon_{nq}, \varepsilon_{eq}$ and δ_{eq}, have to be adapted properly. More details about how to select these factors are presented with the given case studies in Section 7.

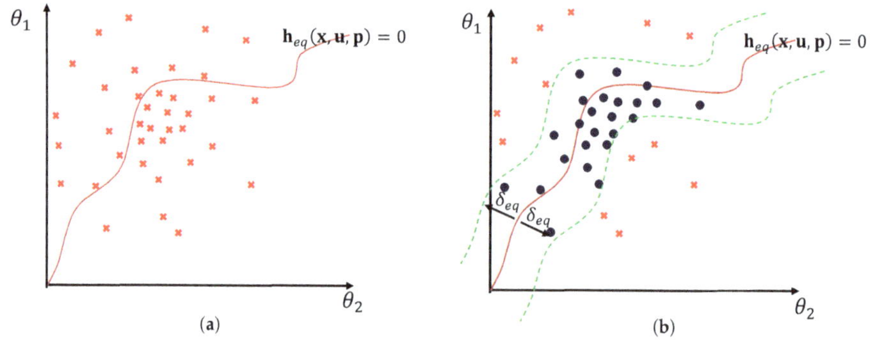

Figure 3. Illustration of *soft* equality constraints $\mathbf{h_{eq}}(\mathbf{x}, \mathbf{u}, \mathbf{p}) = 0$. For the sake of explanation, a two-dimensional random space with uncertain parameters θ_1 and θ_2 is used. Samples satisfying the constraints are shown by blue-filled circles ●, while samples that violate the constraints are shown by red cross ✖. (**a**) The probability of samples that satisfy the equality constraint (red line —) is equal to zero for the continuous random space; (**b**) the equality constraint and its relaxed boundaries (green dashed line - -) with width δ_{eq}. The probability of satisfying the equality constraints is given by the percentage of samples, i.e., ●, which are located within the boundaries.

5. Robust Optimization with the PEM

The final structure to solve the RO problem defined in Problem 1 is summarized in what follows. Note that $F(\cdot)$ in Equations (31) and (32) indicates the CDF of a standard Gaussian distribution. The PEM is used to estimate the relevant statistical moments to include the effect of parameter uncertainties. Equations (26)–(30) are the evaluations of the dynamic system and the constraint functions for all deterministic sample points that are generated from the probability distributions of the uncertain model parameter. Based on the evaluations, Equations (34)–(39) calculate the statistical moments of the objective function and constraints, which are used in Equations (24), (31) and (32). Although Equations (31) and (32) demonstrate the approximation with the fourth moment method, we can easily switch to the second moment method by changing the structure from Equation (20) to Equation (19).

$$\min_{\mathbf{x}(\cdot),\mathbf{u}(\cdot)} \quad \mathbf{E}[M(\mathbf{x}_{t_f})] + \alpha \mathbf{Var}[M(\mathbf{x}_{t_f})], \tag{24}$$

subject to:

$$i = 1, \ldots, 2d^2+1, \qquad m = 1,2,3 \tag{25}$$

$$\boldsymbol{\theta}_i = [\mathbf{p}_i, \mathbf{x}_{0,i}]^T, \mathbf{x}_i = [\mathbf{x}_{\mathbf{d},i}, \mathbf{x}_{\mathbf{a},i}]^T, \mathbf{x}_{\mathbf{d},i}(0) = \mathbf{x}_{0,i}, \mathbf{x}_{t_f,i} = \mathbf{x}_i(t_{final}), \tag{26}$$

$$\dot{\mathbf{x}}_{\mathbf{d},i}(t) = \mathbf{g_d}(\mathbf{x}_i(t), \mathbf{u}(t), \mathbf{p}_i), \quad \mathbf{0} = \mathbf{g_a}(\mathbf{x}_i(t), \mathbf{u}(t), \mathbf{p}_i), \tag{27}$$

$$\overline{h}_{1,i} = -h_{nq}(\mathbf{x}_i(t), \mathbf{u}(t), \mathbf{p}_i) \tag{28}$$

$$\overline{h}_{2,i} = h_{eq}(\mathbf{x}_i(t), \mathbf{u}(t), \mathbf{p}_i) + \delta_{eq} \tag{29}$$

$$\overline{h}_{3,i} = -h_{eq}(\mathbf{x}_i(t), \mathbf{u}(t), \mathbf{p}_i) + \delta_{eq} \tag{30}$$

$$F\left(-\frac{3(\alpha_{\overline{h}_1,4}-1)(\frac{\mu_{\overline{h}_1}}{\sigma_{\overline{h}_1}}) + \alpha_{\overline{h}_1,3}((\frac{\mu_{\overline{h}_1}}{\sigma_{\overline{h}_1}})^2-1)}{\sqrt{(9\alpha_{\overline{h}_1,4} - 5\alpha^2_{\overline{h}_1,3} - 9)(\alpha_{\overline{h}_1,4}-1)}}\right) \leq \varepsilon_{nq} \tag{31}$$

$$F\left(-\frac{3(\alpha_{\overline{h}_2,4}-1)(\frac{\mu_{\overline{h}_2}}{\sigma_{\overline{h}_2}}) + \alpha_{\overline{h}_2,3}((\frac{\mu_{\overline{h}_2}}{\sigma_{\overline{h}_2}})^2-1)}{\sqrt{(9\alpha_{\overline{h}_2,4} - 5\alpha^2_{\overline{h}_2,3} - 9)(\alpha_{\overline{h}_2,4}-1)}}\right) + F\left(-\frac{3(\alpha_{\overline{h}_4,4}-1)(\frac{\mu_{\overline{h}_4}}{\sigma_{\overline{h}_4}}) + \alpha_{\overline{h}_4,3}((\frac{\mu_{\overline{h}_4}}{\sigma_{\overline{h}_4}})^2-1)}{\sqrt{(9\alpha_{\overline{h}_4,4} - 5\alpha^2_{\overline{h}_4,3} - 9)(\alpha_{\overline{h}_4,4}-1)}}\right) \leq \varepsilon_{eq} \tag{32}$$

$$\mathbf{u}_{min} \leq \mathbf{u} \leq \mathbf{u}_{max}, \tag{33}$$

$$\mathbf{E}[M(\mathbf{x}_{t_f})] = w_0 M(\mathbf{x}_{t_f,1}) + w_1 \sum_{i=2}^{2d+1} M(\mathbf{x}_{t_f,i}) + w_2 \sum_{j=2d+2}^{2d^2+1} M(\mathbf{x}_{t_f,j}), \tag{34}$$

$$\begin{aligned}\mathbf{Var}[M(\mathbf{x}_{t_f})] = w_0(M(\mathbf{x}_{t_f,1}) - \mathbf{E}[M(\mathbf{x}_{t_f})])^2 + w_1 \sum_{i=2}^{2d+1} (M(\mathbf{x}_{t_f,i}) - \mathbf{E}[M(\mathbf{x}_{t_f})])^2 \\ + w_2 \sum_{j=2d+2}^{2d^2+1} (M(\mathbf{x}_{t_f,j}) - \mathbf{E}[M(\mathbf{x}_{t_f})])^2,\end{aligned} \tag{35}$$

$$\mu_{\overline{h}_m} = w_0 \overline{h}_{m,1} + w_1 \sum_{i=2}^{2d+1} \overline{h}_{m,i} + w_2 \sum_{j=2d+2}^{2d^2+1} \overline{h}_{m,j}, \tag{36}$$

$$\begin{aligned}\sigma^2_{\overline{h}_m} = w_0(\overline{h}_{m,1} - \mu_{\overline{h}_m})^2 + w_1 \sum_{i=2}^{2d+1} (\overline{h}_{m,i} - \mu_{\overline{h}_m})^2 + \\ w_2 \sum_{j=2d+2}^{2d^2+1} (\overline{h}_{m,j} - \mu_{\overline{h}_m})^2,\end{aligned} \tag{37}$$

$$\alpha_{\overline{h}_m,3} = \frac{w_0(\overline{h}_{m,1} - \mu_{\overline{h}_m})^3 + w_1 \sum_{i=2}^{2d+1} (\overline{h}_{m,i} - \mu_{\overline{h}_m})^3 + w_2 \sum_{j=2d+2}^{2d^2+1} (\overline{h}_{m,j} - \mu_{\overline{h}_m})^3}{\sigma^3_{\overline{h}_m}}, \tag{38}$$

$$\alpha_{\overline{h}_m,4} = \frac{w_0(\overline{h}_{m,1} - \mu_{\overline{h}_m})^4 + w_1 \sum_{i=2}^{2d+1} (\overline{h}_{m,i} - \mu_{\overline{h}_m})^4 + w_2 \sum_{j=2d+2}^{2d^2+1} (\overline{h}_{m,j} - \mu_{\overline{h}_{nq}})^4}{\sigma^4_{\overline{h}_m}}, \tag{39}$$

6. Global Sensitivity Analysis

Before we apply the robust optimization framework, we briefly introduce the idea of global sensitivity analysis (GSA). In general, GSA is not mandatory for the robust optimization framework but provides useful information for analyzing and optimizing complex systems.

GSA is a valuable tool for determining the impact of individual parameters and parameter combinations on the result of a mathematical model for given parameter variations [41,47–51]. Thus, GSA determines the most relevant parameters and parameter uncertainties to be considered in RO. By focusing on the relevant parameters and neglecting the insensitive parameters, we can reduce the complexity of the RO problem considerably.

As most model parameters, which are identified via experimental data, are correlated, the effect of parameter correlation has to be considered in GSA. In this work, we present GSA methods that are capable of problems with independent parameters and problems with dependent parameters. Although parameter dependence is quite common in practical applications, studies of GSA with dependent parameters have been considered only recently; see [25–27]. Moreover, the GSA concepts can be categorized into two types: variance-based methods [22,52,53] and moment-independent methods [54]. For details about the definitions and a critical comparison of these two concepts, the interested reader is referred to [55].

Although the moment-independent method has a more rigorous definition than the variance-based method, the variance-based approach is the standard in GSA, and thus, it is implemented in this work. The structure and types of sensitivity indices used in the variance-based method are illustrated in Figure 4. On the left of Figure 4, the variance-based method defines three types of sensitivity indices for independent parameters. First-order sensitivity indices S_i^{ind} measure the main effect of an individual parameter i on the model output, interaction sensitivity indices $S_{i,j,k,...}^{ind}$ measure the dependence of the effect for two or more parameters, and total sensitivity indices $S_{T_i}^{ind}$ are the sum of the main and interactive effects of parameter i. On the right of Figure 4, new sensitivity indices are derived from the covariance decomposition of the model output for correlated parameters. They have the same types of sensitivity indices as the independent case but for three different groups: structural sensitivity indices S^U, correlative sensitivity indices S^C, and total covariance-based sensitivity indices S^{cov}. Structural sensitivity indices reflect the impact of an individual model parameter or parameter interactions on the model output and are determined by the model structure. They have the same trend as independent sensitivity indices but with different magnitudes. The correlative sensitivity indices exclusively show the impact of parameter correlations. The sum of these indices leads to total covariance-based sensitivity indices S^{cov} that express the overall impact of the correlated parameters. In this work, the first-order sensitivity indices S_i^{ind} for the independent case and the total covariance-based first-order sensitivity indices S_i^{cov} for the correlated case are sufficient, because there are few interactions between the uncertain parameters. Values of the sensitivity indices were utilized as indicators for reducing the complexity of our RO problem as demonstrated in the design of the penicillin fermentation process.

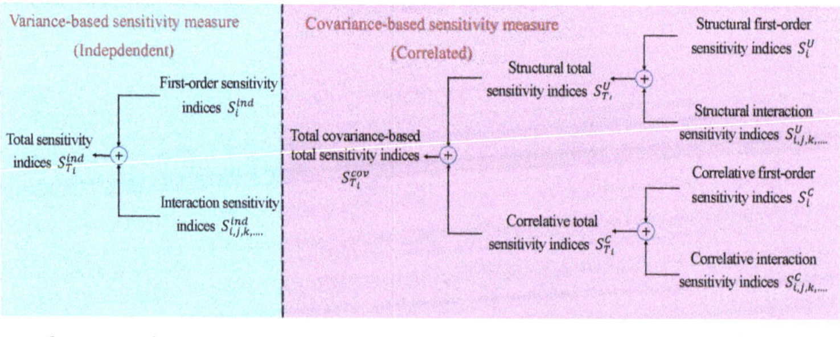

Figure 4. Structure of sensitivity measures for independent (left) and correlated (right) parameters.

7. Case Studies

In this section, we demonstrate the performance of the proposed framework with two case studies. In Case Study 1, we design an optimal jacket temperature profile for a tubular reactor considering two uncertain and correlated model parameters. Additionally, a robust equality constraint for the product temperature at the reactor outlet is assumed to incorporate process intensification aspects in the design problem. In Case Study 2, a penicillin fermentation process is analyzed as it is of interest in the pharmaceutical industry. A fed-batch bioreactor model is used to design an optimal feeding profile under parameter uncertainties. GSA is applied to determine the influence of parameter uncertainties on the process states and to offer a more tractable problem, i.e., a reduced number of uncertain model parameters, which have to be considered in the robust process design.

GSA and the RO problem were solved in MATLAB®(Version 2017b, The MathWorks Inc., Natick, MA, USA). Parameter sensitivities for the independent case were calculated with UQLAB (Version 1.0, ETH Zurich, Switzerland) [56]. The RO problem for the first case study was solved with the MATLAB function *fmincon*, while the RO problem for the second case study, which is more complex, was solved by the simultaneous approach [57] and implemented in the symbolic framework CasADi (Verion 3.3.0, KU Leuven, Belgium) for numerical optimization [58] using the NLP solver IPOPT [59] and the MA57 linear solver [60].

7.1. Case Study 1: A Jacket Tubular Reactor

Here, the design of a tubular reactor, where an irreversible first-order reaction Equation (40) takes place, is considered the first benchmark case study.

$$A \rightarrow B + C. \tag{40}$$

The reactor, which is operated under the steady-state condition, is described by the following governing Equations [31]:

$$\frac{d\mathbf{x}_1}{dz} = \frac{\alpha_{kin}}{v}(1-\mathbf{x}_1)e^{\frac{\gamma \mathbf{x}_2}{1+\mathbf{x}_2}}, \tag{41}$$

$$\frac{d\mathbf{x}_2}{dz} = \frac{\alpha_{kin}\delta}{v}(1-\mathbf{x}_1)e^{\frac{\gamma \mathbf{x}_2}{1+\mathbf{x}_2}} + \frac{\beta}{v}(\mathbf{u}-\mathbf{x}_2), \tag{42}$$

where z is the relative position along the reactor, $0 \leq z \leq 1$. The states $\mathbf{x}_1$ and $\mathbf{x}_2$ are the dimensionless forms of the reactant concentration of A and the reactor temperature, respectively. The jacket temperature is the control input given in its dimensionless form $\mathbf{u} = (T_j - T_{in})/T_{in}$ and is adjusted to meet the desired performance and robustness requirements. The control input is discretized into 25 equidistant elements constrained by 280 K and 400 K. The kinetic coefficient α_{kin} and the heat transfer coefficient β are assumed to be uncertain, i.e., they follow a Gaussian distribution with a standard deviation of 10 % of their nominal values. The implemented parameter values and operating conditions are summarized in Table 1. For additional details of the proposed reactor model, the interested reader is referred to [31]. The conversion of the reactor C_f, as well as the reactor temperature T_r, can be calculated from their dimensionless form via:

$$C_f(z=1) = x_1(z=1), \tag{43}$$

$$T_r(z) = x_2(z) \times T_{in} + T_{in}. \tag{44}$$

In this case study, we aim to maximize the final conversion of reactant A while fulfilling the given constraints on the reactor temperature. In particular, an upper boundary is added to the reactor temperature to prevent undesired side reactions. The results of the deterministic optimal design are depicted on the left of Figure 5. As we can see, the reactor temperature increases rapidly to the upper boundary to ensure the maximum reaction rate and final conversion of 0.996, respectively. However, numerous violations of the temperature boundary occur when the parameter uncertainties are taken into account. In contrast to the deterministic process design, a robust optimal design that includes parameter uncertainties is conducted next. Here, a weight factor $\alpha = 3$ and a tolerance value $\varepsilon_{nq} = 1\%$ are used for the robust objective and inequality constraints. Please note that the weight factor α indicates the amount of trade-off between process performance and robustness of objective function, and is selected based on our previous studies [6]. The tolerance value $\varepsilon_{np} = 1\%$ means the robust solution has to guarantee that the violation probability of inequality constraints should not be larger than 1%, and could be changed depending on robustness required for the inequality constraint. The robust inequality constraints are approximated with the second moment method as in Equation (19). The corresponding results are given on the right of Figure 5. As we can see from the results, the jacket temperature profile is different from the nominal design, especially from location 0.3 to the end of the reactor. Moreover, the reactor temperature profile of the robust design remains below its upper limit with a probability of 99% with the loss in the reactor performance; i.e., final conversion decreases to 0.985.

Table 1. Parameters for the tubular reactor model.

Parameters	Unit	Nominal Value	Uncertainty
$x_1(0)$	-	0	-
$x_2(0)$	-	0	-
α_{kin}	s^{-1}	0.058	$\mathcal{N}(0.058, 0.0058^2)$
β	s^{-1}	0.2	$\mathcal{N}(0.2, 0.02^2)$
v	$m\,s^{-1}$	0.1	-
γ	-	16.66	-
δ	-	0.25	-

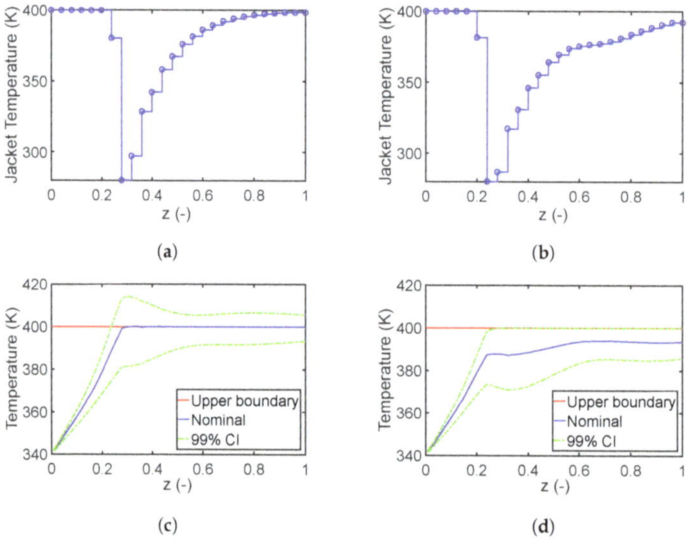

Figure 5. Results for nominal design (left) and robust design (right). (**a**,**b**) are the optimal profiles of the jacket temperature; (**c**,**d**) are the evolution of the reactor temperature and the 99% confidence interval (CI). The mean and standard deviation of the conversion of reactant A have values of [0.996, 0.004] and [0.985, 0.010] for the nominal and robust design, respectively.

7.1.1. Robust Design With Parameter Correlation

Next, we investigate the influence of parameter correlation on the robust process design. We assign the two uncertain parameters α_{kin} and β with the marginal distributions shown in Table 1 and the additional Pearson correlation coefficient of 0.8. Deterministic sample points for the correlated parameters are generated with Algorithm 1 of the modified PEM. The structure of the RO problem is similar to that for independent parameters with a weight factor $\alpha = 3$ and a tolerance value $\varepsilon_{nq} = 1\%$. Here, too, the second moment method is applied. In Figure 6, results for the optimal design with parameter correlation are given. As we can see, the profile of the jacket temperature has considerable differences compared to the nominal case; see Figure 5. Especially, the drop in the jacket temperature between position $z = 0.5$ and $z = 0.8$ results from the parameter correlation effect.

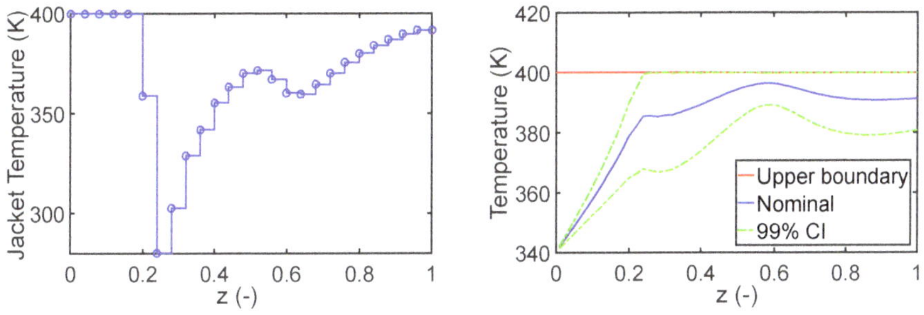

(a) (b)

Figure 6. Results for robust design with parameter correlation: (**a**) the optimal jacket temperature profile and (**b**) the reactor temperature and its 99% confidence interval (CI). The mean and standard deviation of the conversion of reactant A are 0.986 and 0.008, respectively.

7.1.2. Performance of the Fourth Moment Method

Thus far, only the second moment method has been used to approximate the robust inequality constraints. The resulting confidence intervals of the reactor temperature are illustrated with green dashed curves in Figures 5d and 6b. We can observe that the upper boundary of the confidence intervals are consistent with the upper limit of the reactor temperature once they approach it. However, the confidence intervals are approximated by taking into account only the first and second statistical moments and are insufficient if the probability distribution of the reactor temperature is non-Gaussian. Reference values based on MC simulations with 10,000 sample points are summarized in Table 2. In the case of the second moment method, the violation probabilities are 4.7% and 3.6%, respectively, which exceeds the tolerance value $\varepsilon_{nq} = 1\%$. The reason for this mismatch is mainly due to the approximation error of the robust inequality constraints while considering only the first two statistical moments.

Table 2. The number of constraint violations from 10,000 Monte Carlo simulations, where the robust inequality constraints are approximated by the second and fourth moment methods for process designs with independent and correlated parameters.

	Second Moment Method		**Fourth Moment Method**	
	Independent	Correlated	Independent	Correlated
Number of violations	470	357	440	385
Probability	0.047	0.036	0.044	0.039

As discussed in Section 4.2, the fourth moment method uses more statistical information than the second moment method and has a lower approximation error. The same RO problem is solved again with the fourth moment approach, and the violation probabilities are estimated and listed in the right of Table 2. However, the expected improvement could not be validated. In fact, the violation probability for the correlated scenario increases in case of the fourth moment method. The reason for this unexpected performance is mainly due to the estimation error of higher order statistical moments. When we compare the first four statistical moments estimated by the PEM and the MC simulations, we can see in Figure 7 that the PEM provides useful approximations for the first and second moments and deteriorates considerably for the higher order moments. As has been mentioned, the PEM is accurate if the system can be accurately approximated with the sum of monomials up to order of five, and as such its accuracy deteriorates with the increasing order of the statistical moments.

The comparison indicates that the fourth moment method might not be suitable for the PEM-based robust optimization framework, especially for practical applications where the systems might be strongly nonlinear and complex. Please note that for calculating the n-th statistical moment, not only the function $k(\xi)$ but also $k(\xi)^n$ has to be approximated, which is challenging for all sample-based approximation schemes, including MC simulations [2]. The fourth moment approach, in turn, is still a promising way to approximate probability distributions if the higher order moments can be estimated accurately, e.g., using polynomial chaos expansion or Gaussian mixture density approximation [61]. Based on this finding, in the following section, we exclusively implement the second moment method.

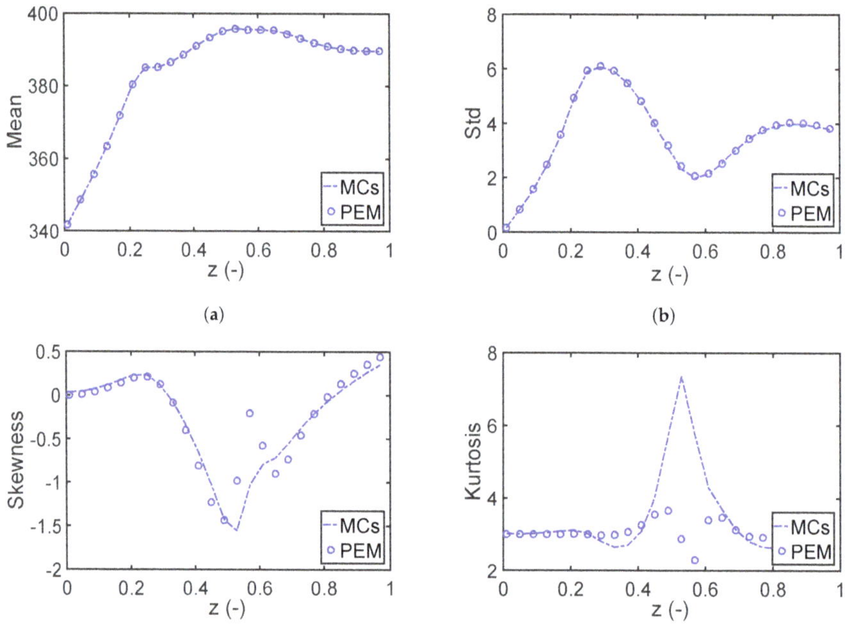

(a) (b)

(c) (d)

Figure 7. A comparison of the first four statistical moments, (**a**) mean value, (**b**) standard deviation, (**c**) skewness, and (**d**) kurtosis, estimated with the point estimate method (PEM) and Monte Carlo (MC) simulations for the reactor temperature in Case Study 1.

Alternatively, one might adjust the tolerance value for the robust inequality constraints to mitigate the effect of approximation errors when using the second moment method. The violation probabilities of the inequality constraints for the robust design with different tolerance values are given in Figure 8. As we can see, the probability can achieve 0.01 by setting the tolerance value to 0.002 for the independent and correlated cases, while the average conversion of reactant A was slightly impacted by changing the tolerance value; see Figure 8.

7.1.3. Impact of Robust Equality Constraints

Here, we would like to investigate the effect of robust equality constraints that might result from process specifications. The design of continuous processes follows the trend of process integration and intensification to reduce energy costs and raw material. For instance, to avoid extra cooling expenses for a downstream process, we can integrate the heat management into the reactor design directly. To this end, a terminal equality constraint is added to lower the outlet temperature to the value of the inlet temperature:

$$|T_r(z=1) - T_r(z=0)| = 0. \tag{45}$$

With this additional *soft* equality constraint, there exists a trade-off between maximizing the reactant conversion while minimizing the temperature difference. First, the results of the reactor design where we neglected parameter uncertainties are given in Figure 9a. The jacket temperature drops sharply to its lower limit for the second half of the reactor, and the outlet temperature returns exactly to 340 K. Consequently, the reactant conversion decreases with 2% compared to the nominal design without the equality constraint (Figure 5). Next, the effect of parameter uncertainties on the nominal design is illustrated in Figure 9b with the green dotted line. Because of limited space, we mainly consider the case where uncertain parameters are correlated. In this case, a strong violation of inequality and equality constraints exists and has to be tackled properly. The robust optimization framework proposed in Section 5 is used to solve this problem. An identical setting ($\alpha = 3$ and $\varepsilon_{nq} = 1\%$) is used for the objective function and inequality constraints here. Different scenarios with different relaxation factors δ_{eq} and tolerance factors ε_{eq} are used to demonstrate the effect of robust equality constraints on the process performance. Values for the relaxation factors and results are summarized in Figure 10. We can see that the probability distribution of the outlet temperature narrows quickly once we reduce the relaxed region and violation probability, while the performance of the reactor (the reactant conversion) deteriorates considerably. The process engineer has to decide on the trade-off between product performance and energy expense. Note that the robust inequality constraints in these scenarios are always satisfied, and thus, are not explicitly shown here.

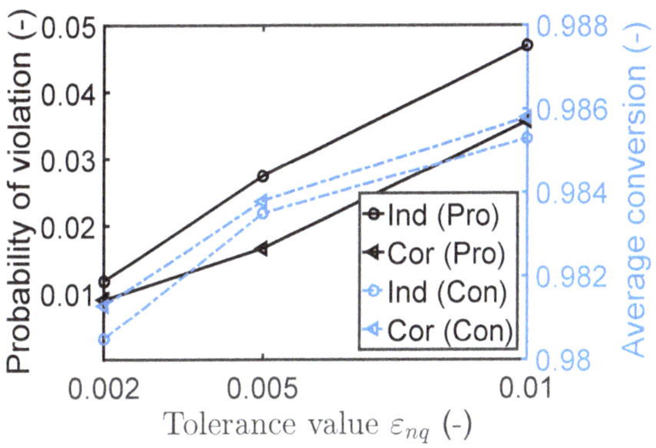

Figure 8. The violation probability of the reactor temperature (Pro) and the average conversion of the reactant A (Con) for process designs with different tolerance values. Ind and Cor indicate the results for the independent and correlated scenarios.

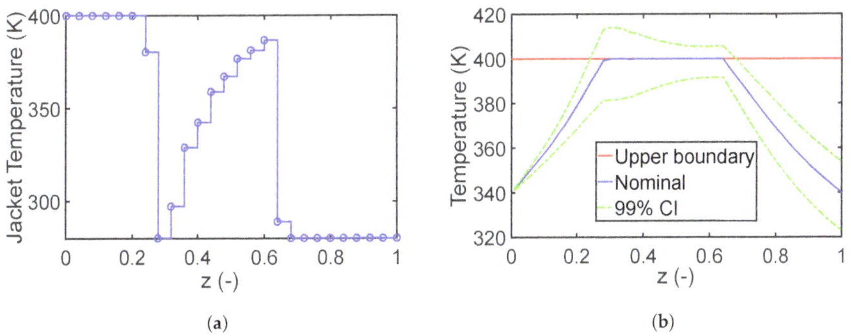

Figure 9. Results for the nominal design with terminal equality constraints: (**a**) the optimal jacket temperature profile and (**b**) the reactor temperature with its 99% confidence interval (CI). The mean and standard deviation of the conversion of reactant A are 0.980 and 0.016, respectively.

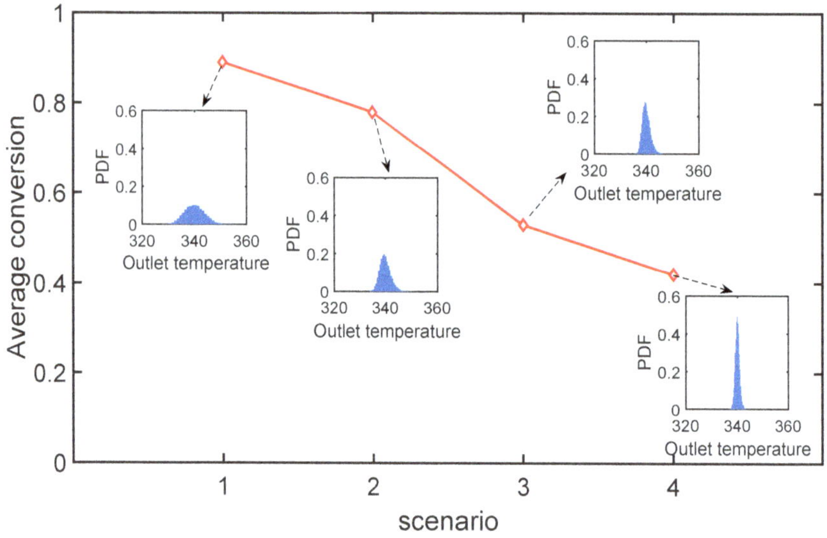

Figure 10. The average conversion of reactant A and the probability density function of the outlet temperature for four scenarios that have different relaxation factors δ_{eq} and tolerance factors ε_{eq}. 1: $\delta_{eq} = 5, \varepsilon_{eq} = 10\%$, 2: $\delta_{eq} = 5, \varepsilon_{eq} = 1\%$, 3: $\delta_{eq} = 2, \varepsilon_{eq} = 10\%$, 4: $\delta_{eq} = 2, \varepsilon_{eq} = 1\%$.

7.2. Case Study 2: Fed-Batch Bioreactor for Fermentation of Penicillin

The performance of the PEM-based robust optimization framework is also demonstrated with a fermentation process as illustrated in Figure 11. Fermentation processes have received great interest in the pharmaceutical industry, and in this study, we try to optimize the penicillin fermentation [62]. To this end, we design a feeding substrate profile that ensures the optimal performance and robustness of the bioreactor. A fed-batch reactor model is used based on the following assumptions: (1) ideal mixing of all components in the bioreactor; (2) isothermal condition in the reactor; and (3) the effect of the oxygen transfer can be neglected by considering an upper limitation on the biomass and substrate concentrations. The mathematical model for the fermentation process reads as:

$$\frac{dX}{dt} = \mu X - \frac{F}{V}X \tag{46}$$

$$\frac{dS}{dt} = -\frac{\mu}{Y_x}X - \frac{\theta_p}{Y_p}X - m_x X + \frac{F}{V}(S_f - S) \tag{47}$$

$$\frac{dP}{dt} = \theta_p X - KP - \frac{F}{V}P \tag{48}$$

$$\frac{dV}{dt} = F, \tag{49}$$

where the state variables, X, S, P and V, indicate the concentrations of the biomass, substrate, product and volume of components in the reactor, respectively. The feeding stream of the substrate has a constant concentration S_f and a time-dependent flow rate F. The specific growth rate of the biomass μ and the product θ_p is represented by the substrate inhibition kinetic of the following form:

$$\mu = \frac{\mu_m S}{S + K_x X} \tag{50}$$

$$\theta_p = \frac{\theta_m S}{S + K_p + S^2/K_i}. \tag{51}$$

The initial conditions of the state variables and the nominal value of the other kinetic parameters are summarized in Table 3. Further details about the model are given in [62].

Table 3. Nominal values of the model parameters and the initial conditions for the fed-batch model.

Parameters	Unit	Nominal Value	Parameters	Unit	Nominal Value
μ_m	1/h	0.11	m_x	1/h	0.029
K_x	-	0.006	S_f	g/L	400
θ_m	1/h	0.004	t	h	0–80
K_p	g/L	0.0001	$X(0)$	g/L	1
K_i	g/L	0.1	$S(0)$	g/L	0.5
K	1/h	0.01	$P(0)$	g/L	0
Y_x	-	0.47	$V(0)$	L	250
Y_p	-	1.2			

First, the process is optimized assuming that all parameters are estimated precisely; i.e., parameter uncertainties are neglected. The goal is to maximize the final concentration of product P within a given time range while the concentration of biomass X and substrate S should be below 40 g/L (limited by the oxygen transfer capacity) and 0.5 g/L (to avoid side reactions) for the entire time horizon, respectively [63]. The control variable F is parametrized with 100 elements, which are bounded within the range of [0, 10]. The resulting dynamic optimization problem is solved with the nominal value of all parameters, and the results are shown in Figure 12. Here, the feed rate of the substrate is adjusted to keep the substrate concentration equal to 0.5 g/L at which the maximum growth rate of the biomass is achieved at the beginning. After the biomass concentration reaches its upper limit, the substrate concentration drops nearly to zero to cease the self-reproduction of the biomass. Moreover, the substrate is fed at low rate that is consumed by the biomass to produce the desired product.

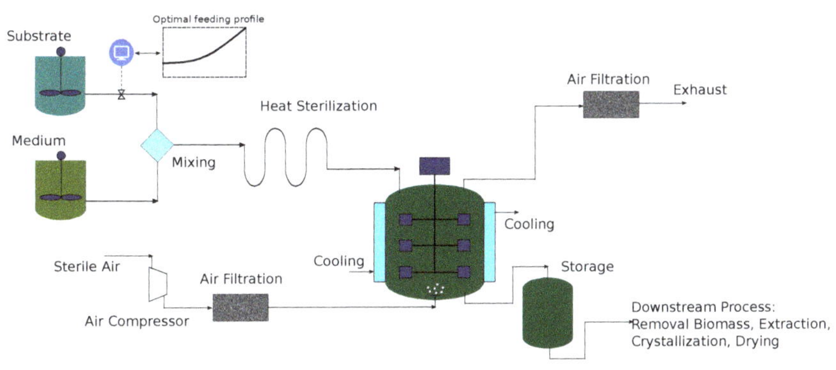

Figure 11. Scheme of a fermentation process with a fed-batch bioreactor.

However, due to imperfect measurement data and model simplifications, the estimates of the model parameters may have a considerable error as well as being correlated. Based on the results given in [64], we assign the nine parameters with a multivariate normal distribution, where their marginal distributions have mean values equal to the nominal values and standard deviations equal to 10% of the nominal values. To investigate the effect of parameter correlations, two scenarios are analyzed: (1) the parameter correlations are neglected, and the correlation matrix Σ is set to the identity matrix; (2) the correlation coefficients of μ_m, θ_m, Y_x, and m_x in Σ are set to 0.95. The effect of imprecise model parameters on the process performance is also shown in Figure 12 with the blue dotted lines. Strong violation of the constraints and large variation of the product quality are observed, and thus, the parameter uncertainty has to be considered in the process design for robustness. Please note that the negative confidence interval (CI) of the substrate concentration stems from the assumption that the CIs are symmetric and directly derived from the mean and variance of the states.

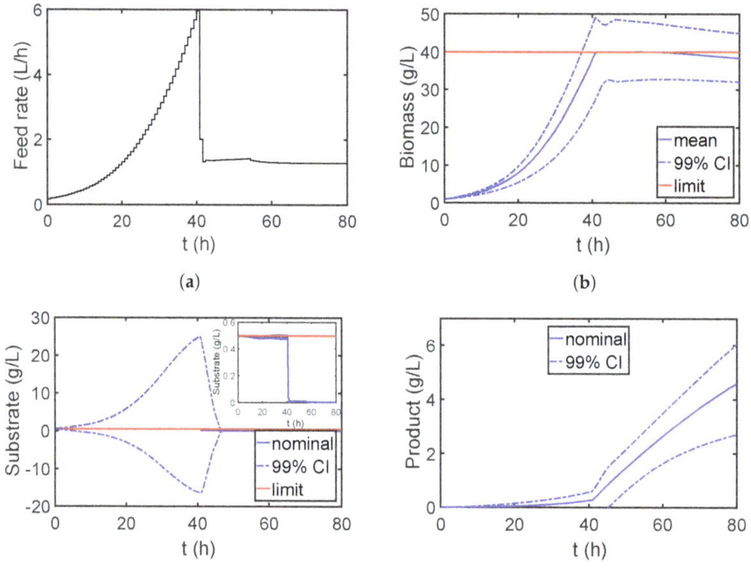

(a) (b) (c) (d)

Figure 12. (**a**) Feeding profile; evolution of the (**b**) biomass; (**c**) substrate and (**d**) product obtained from the nominal design, where the parameter uncertainties are neglected. In turn, the blue dotted lines illustrate the effect of the parameter uncertainties.

7.2.1. Global Sensitivity Analysis

Before solving the RO problem for the fermentation process, we want to decrease its computational cost by deciding which parameters are not relevant and can be neglected in the robust process design. Thus, the corresponding time-dependent sensitivity indices of the parameters are calculated for the biomass and substrate concentrations in addition to the product concentration at the final time point, i.e., for those quantities involved in either the objective function or the constraints of the optimization problem. Figures 13a,c and 14a show the sensitivity results for the independent case. As we can see, the biomass and product concentrations are strongly affected by parameters μ_m, θ_m, Y_x, and m_x, while the other parameters have a minor impact. Moreover, by summing up the first-order sensitivity indices, the interaction among the parameters are negligible. Next, we calculate the correlative (S^C) and total covariance-based (S^{cov}) first-order sensitivity regarding the biomass, substrate, and product concentration; see Figures 13b,d and 14b. Here, we do not show the results for the structural sensitivity indices and all the total sensitivity indices. The reason is that the model structure does not change with the existence of parameter correlations, and thus, the structural sensitivity indices and parameter interactions are similar to those for the independent case. Nevertheless, an evident effect of parameter correlations on the sensitivity analysis result can be observed: The parameter sensitivities have a completely different trend compared to the independent case. The sensitivity results from the correlated case also suggest considering the uncertainties and correlations from μ_m, θ_m, Y_x, and m_x for the RO problem. By using the information from the sensitivity analysis, we can significantly reduce the number of required PEM points for the RO problem. The number of model evaluations for each optimization iteration decreases from $2 \times 9^2 + 1 = 163$ to $2 \times 4^2 + 1 = 33$ for the independent and correlated cases. The performance of the RO with parameter uncertainties of appreciable sensitivities is studied in the following section.

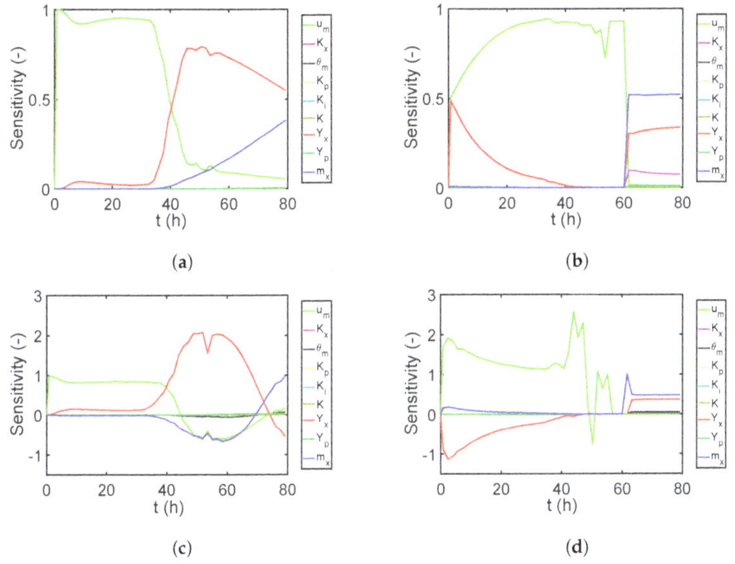

Figure 13. Sensitivity results of the nine parameters on the biomass and substrate concentrations for the independent case: (**a**) first-order sensitivity indices for the concentration of biomass X; (**b**) first-order sensitivity indices for the concentration of substrate S; and correlated case (**c**) total covariance-based first-order sensitivity indices for biomass X; (**d**) total covariance-based first-order sensitivity indices for substrate S .

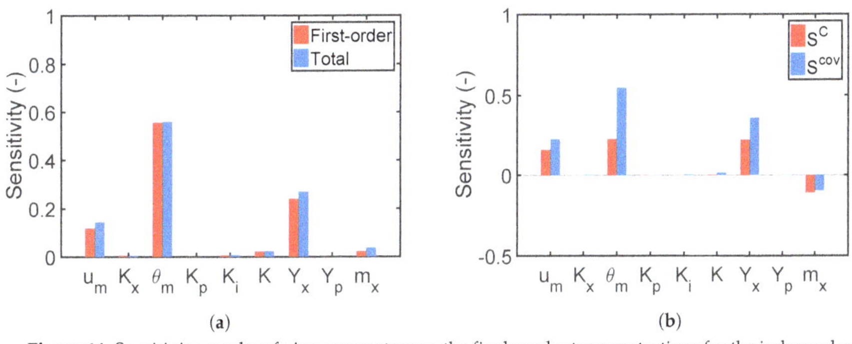

Figure 14. Sensitivity results of nine parameters on the final product concentrations for the independent (**a**) and correlated (**b**) case.

7.2.2. Robust Optimization

The RO is solved with the framework proposed in Section 5. To this end, a weight factor $\alpha = 3$ and a tolerance value $\varepsilon_{nq} = 1\%$ are used for the robust objective and inequality constraints. The PEM points for RO are generated only for parameters with appreciable sensitivities, i.e., four parameters are considered. First, the RO is solved for the simplifying assumption of the independent parameters. The evolution of the mean and 99% CIs for the biomass and the substrate are illustrated in Figure 15. Please note that the CIs in all the plots are quantified with considering the uncertainties from all nine parameters. As we can see from Figure 15, the biomass grows rapidly until its CI approaches the upper boundary to maximize the productivity, while the CI of the substrate remains at its upper boundary at the beginning and decreases to a low value to activate the production phase. However, the result of the RO ignoring parameter dependencies is too conservative. The effect of parameter dependencies is shown in Figure 16 for the previous optimized setting, i.e., assuming independent parameters. The shape of the CIs of the biomass and the substrate are quite different from those in Figure 15 and do not reach their upper boundaries, which leaves some space for improvement. Therefore, we repeat the RO considering the parameter dependencies accordingly and show the results in Figure 17. As we can see, the CIs of both biomass and substrate concentration reach the upper boundaries and are less conservative compared to the results in Figure 16. The optimized feeding profile of the substrate for the independent and correlated cases are compared in Figure 18a. The substrate for the correlated case is fed with a higher rate and descended a bit earlier than that for the independent scenario. The PDFs of the product concentrations at the final time point shown in Figures 16 and 17 are compared in more detail in Figure 18b. The product concentration is improved considerably as the dashed curve, which represents the parameter dependency case, is a bit narrowed and shifted to higher concentrations.

As mentioned above, the negative CIs of the substrate concentration in all the figures are due to the assumption of symmetric distributions of the states. This also indicates that the CIs might not be accurate, and thus, we validate them by checking the number of constraint violations with 10,000 Monte Carlo simulation for the independent and correlated case, where the corresponding optimal feeding profiles are applied. The results are listed in Table 4. As we can see from the second row, the violation frequencies are higher than expected, $\varepsilon_{nq} = 1\% = \frac{100}{10{,}000}$, especially for the substrate concentration. Although the violation frequencies might be acceptable for practical applications, we can improve the RO credibility by using a smaller tolerance factor as introduced in Section 7.1.2. Corresponding results are shown in the third row of Table 4. All violation numbers are improved, while we slightly lower the reactor performance regarding the penicillin productivity.

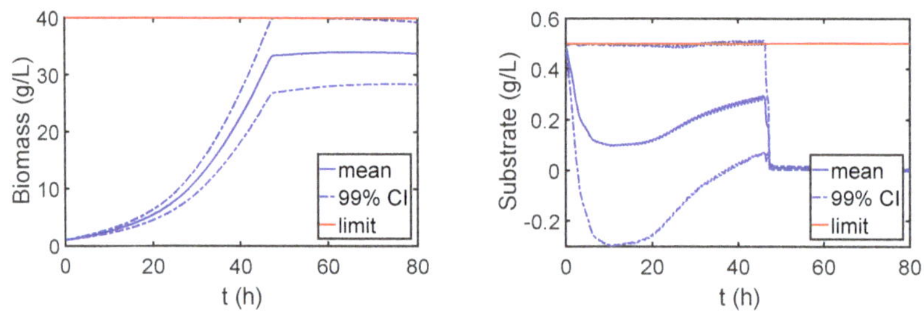

(**a**) (**b**)

Figure 15. Evolution of the mean and 99% confidence interval (CI) of the (**a**) biomass and (**b**) substrate concentrations for the robust design of the fed-batch bioreactor, where the uncertain parameters are independent. The feeding profile from the robust design with independent uncertain parameters is applied.

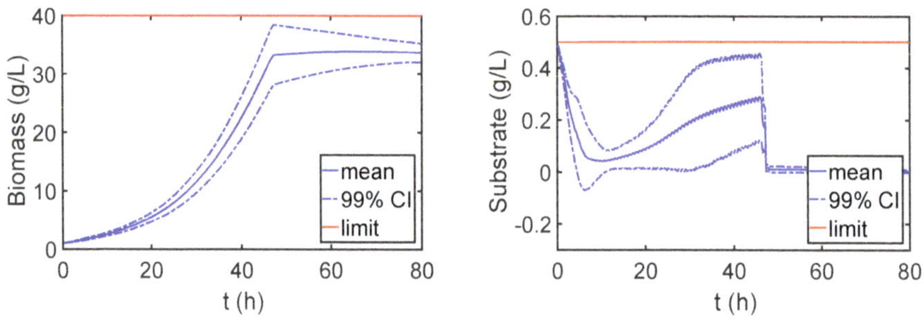

(**a**) (**b**)

Figure 16. Evolution of the mean and 99% confidence interval (CI) of the (**a**) biomass and (**b**) substrate concentrations for the robust design of the fed-batch bioreactor, where the uncertain parameters are correlated. The feeding profile from the robust design with independent uncertain parameters is applied.

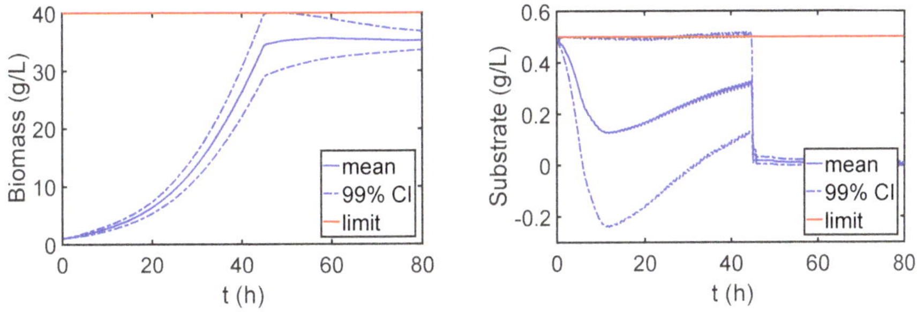

(**a**) (**b**)

Figure 17. Evolution of the mean and 99% confidence interval (CI) of the (**a**) biomass and (**b**) substrate concentrations for the robust design of the fed-batch bioreactor, where the uncertain parameters are correlated. The feeding profile from the robust design with correlated uncertain parameters is applied.

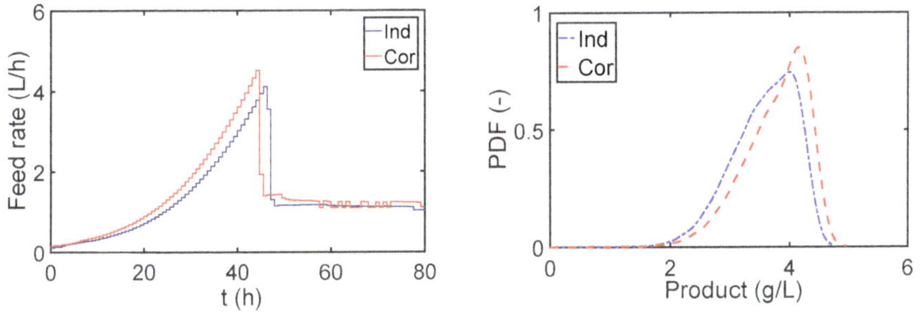

(**a**) (**b**)

Figure 18. Results for the robust design of the fed-batch bioreactor, where the uncertain parameters are either independent or correlated. (**a**) control sequence for substrate feeding; and (**b**) final concentration of the product, respectively.

Table 4. The number of constraint violations from 10,000 Monte Carlo simulations, where the tolerance factor $\varepsilon_{nq} = 1\%$ and $\varepsilon_{nq} = 0.14\%$ for both designs with independent and correlated parameters. The performance indicates the mean value of the production concentration at the end.

		Independent	**Correlated**
$\varepsilon_{nq} = 1\%$	X	146	35
	S	572	554
performance		3.63	3.76
$\varepsilon_{nq} = 0.14\%$	X	19	2
	S	378	369
performance		3.53	3.67

8. Conclusions

In this work, we proposed a new framework for solving robust optimization problems using the point estimate method. Here, a sampling strategy derived from an isoprobabilistic transformation was used to include parameter dependencies and *soft* equality constraints of practical relevance. In parallel, we also analyzed methods including fourth-order statistical moments to approximate robust equality and inequality constraints. To include only the most relevant model parameters and to reduce the computation costs, we also calculated the global parameter sensitivities before the robustification step.

Two case studies, which include chemical and biological production processes, were used to demonstrate the performance of the proposed framework. The first case study attempts to maximize the conversion of a reactant while simultaneously satisfying the constraints on the reactor temperature of a tubular reactor. The proposed method addresses the trade-off between performance and robustness for the reactor under parameter uncertainties. We observed an evident influence of parameter correlation on the designed control profile and confidence intervals of the system states. Performances of the second and fourth moment methods for approximating the robust inequality constraints were also examined. The fourth moment method has a more rigorous structure compared to the second moment approach. However, the performance of the fourth moment method is limited by the accuracy of the PEM. Thus, we concluded that the second moment method might be more favorable in this particular case. Furthermore, the approximation error could be compensated by using more conservative tolerance values, which resulted in slight deterioration of the reactor performance. To save energy costs, we also added an equality constraint to the outlet temperature. The robust equality constraint had to be relaxed deliberately to be solvable. The process performance deteriorated dramatically with lower relaxation factors. The second example is the optimal design of a bioreactor for a penicillin fermentation process. Global sensitivity analysis was used to determine the relevant parameters and to ease the computational expense of the robustification framework. This is extremely useful for large-scale problems with a high number of uncertain parameters. Moreover, the effect of parameter correlations on the robust process design was also observed. Here, the PEM still performs reasonably well and retains a relatively low computational cost.

In conclusion, the proposed framework provides a comprehensive strategy for robust optimization problems and covers features that have not been considered in previous works. It is able to achieve suitable robust design in the absence and presence of parameter correlations at low computational costs. As discussed, the PEM might fail in estimating higher order statistical moments, especially for systems with strong nonlinearities. This is also the main reason why the performance of the fourth moment method did not provide the expected improvement in robustification. Alternatively, the accuracy of the PEM can be increased using extended sample-generating rules, i.e., higher sample number results in more precise approximation at the cost of efficiency, or different methods for uncertainty quantification might be studied. Future work will focus on this issue.

Author Contributions: X.X. and R.S. designed the study. X.X. performed simulations and prepared the draft. R.S. and U.K provided feedback to the content and participated in writing the manuscript.

Funding: This research was partially funded by MWK Niedersachsen under grant "Promotionsprogramm μ-Props".

Acknowledgments: We acknowledge support by the German Research Foundation and the Open Access Publication Funds of the Technische Universität Braunschweig. Funding of the "Promotionsprogramm μ-Props" for Xiangzhong Xie by MWK Niedersachsen is gratefully acknowledged.

Conflicts of Interest: The authors declare no conflict of interest.

References

1. Biegler, L.T. *Nonlinear Programming: Concepts, Algorithms, and Applications To Chemical Processes;* SIAM: Philadelphia, PA, USA, 2010.
2. Schenkendorf, R. Optimal Experimental Design for Paramter Identification and Model Selection. Ph.D. Thesis, Otto-von-Guericke-Universität Magdeburg, Magdeburg, Germany, 2014.
3. Mortier, S.T.F.; Van Bockstal, P.J.; Corver, J.; Nopens, I.; Gernaey, K.V.; De Beer, T. Uncertainty analysis as essential step in the establishment of the dynamic Design Space of primary drying during freeze-drying. *Eur. J. Pharm. Biopharm.* **2016**, *103*, 71–83. [CrossRef] [PubMed]
4. Taguchi, G.; Clausing, D. Robust quality. *Harv. Bus. Rev.* **1990**, *68*, 65–75.
5. Vallerio, M.; Telen, D.; Cabianca, L.; Manenti, F.; Van Impe, J.; Logist, F. Robust multi-objective dynamic optimization of chemical processes using the Sigma Point method. *Chem. Eng. Sci.* **2016**, *140*, 201–216. [CrossRef]
6. Xie, X.; Schenkendorf, R.; Krewer, U. Robust design of chemical processes based on a one-shot sparse polynomial chaos expansion concept. *Comput. Aided Chem. Eng.* **2017**, *40*, 613–618.

7. Nagy, Z.K.; Braatz, R.D. Worst-case and distributional robustness analysis of finite-time control trajectories for nonlinear distributed parameter systems. *IEEE Trans. Control Syst. Technol.* **2003**, *11*, 694–704. [CrossRef]
8. Ghaoui, L.E.; Oks, M.; Oustry, F. Worst-case value-at-risk and robust portfolio optimization: A conic programming approach. *Oper. Res.* **2003**, *51*, 543–556. [CrossRef]
9. Janak, S.L.; Lin, X.; Floudas, C.A. A new robust optimization approach for scheduling under uncertainty: II. Uncertainty with known probability distribution. *Comput. Chem. Eng.* **2007**, *31*, 171–195. [CrossRef]
10. Venter, G.; Haftka, R. Using response surface approximations in fuzzy set based design optimization. *Struct. Multidiscipl. Optim.* **1999**, *18*, 218–227. [CrossRef]
11. Beyer, H.G.; Sendhoff, B. Robust optimization–a comprehensive survey. *Comput. Methods Appl. Mech. Eng.* **2007**, *196*, 3190–3218. [CrossRef]
12. Smith, R.C. *Uncertainty Quantification: Theory, Implementation, And Applications*; SIAM: Philadelphia, PA, USA, 2013; Volume 12.
13. Shi, J.; Biegler, L.T.; Hamdan, I.; Wassick, J. Optimization of grade transitions in polyethylene solution polymerization process under uncertainty. *Comput. Chem. Eng.* **2016**, *95*, 260–279. [CrossRef]
14. Wiener, N. The homogeneous chaos. *Am. J. Math.* **1938**, *60*, 897–936. [CrossRef]
15. Xiu, D.; Karniadakis, G.E. The Wiener–Askey polynomial chaos for stochastic differential equations. *SIAM J. Sci. Comput.* **2002**, *24*, 619–644. [CrossRef]
16. Mesbah, A.; Streif, S. A probabilistic approach to robust optimal experiment design with chance constraints. *IFAC-PapersOnLine* **2015**, *48*, 100–105. [CrossRef]
17. Nimmegeers, P.; Telen, D.; Logist, F.; Van Impe, J. Dynamic optimization of biological networks under parametric uncertainty. *BMC Syst. Biol.* **2016**, *10*, 86. [CrossRef] [PubMed]
18. Paulson, J.A.; Mesbah, A. An efficient method for stochastic optimal control with joint chance constraints for nonlinear systems. *Int. J. Robust Nonlinear Control* **2017**. [CrossRef]
19. Golub, G.H.; Welsch, J.H. Calculation of Gauss quadrature rules. *Math. Comput.* **1969**, *23*, 221–230. [CrossRef]
20. Xie, X.; Krewer, U.; Schenkendorf, R. Robust Optimization of Dynamical Systems with Correlated Random Variables using the Point Estimate Method. *IFAC-PapersOnLine* **2018**, *51*, 427–432. [CrossRef]
21. Freund, H.; Maußner, J. Optimization Under Uncertainty in Chemical Engineering: Comparative Evaluation of Unscented Transformation Methods and Cubature Rules. *Chem. Eng. Sci.* **2018**, *183*, 329–345.
22. Saltelli, A.; Chan, K.; Scott, E.M. *Sensitivity Analysis*; Wiley: New York, NY, USA, 2000; Volume 1.
23. Reizman, B.J.; Jensen, K.F. An automated continuous-flow platform for the estimation of multistep reaction kinetics. *Org. Process Res. Dev.* **2012**, *16*, 1770–1782. [CrossRef]
24. Sudret, B.; Caniou, Y. Analysis of covariance (ANCOVA) using polynomial chaos expansions. In Proceedings of the 11th International Conference on Structural Safety & Reliability, New York, NY, USA, 16–20 June 2013.
25. Valkó, É.; Varga, T.; Tomlin, A.; Busai, Á.; Turányi, T. Investigation of the effect of correlated uncertain rate parameters via the calculation of global and local sensitivity indices. *J. Math. Chem.* **2018**, *56*, 864–889. [CrossRef]
26. López-Benito, A.; Bolado-Lavín, R. A case study on global sensitivity analysis with dependent inputs: The natural gas transmission model. *Reliab. Eng. Syst. Saf.* **2017**, *165*, 11–21. [CrossRef]
27. Valkó, É.; Varga, T.; Tomlin, A.; Turányi, T. Investigation of the effect of correlated uncertain rate parameters on a model of hydrogen combustion using a generalized HDMR method. *Proc. Combust. Inst.* **2017**, *36*, 681–689. [CrossRef]
28. Xie, X.; Ohs, R.; Spieß, A.; Krewer, U.; Schenkendorf, R. Moment-Independent Sensitivity Analysis of Enzyme-Catalyzed Reactions with Correlated Model Parameters. *IFAC-PapersOnLine* **2018**, *51*, 753–758. [CrossRef]
29. Lebrun, R.; Dutfoy, A. Do Rosenblatt and Nataf isoprobabilistic transformations really differ? *Probab. Eng. Mech.* **2009**, *24*, 577–584. [CrossRef]
30. Logist, F.; Smets, I.; Van Impe, J. Derivation of generic optimal reference temperature profiles for steady-state exothermic jacketed tubular reactors. *J. Process Control* **2008**, *18*, 92–104. [CrossRef]
31. Telen, D.; Vallerio, M.; Cabianca, L.; Houska, B.; Van Impe, J.; Logist, F. Approximate robust optimization of nonlinear systems under parametric uncertainty and process noise. *J. Process Control* **2015**, *33*, 140–154. [CrossRef]
32. Zhao, Y.G.; Ono, T. Moment methods for structural reliability. *Struct. Saf.* **2001**, *23*, 47–75. [CrossRef]
33. Chaloner, K.; Verdinelli, I. Bayesian experimental design: A review. *Stat. Sci.* **1995**, *10*, 273–304. [CrossRef]

34. Lerner, U.N. Hybrid Bayesian Networks for Reasoning About Complex Systems. Ph.D. Thesis, Stanford University Stanford, Stanford, CA, USA, 2002.
35. Schenkendorf, R. A general framework for uncertainty propagation based on point estimate methods. In Proceedings of the Second european conference of the prognostics and health management society, phme14, Nantes, France, 8–10 July 2014.
36. Maußner, J.; Freund, H. Optimization under uncertainty in chemical engineering: Comparative evaluation of unscented transformation methods and cubature rules. *Chem. Eng. Sci.* **2018**, *183*, 329–345. [CrossRef]
37. Julier, S.J.; Uhlmann, J.K. *A General Method for Approximating Nonlinear Transformations Of Probability Distributions*; Robotics Research Group, Department of Engineering Science, University of Oxford: Oxford, UK, 1996.
38. Schenkendorf, R.; Groos, J.C. Global sensitivity analysis applied to model inversion problems: A contribution to rail condition monitoring. *Int. J. Progn. Health Manag.* **2015**, *6*, 1–14.
39. Schenkendorf, R.; Mangold, M. Qualitative and quantitative optimal experimental design for parameter identification of a map kinase model. *IFAC Proc. Vol.* **2011**, *44*, 11666–11671. [CrossRef]
40. Telen, D.; Logist, F.; Van Derlinden, E.; Van Impe, J.F. Robust optimal experiment design: A multi-objective approach. *IFAC Proc. Vol.* **2012**, *45*, 689–694. [CrossRef]
41. Schenkendorf, R.; Xie, X.; Rehbein, M.; Scholl, S.; Krewer, U. The Impact of Global Sensitivities and Design Measures in Model-Based Optimal Experimental Design. *Processes* **2018**, *6*, 27. [CrossRef]
42. Lebrun, R.; Dutfoy, A. A generalization of the Nataf transformation to distributions with elliptical copula. *Probab. Eng. Mech.* **2009**, *24*, 172–178. [CrossRef]
43. Rangavajhala, S.; Mullur, A.; Messac, A. The challenge of equality constraints in robust design optimization: Examination and new approach. *Struct. Multidiscipl. Optim.* **2007**, *34*, 381–401. [CrossRef]
44. Ostrovsky, G.; Ziyatdinov, N.; Lapteva, T. Optimal design of chemical processes with chance constraints. *Comput. Chem. Eng.* **2013**, *59*, 74–88. [CrossRef]
45. Kolmogorov, A.N. *Foundations of the Theory of Probability: Second English Edition*; Courier Dover Publications: Mineola, NY, USA, 2018.
46. Smola, A.J.; Schölkopf, B. A tutorial on support vector regression. *Stat. Comput.* **2004**, *14*, 199–222. [CrossRef]
47. Boukouvala, F.; Niotis, V.; Ramachandran, R.; Muzzio, F.J.; Ierapetritou, M.G. An integrated approach for dynamic flowsheet modeling and sensitivity analysis of a continuous tablet manufacturing process. *Comput. Chem. Eng.* **2012**, *42*, 30–47. [CrossRef]
48. Rehrl, J.; Gruber, A.; Khinast, J.G.; Horn, M. Sensitivity analysis of a pharmaceutical tablet production process from the control engineering perspective. *Int. J. Pharm.* **2017**, *517*, 373–382. [CrossRef] [PubMed]
49. Kiparissides, A.; Kucherenko, S.; Mantalaris, A.; Pistikopoulos, E. Global sensitivity analysis challenges in biological systems modeling. *Ind. Eng. Chem. Res.* **2009**, *48*, 7168–7180. [CrossRef]
50. Wang, Z.; Ierapetritou, M. Global sensitivity, feasibility, and flexibility analysis of continuous pharmaceutical manufacturing processes. *Comput. Aided Chem. Eng.* **2018**, *41*, 189–213.
51. Lin, N.; Xie, X.; Schenkendorf, R.; Krewer, U. Efficient global sensitivity analysis of 3D multiphysics model for Li-ion batteries. *J. Electrochem. Soc.* **2018**, *165*, A1169–A1183. [CrossRef]
52. Li, G.; Rabitz, H.; Yelvington, P.E.; Oluwole, O.O.; Bacon, F.; Kolb, C.E.; Schoendorf, J. Global sensitivity analysis for systems with independent and/or correlated inputs. *J. Phys. Chem. A* **2010**, *114*, 6022–6032. [CrossRef] [PubMed]
53. Mara, T.A.; Tarantola, S.; Annoni, P. Non-parametric methods for global sensitivity analysis of model output with dependent inputs. *Environ. Model. Softw.* **2015**, *72*, 173–183. [CrossRef]
54. Borgonovo, E. A new uncertainty importance measure. *Reliab. Eng. Syst. Saf.* **2007**, *92*, 771–784. [CrossRef]
55. Xie, X.; Schenkendorf, R.; Krewer, U. Efficient sensitivity analysis and interpretation of parameter correlations in chemical engineering. *Reliab. Eng. Syst. Saf.* **2018**. [CrossRef]
56. Marelli, S.; Sudret, B. UQLab: A framework for uncertainty quantification in Matlab. In Proceedings of the Second International Conference on Vulnerability and Risk Analysis and Management (ICVRAM) and the Sixth International Symposium on Uncertainty, Modeling, and Analysis (ISUMA), Liverpool, UK, 13–16 July 2014; pp. 2554–2563.
57. Biegler, L.T. An overview of simultaneous strategies for dynamic optimization. *Chem. Eng. Process. Process Intensif.* **2007**, *46*, 1043–1053. [CrossRef]
58. Andersson, J.; Åkesson, J.; Diehl, M. CasADi: A symbolic package for automatic differentiation and optimal control. In *Recent Advances in Algorithmic Differentiation*; Springer: New York, NY, USA, 2012; pp. 297–307.

59. Wächter, A.; Biegler, L.T. On the implementation of an interior-point filter line-search algorithm for large-scale nonlinear programming. *Math Program* **2006**, *106*, 25–57. [CrossRef]
60. Duff, I.S. MA57—A code for the solution of sparse symmetric definite and indefinite systems. *ACM Trans. Math. Softw. (TOMS)* **2004**, *30*, 118–144. [CrossRef]
61. Rossner, N.; Heine, T.; King, R. Quality-by-design using a gaussian mixture density approximation of biological uncertainties. *IFAC Proc. Vol.* **2010**, *43*, 7–12. [CrossRef]
62. Bajpai, R.; Reuss, M. A mechanistic model for penicillin production. *J. Chem. Technol. Biotechnol.* **1980**, *30*, 332–344. [CrossRef]
63. San, K.Y.; Stephanopoulos, G. Optimization of fed-batch penicillin fermentation: A case of singular optimal control with state constraints. *Biotechnol. Bioeng.* **1989**, *34*, 72–78. [CrossRef] [PubMed]
64. Chu, Y.; Hahn, J. Necessary condition for applying experimental design criteria to global sensitivity analysis results. *Comput. Chem. Eng.* **2013**, *48*, 280–292. [CrossRef]

processes

MDPI

Article

A Systematic Grey-Box Modeling Methodology via Data Reconciliation and SOS Constrained Regression

José Luis Pitarch [1,*], Antonio Sala [2] and César de Prada [1,3]

1 Systems Engineering and Automatic Control department, EII, Universidad de Valladolid, C/Real de Burgos s/n, 47011 Valladolid, Spain; prada@autom.uva.es
2 Instituto Universitario de Automática e Informática Industrial (ai2), Universitat Politècnica de Valencia, Camino de Vera S/N, 46022 Valencia, Spain; asala@isa.upv.es
3 Institute of Sustainable Processes (IPS), Universidad de Valladolid, C/Real de Burgos s/n, 47011 Valladolid, Spain
* Correspondence: jose.pitarch@autom.uva.es

Received: 26 February 2019; Accepted: 20 March 2019; Published: 23 March 2019

Abstract: Developing the so-called grey box or hybrid models of limited complexity for process systems is the cornerstone in advanced control and real-time optimization routines. These models must be based on fundamental principles and customized with sub-models obtained from process experimental data. This allows the engineer to transfer the available process knowledge into a model. However, there is still a lack of a flexible but systematic methodology for grey-box modeling which ensures certain coherence of the experimental sub-models with the process physics. This paper proposes such a methodology based in data reconciliation (DR) and polynomial constrained regression. A nonlinear optimization of limited complexity is to be solved in the DR stage, whereas the proposed constrained regression is based in sum-of-squares (SOS) convex programming. It is shown how several desirable features on the polynomial regressors can be naturally enforced in this optimization framework. The goodnesses of the proposed methodology are illustrated through: (1) an academic example and (2) an industrial evaporation plant with real experimental data.

Keywords: grey-box model; machine learning; SOS programming; process modeling

---

1. Introduction

Due to the increasing levels of digitalization, motivated by the ideas of the so-called Industry 4.0 [1], both academic researchers and technological companies search for methods to transform raw data in useful information. This information is expected to significantly impact in the decision-making procedures at all levels in a factory: from operation and maintenance to production scheduling and supply chain.

The process industries are not alien to this digital transformation, although the challenges to face are slightly different from the ones in other sectors. On the one hand, their activity is based in complex plants formed by very heterogeneous (usually expensive) equipment, performing complex processes such as (bio)chemical reactions, phase transformations, etc. On the other hand, their markets are quite constrained in terms of raw materials or product demands, whereas environmental regulations become tighter every year. In this context, improved efficiency thanks to smart-production systems can be achieved by: (1) transforming data in reliable information and use such information to optimize the operation [2] and (2) improve the coordination of tasks in a plant [3] and between plants [4]—in summary, to equip people (operators, engineers, managers, etc.) with support systems in order to make better tactical decisions.

Suitable models of a different nature are key to making better decisions through the above stages. Current trends encourage the use of pure data-driven approaches coming from the framework of artificial intelligence and big data (e.g., artificial neural networks [5] and machine learning [6]). These techniques are rather systematic and do not require a deep knowledge on the systems where applied. Nevertheless, the process industry is not characterized by a scarce knowledge on the involved physicochemical processes. Indeed, detailed models for some equipment/plants have already existed for the last two decades (e.g., distillation columns [7]). Therefore, throwing out all this deep knowledge and just relying on the decisions inferred by data-driven machines would be risky.

These high-fidelity models have normally been used in offline simulations, for making decisions about the process design. This drawback is due to the usually high computational complexity and the relatively limited degrees of freedom to fit the actual plants. Therefore, there is still a lack of suitable models able for online prediction at almost all tactical levels of the automation pyramid: from real-time predictive simulation and optimization [8] to production planning and scheduling [3]. Therefore, the concept of plant *digital twin* contains a (virtual replica of actual assets that matches their behavior in real time), playing an important role in decision-support systems.

Consequently, many people in the process control community have been devoting efforts during the last decade to develop efficient and reliable models to support operators and managers in their decisions [9,10]. The preferred option is building models that combine as much physical information as possible/acceptable with relationships obtained from experimental data collected from the plant [11]. In this way, these hybrid or *grey-box models* get a high level of matching with the actual plant and, importantly, they get improved prediction capabilities as their outputs will at least fulfil the basic physical laws considered.

There are many good reviews and publications on process modelling, both covering first principles [12] and data-based approaches [13], but, in the authors' opinion, there is still a lack of methodology for the systematic development of grey-box models. In addition, several different approaches have been proposed in the literature to identify the "black part" of the grey-box model from input–output data. Among them, least-squares (LS) regression with regularization in the model coefficients [14,15] is one of the most used. Nonetheless, although the obtained models with this family of methods are quite balanced in terms of fitness to data and model complexity, the guarantees of physical coherence are under discussion, as they mainly depend on the quality and quantity of the collected data for regression.

This paper proposes a two-stage methodology which combines robust data reconciliation [16] with improved constrained regression. In the first stage, one gets estimations for all process variables that are coherent with some basic physical laws. Then, in a second stage of experimental customization, sophisticated constrained regression is used to get reliable experimental relationships among variables (that are not necessarily physical inputs and outputs measured in the plant), which will complete the first-principles backbone [17].

In this context, the authors of [18,19] already proposed a useful concept for black-box modeling: a machine-learning approach which automatically selects the suitable model complexity among a set of basis functions by balancing some model-complexity criteria with the fitness to regression data. Thus, this approach can be used in the second stage (constrained regression) of our proposed grey-box model building methodology [17]. The goal in this stage is to include as much process knowledge as possible (bounds on the model response, valid input domain, monotonic responses, maximum slopes and curvatures, etc.) as constraints in the regression. However, as these types of constraints on the model need to be enforced on infinitely many points belonging to the input–output domain, the regression becomes a semi-infinite programming problem [20] where a set of finite decision variables (the model parameters) but an infinite set of constraints arise. To tackle this problem numerically, the authors of [18] break down the problem into two parts: first, a relaxation of the original problem over a finite subset $x \in \mathcal{X}^*$ of the inputs domain $\mathcal{X}^* \subset \mathcal{X}$ (typically the points in the regression dataset) is solved via mixed-integer programming (MIP). Once a solution (i.e., values of the n model coefficients $\beta \in \mathbb{R}^n$) for this problem is gathered, a subsequent maximum-violation problem needs to be solved—that is, basically, a maximization of the constraints violation over all $x \in \mathcal{X}$, with the model fixed from the previous stage. If the constraints are violated at some point, this is added to the regression dataset and the procedure repeats until no constraint violation is detected. In the general case, this procedure involves solving nonlinear optimization problems (except the MIP one if candidate basis functions and constraints are chosen to be linear in decision variables). Moreover, the problem of finding the point where maximum constraint violation takes place is generally *nonconvex*. Therefore, a *global* optimizer is required to guarantee that the best fit fulfilling constraints have been found. Altogether, this means that the constrained-regression problem can be very time-consuming and computationally demanding.

To overcome this issue, in this paper, we propose casting the constrained-regression problem as a sum-of-squares (SOS) polynomial programming one, a technique that emerged ten years ago as the generalization of the semidefinite programming to the polynomial optimization over semi-algebraic sets [21]. The great advantage of SOS programming is the ability of guaranteeing constraint satisfaction for all $x \in \mathcal{X}$ (infinitely-constrained problem) without the need for fine-sampling datasets and via *convex optimization*. Although SOS programming is quite popular now within the automatic-control community, it has not penetrated too much into other fields of application. In particular, the authors only know one work on SOS programming applied to constrained regression [22], where explicit equilibrium approximations of fast-reacting species are sought via polynomial regressors. This work is particularly interesting because its authors outlined ideas similar to ours about grey-box modeling: they searched for reduced-order representations of kinetic networks which were physically consistent. In this paper, such an initial approach is extended to pose a constrained regression problem with guaranteed satisfaction of more advanced constraints than just model positivity, e.g., boundary constraints and limits on the model (partial) derivatives.

The rest of the article is organized as follows: Section 2 presents the problem formulation and its context in a formal way. The necessary definitions and preliminary results supporting the methodology and/or the examples are summarized in Section 3. Subsequently, Section 4 presents our proposed grey-box modeling methodology and Section 5 goes deeper into the SOS constrained regression. The benefits of the proposal are illustrated in Section 6 with two examples, one academic and another based on an industrial case study. Finally, the results are discussed in the last section, providing final remarks as well as an overview for possible extensions of the method.

2. Problem Statement

Let us assume that some first-principles equations of a process are available:

$$\frac{\mathrm{d}x}{\mathrm{d}t} = f(x(t), u(t), z(t), \theta), \quad h(x(t), u(t), z(t), \theta) = 0, \tag{1}$$

where $x \in \mathbb{R}^n$ is the state vector, $u \in \mathbb{R}^m$ are *known* process inputs (manipulated variables or measured disturbances taking arbitrary values independently of the rest of the variables), $z \in \mathbb{R}^q$ are algebraic variables (with not-yet-fixed roles: some of them may be arbitrary inputs, some others may be functions of other variables, as discussed below), $\theta \in \mathbb{R}^p$ are model parameters (assumed to be constant) and $f(\cdot) \in \mathbb{R}^n$, $h(\cdot) \in \mathbb{R}^l$ can be nonlinear functions of their arguments.

Let us also assume that the above model is "incomplete", meaning that the system (1) is not fully determined by only the inputs u. Formally, this means that there are $q - n - l - m > 0$ variables $z^* \subset z$ assumed arbitrarily time-varying. However, some of them are not *unknown inputs* actually, but must be a function of other variables $z^*(x, u, z)$, representing the not well-known parts of the process. Therefore, assuming no significant unmodeled dynamics, let us assert that some additional equations $r(x, u, z) = 0$ need to be identified from process experimental data. Note that, although the first-principle model was "complete", it incorporates parameters θ whose value is not perfectly known. Therefore, stages of data collection and parameter identification are always present in practice, so that the model outputs fit those of the plant.

The classical way to approach this problem is to set up a certain functional structure for the equations $r(x, u, z, \theta_z) = 0$ searched with some parameters θ_z left for identification, and then formulate the following least-squares (LS) constrained (nonlinear) regression problem with N data samples collected at time instants $t_1, t_2, \ldots, t_N$ [23]:

$$\begin{aligned} &\min_{\theta, x_0} \sum_{t=t_1}^{t_N} \left\| \left(\hat{y}(t) - y(t)\right)/\sigma \right\|_2^2, \\ &\text{s.t.: } \frac{\mathrm{d}x}{\mathrm{d}t} = f(x(t), u(t), z(t), \theta), \quad x(0) = x_0, \\ &\quad h(x(t), u(t), z(t), \theta) = 0, \quad r(x(t), u(t), z(t), \theta_z) = 0, \end{aligned} \tag{2}$$

where $\hat{y}$ are the process measured outputs, normalized by their respective mean or standard deviation σ, and $y = c(x, z)$ are their corresponding model predictions.

Note that, in many cases, x_0 may itself be part of the adjustable parameters (initial condition fitting).

This approach assumes that the chosen structure (implicitly in whichever equations are asserted in the expression of $r(\cdot) = 0$) for $z^*(x, u, z)$ is correct and the parameters are being estimated in the right way. Therefore, a proper selection of this structure is key to obtaining a good fitness with the real plant. However, the above assumptions and expectations may fail due to the following reasons:

- Proposing a good candidate structure often implies certain knowledge of the interactions and phenomena taking place in the process, which are normally too complex to model or are not well understood.
- Some parameters θ_z may not be identifiable within reasonable precision with a scarce set of measured variables y.

Furthermore, as problem (2) is normally a nonlinear dynamic optimization one, the computational demands can exploit rapidly with the size of the dataset N, with the complexity of the model equations and the time scales of the involved dynamics [24]. Therefore, it is not recommended to unnecessarily go deep into the physical phenomena as long as the final aim does not require it, e.g., different model requirements arise for the development of digital twins than for the ones for control or optimization purposes.

Hence, as the initially proposed structure $r(\cdot)$ will likely not be correct in a complex identification setup, the fit will need to be repeated with a modified candidate structure, following a very time-consuming trial-and-error procedure, without any guarantee of optimality of course.

In these cases, it may be sensible to combine what is known with certainty about the model, such as mass or energy balances, with data-driven equations obtained from measurements, representing those parts of the process which are unknown or complex to model. This results in a *grey box* model being a mixture of first-principles and data-driven equations [25].

Pursued Goal

As these grey-box models are built to be used for interpolation and extrapolation in control and optimization routines, the data-driven parts must be in coherence with the process physics [26]. Hence, some properties on z^* and/or in their derivatives (bounds, monotony, curvature, convexity, etc.) would like to be ensured, not only in the regression data but in the entire expected region of operation.

Machine learning is thus recalled to identify such "black parts" with suitable complexity, but, as these constraints on the model outputs need to be enforced on *infinitely many points* belonging to the input–output variables domain, the constrained regression becomes a semi-infinite programming problem [20] where there is a set of finite decision variables (model parameters) but an *infinite* set of constraints. To tackle this problem numerically, the authors in [18] developed the tool ALAMO which performs a kind of two-stage procedure, where, in the first stage, there is a relaxation of the original fitting problem over a finite subset of the input variables (any variable in (1) considered to be input into the black-box submodel $z^*(x, u, z)$) is solved via mixed-integer linear programming (MILP) for suitable model feature selection. Then, once optimal values for the model parameters are gathered, a second stage of validation is performed. This step consists of solving a maximum-violation problem that is basically a nonlinear maximization of the constraint violation over the whole input region, with the model fixed from the previous stage. Then, if such maximum violation is not zero, the point where it happens is added to the regression dataset and the procedure repeats. This can be a very time-consuming process involving the resolution of several mixed-integer and nonconvex optimization problems, depending on how many re-samples are required to ensure constraint satisfaction in the whole operating region.

In this paper, we propose an alternative way to efficiently tackle the grey-box modeling problem via data reconciliation and regression sub-models based on polynomials with guaranteed constraint satisfaction.

3. Materials and Methods

The following notation and previous results will be used in the proposed methodology.

3.1. Dynamic Data Reconciliation

Data reconciliation (DR) is a well-known technique to provide a set of estimated values for process variables over time that are as close as possible to the measured values inferred by sensors, but that are coherent with the process underlying physics: fulfilling some basic first-principle laws such as mass and energy balances [27]. The approach is based on the assumption that redundant information (duplicated sensors and/or existence of extra algebraic constraints among variables) is available. Hence, an optimization problem is set up to minimize some weighted sum of the deviations between the values measured over a past time horizon H until the current time t_c and their corresponding estimations (decision variables), subject to the (usually nonlinear) model equations plus any other inequality constraint to bound the unmeasured internal variables of the model (see Figure 1).

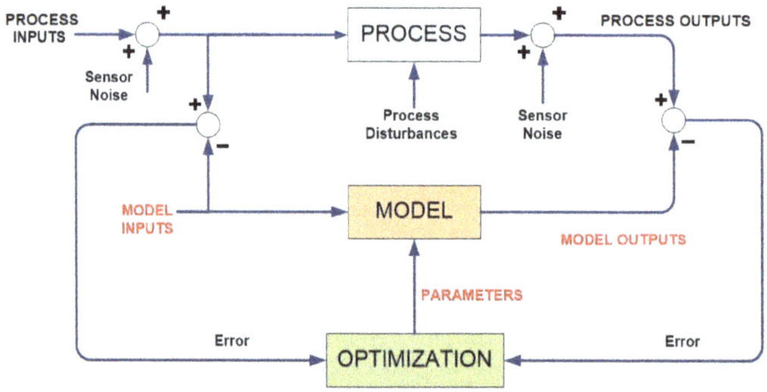

Figure 1. Standard DR scheme (decision variables highlighted in red).

The main obstacle of DR in industrial plants is the often scarce number of sensors, so that it is difficult to provide an acceptable level of redundancy with the collected data. Hence, in many cases, the approach is only able to calculate the unknown variables in the model, from perhaps some corrupted measurements that lead to incorrect estimations. In order to palliate this limitation, two courses of action are explored: (a) artificially increasing the system redundancy and (b) reducing the influence of gross errors in the measurements. Both aspects are covered in the following dynamic DR formulation:

$$\begin{aligned} \min_{u,z,w,\theta} & \int_{t_c-H}^{t_c} \left(K^2 \sum_{i=1}^{s} \left(\frac{|\epsilon_i(t)|}{K} - \log\left(1 + \frac{|\epsilon_i(t)|}{K}\right)\right) + \sum_{j=1}^{r} w_j(t)^2 \right) \mathrm{d}t, \\ \text{s.t.:} & \frac{\mathrm{d}x}{\mathrm{d}t} = f(x(t), u(t), z(t), \theta), \quad x(t_c - H) = x_0, \\ & \frac{\mathrm{d}z^*}{\mathrm{d}t} = \omega_c \cdot z^*(t) + \kappa w(t), \quad z^*(t_c - H) = z_0^*, \\ & h(x(t), u(t), z(t), \theta) = 0, \quad g(x(t), u(t), z(t), \theta) \geq 0. \end{aligned} \tag{3}$$

In this (nonlinear) dynamic optimization problem:

- $\epsilon_i := (y_i - \hat{y}_i)/\sigma_i$, $\hat{y}$ being the process measured variables with $y = c(x, u, z) \in \mathbb{R}^s$ their analogies in the model, and σ are the sensors' standard deviations.
- $f(\cdot), h(\cdot), g(\cdot), c(\cdot)$ are vectors of possibly nonlinear functions comprising the model equations (f and h), the measured outputs (vector c) and additional constraints such as upper and lower bounds in some variables or/and their variation over time (vector g).
- z^* are the *free* model variables whose value will be estimated by the DR. These z^* are supposed to vary conforming a wide-sense stationary process w whose power spectral density is limited by bandwidths $\omega_c \in \mathbb{R}^+$. Bandwidths ω_c and gains κ can be set according to an engineering guess on the variation of the mean values of θ and via the sensitivity matrix of θ in y as proposed in ([28] Chap. 3), respectively. For instance, a limit case of $\omega_c \to 0$ and $\kappa \to 0$ would represent a constant parameter.
- $K \in \mathbb{R}^+$ is a user-defined parameter to tune the slope of the fair estimator [16], i.e., the insensitivity to outliers.

The initial states x_0 and z_0^* at $t = t_c - H$ may be either assumed known from the estimations provided at the previous reconciliation run, or also left as decision variables (with possible addition of some pondering of their deviations w.r.t. such previous estimations to the cost index).

Remark 1. *Note that inclusion of x_0 and z_0^* carry out all the system information from the past, thus avoiding the need of solving (3) for large H. In fact, z_0^* can be interpreted as "virtual measurements" for the unknown variables, increasing thus the system redundancy ([28] Chap. 3).*

3.2. Sum-of-Squares Programming

Sum-of-Squares (SOS) programming will lie at the heart of the proposed methodologies in this work. This section briefly reviews the basic concepts in it.

A multivariate real polynomial p in variables $x = (x_1, \ldots, x_n)$ and coordinate degree $d = (i_1, \ldots, i_n)$ is a linear combination of monomials in x with coefficients $c_i \in \mathbb{R}$

$$p(c, x) = \sum_{i \leq d} c_i \cdot x^i, \quad x^i = x_1^{i_1} \cdots x_n^{i_n}$$

and will be denoted by $p(c, x) \in \mathcal{R}_x$.

Definition 1 (SOS polynomials). *An even-degree polynomial $p(c, x)$ is said to be SOS if it can be decomposed as a sum of squared polynomials $p(c, x) = \sum_i g_i(a, x)^2$, or, equivalently, iff $\exists\ Q(c) \succeq 0 \mid p(c, x) = z^T(x) Q(c) z(x)$, with $z(x)$ being a vector of monomials in x [29].*

Matrix Q is called the *Gram Matrix*, and checking if any $Q(c) \succeq 0$ (i.e., Q positive semidefinite) exists for a given p is a linear matrix inequality (LMI) problem [30]. In this way, checking if a polynomial $p(c, x)$ is SOS can be efficiently done via semidefinite (i.e., convex) programming (SDP) solvers [31].

The set of SOS polynomials in variables x will be denoted by Σ_x. E.g., stating that a polynomial $p(c,x)$, being c adjustable parameters, is SOS will be represented as $p(c,x) \in \Sigma_x$. Note that, evidently, all SOS polynomials are non-negative, but the inverse is not true [32].

Definition 2 (SOS polynomial matrices). *Let $F(c,x) \in \mathcal{R}_x^m$ be an $m \times m$ symmetric matrix of real polynomials in x. Then, $F(c,x)$ is an SOS polynomial matrix if it can be decomposed as $F(c,x) = H^T(a,x)H(a,x)$ or, equivalently, if $y^T F(c,x) y \in \Sigma_{x,y}$ [33].*

An $m \times m$ SOS polynomial matrix F in variables x will be denoted by $F(c,x) \in \Sigma_x^m$. Analogously to SOS polynomials, if F is SOS, then $F(c,x) \succeq 0 \forall x$.

SOS Optimization

In the same way as certifying that a polynomial (matrix) is SOS, the minimization of a linear objective in decision variables β subject to some affine-in-β SOS constraints $F(\beta,x) \in \Sigma_x^m$ or positive-definiteness constraints $M(\beta) \succeq 0$ can be cast as an SDP problem. *Local* certificates of positivity on semialgebraic sets can be checked via the Positivstellensatz theorem [34]. The following lemmas are particular versions of such general result [35].

Lemma 1. *Consider a region $\Omega(x)$ defined by polynomial boundaries as follows:*

$$\Omega(x) := \{x \mid g_1(x) \geq 0, \ldots, g_q(x) \geq (0), k_1(x) = 0, \ldots, k_e(x) = 0\}$$

If polynomial multipliers $s_i(a_i,x) \in \Sigma_x$ and $v_j(b_j,x) \in \mathcal{R}_x$ are found to be fulfilling

$$p(c,x) - \sum_{i=1}^{q} s_i(a_i,x) g_i(x) + \sum_{j=1}^{e} v_j(b_j,x) k_j(x) \in \Sigma_x, \tag{4}$$

then $p(c,x)$ is locally greater or equal to zero in $\Omega(x)$. Note that $p(c,x)$ can have an arbitrary (not necessarily even) degree, as long as $\deg(s_i \cdot g_i)$ and $\deg(v_j \cdot k_j)$ are even and greater than $\deg(p)$.

Lemma 2. *A symmetric polynomial matrix $F(c,x) \in \mathcal{R}_x^m$ is locally positive semidefinite in $\Omega(x)$ if there exist polynomial matrices $S_i(a_i,x) \in \Sigma_x^m$ and $V_j(b_j,x) \in \mathcal{R}_x^m$ verifying:*

$$F(c,x) - \sum_{i=1}^{q} S_i(a_i,x) g_i(x) + \sum_{j=1}^{e} V_j(b_j,x) k_j(x) \in \Sigma_x^m. \tag{5}$$

By the previous discussion, checking the matrix condition (5) can be done via SDP optimization algorithms and SOS decomposition [31].

Lemma 3. *The set of (polynomial) matrix inequalities nonlinear in decision variables $\beta := \{a,b,c\}$*

$$R(c,x) \succ 0, \qquad Q(a,x) - S(b,x)^T R(c,x)^{-1} S(b,x) \succ 0 \tag{6}$$

with $Q(a,x) = Q(a,x)^T$ and $R(c,x) = R(c,x)^T$, is equivalent to the following matrix expression:

$$M(\beta,x) = \begin{bmatrix} Q(a,x) & S(b,x)^T \\ S(b,x) & R(c,x) \end{bmatrix} \succ 0. \tag{7}$$

This result is the direct extension of the well-known Schur Complement in the LMI framework [36] to the polynomial case. Condition (7) can be (conservatively) checked via SOS programming, as previously discussed in Lemma 2.

3.3. Polynomial Regression with Regularization

Our methodology proposal in this work will be compared to standard regularized regression [14,15], whose basic ideas are briefly summarized next.

Assume that a normalized (zero mean and $\sigma = 1$) set of input–output data $\mathcal{X}_T\{X, Y\}$ for regression is available, where matrices X, Y have the N samples over time in columns, for the respective n_i input and n_o output variables in rows. Consider the candidate models for regression to be polynomials $p(c, x) \in \mathcal{R}_x^{n_o \times 1}$ of coordinate degree less than d in the inputs. Abusing notation, $P(c, X) \in \mathbb{R}^{n_o \times N}$ will represent the matrix resulting from evaluate $p(c, x)$ at the sampled points X.

Though polynomials are flexible candidate models, its use in machine-learning approaches is often limited to degrees $d \leq 3$ because they are very susceptible to overfitting, especially with a small number of samples. In order to palliate this drawback, a suitable *regularization* on the coefficients c of the high-degree monomials can be used, hence balancing the fitness to the training data with model complexity:

$$\min_{c} \quad \|Y - p(c, X)\|_l + \gamma \left\|\Gamma \cdot c^T\right\|_l, \tag{8}$$

where $\Gamma \in \mathbb{R}^{C_{n_i+d,n_i}}$ is a metaparameter matrix (usually diagonal) defining the regularization in each coefficient of c (i.e., its weighting structure in the objective function) and $\gamma \in \mathbb{R}^+$ is a tuning parameter to optimize training versus validation fit—see the next paragraph. Note that fitting errors as well as the regularization term may be formulated in any l-norm, typically the absolute ($l = 1$) or quadratic error ($l = 2$). In fact, the inclusion of bandwidth limits ω_c and random inputs w in (3) can also be understood as a type of regularization in a dynamic framework.

Of course, a further stage of cross validation of the "trained" model against a different dataset $\mathcal{X}_V$ (or leave-one-out validation if few data are collected) is required. Thus, given a metaparameter Γ fixed a priori, the procedure to get the polynomial model which best fits the experimental data is solving (8) performing an exploration in γ (note that the evolution of the fitting error with γ can be non-monotonic, so bisection algorithms do not apply) and choosing the model which minimizes any desired weighted combination of the training and validation errors.

4. Proposed Modeling Methodology

Instead of a priori fixing a certain structure for the unknown equations $r(x, u, z, \theta_z)$ and solving (2) or, directly by brute force using a machine-learning approach to find a complete surrogate model $y = p(u, z)$ for the whole plant or individual equipment [37], we propose following the two-stage approach for grey-box modeling:

1. **Estimation.** With the partial model (1), use data reconciliation (3) to get coherent estimations over time of all variables x, u, z and parameters θ from process data.
2. **Regression.** Identify relationships between variables z^* with any x, u and/or z, and formulate a constrained regression problem to obtain algebraic equations $r(x, u, z) = 0$. Finally, these equations are added to the first-principles ones (1) in order to get a complete model of the process.

Stage 1 typically involves solving a nonlinear dynamic optimization problem, whose resolution can be done either via sequential or simultaneous approaches [24]: Depending on the problem structure, a combination of a dynamic simulator (e.g., IDAS, CVODES, etc. [38]) with an NLP optimization algorithm (rSQP like SNOPT [39] or an evolutionary one like spMODE [40]) can be a good choice, but modern optimization environments including algorithmic differentiation (like CasADi [41] or Pyomo [42]) offer excellent features in simultaneous (sparse) optimization problems, including automatic discretization of the system dynamics by orthogonal collocation, that facilitate the use of efficient interior-point NLP codes (e.g., IPOPT [43]). The outputs of this stage are coherent variables and parameter estimations according to the known physics of the process, including the estimations for the unknown inputs $z*$ whose hidden relations with other variables will be sought in Stage 2.

For Stage 2, different approaches from machine learning can be used. However, as mentioned in Section 2, not all can take advantage of the partial knowledge that one may have about z^*. Therefore, extra (local or global) conditions on the regression models are to be enforced in order to guarantee reliable interpolation, and also extrapolation to allow z^* taking values outside the range where experimental data was collected.

Although this concept is not novel [26], modern machine learning tools generalize the resolution of this constrained-regression problem. For instance, mixed-integer programming (MIP) and global optimization methods (e.g., BARON [44]) are employed to automatically select among a set of user-provided potential basis functions, a linear combination of those that provide the best fit taking into account such extra constraints to guarantee physical coherence. As briefly mentioned in Section 2, algebraic modelling environments like ALAMO offer a good support for this task using MIP solvers and adaptive-sampling procedures. However, their computational demands are high, even in the case where the MIP problem is restricted to be linear in decision variables.

Instead of the "ALAMO approach", an alternative way for solving Stage 2 via SOS constrained regression is proposed next. In this approach, the potential set of basis functions for regression are limited to be polynomial, but the resulting optimization problem is convex and extra constraints on the model response and/or in its derivatives are naturally enforced with full guarantee of satisfaction within a desired input–output region, no matter how many samples are to be fitted or which region was covered by the experiments. In this way, high-order polynomial regressors can be used with guarantees of well-behaved resulting function approximators, compared to most options in prior literature.

5. SOS Constrained Regression

Assume that a given dataset of N sampled (or estimated) values of some output variables (those z^* in Stage 2, Section 4) $Y \in \mathbb{R}^{n_o \times N}$ and some (x, u, z) inputs $X \in \mathbb{R}^{n_i \times N}$ is available. Abusing notation for simplicity, in this section, it is assumed that x represents any set of input variables x, u, z in Stage 2, Section 4. Thus, the problem to solve is building a polynomial model of coordinate degree at most d

$$z^* = p(c, x) \in \mathcal{R}_x^{n_o \times 1}, \qquad c \in \mathbb{R}^{n_o \times C_{n_i+d, n_i}}, \tag{9}$$

with the monomial coefficients c being parameters for regression, such that a measure of the error $\mathcal{E}$ (e.g., $\mathcal{L}_1$-regularized or least-squares) w.r.t. the data being minimized over a set of constraints on the model, locally defined in the parameter $c \in \mathcal{P}$ and input $x \in \mathcal{X}$ spaces:

$$\min_{c} \mathcal{E} := \|Y - P(c, X)\|_l, \tag{10}$$

$$\text{s.t.: } \Omega(\mathcal{X}) := \{c \in \mathcal{P} \mid g(c, x) \geq 0 \; \forall x \in \mathcal{X}\}. \tag{11}$$

The vector function $g(\cdot)$ here represent a general set of *polynomial* constraints to (locally) specify some desired robust features on the model response. Thus, Ref. (11) may range from standard (polynomial) bounds on z^* ensuring, for instance, non-negativity in $x \in \mathcal{X}$, to more complex bounds on its derivatives.

In this way, Refs. (10) and (11) are a semi-infinite constrained optimization problem, but it can be cast as a convex SOS problem if polynomials p and g are affine in decision variables c, $\mathcal{E}$ is linear in c and the region $\mathcal{X}$ is defined by polynomial boundaries on x. Details are given next for each of the entities involved in the above constrained regression problem.

Objective function. Note that $P(c, X)$ in (10) can be written as $P(c, X) = c \cdot F(X)^T$, where $F(X) \in \mathbb{R}^{C_{n_i+d, n_i} \times N}$ is the Vandermonde matrix containing all the monomials up to degree d evaluated at the sample points X. Then, as usually $N >> C_{n_i+d,n_i}$, the *economic* singular value decomposition $F(X) = S_1 V_1 D$ can be used to reduce the size of (10) [22]:

$$\mathcal{E} := \|Y - P(c,X)\|_l = \|YS_1 - cDV_1\|_l \,. \tag{12}$$

Now, the more common regressors based on the $\mathcal{L}_1$ and $\mathcal{L}_2^2$ norms (absolute error and least squares respectively) can be reformulated for SDP optimization as follows:

1. $\|YS_1 - cDV_1\|_1$ is enforced by:

$$\begin{aligned} &\min_{c,\tau} \sum_{i=1}^{n_o} \tau_i \\ &\text{s.t.: } \tau - YS_1 + cDV_1 \geq 0, \quad YS_1 - cDV_1 + \tau \geq 0, \quad \tau \in \mathbb{R}_+^{n_o}. \end{aligned} \tag{13}$$

2. Using Lemma 3, $\|YS_1 - cDV_1\|_2^2$ is enforced by:

$$\begin{aligned} &\min_{c,\tau} \tau \\ &\text{s.t.: } \begin{bmatrix} \tau & YS_1 - cDV_1 \\ S_1^T Y^T - V_1 D^T c^T & I \end{bmatrix} \succeq 0. \end{aligned} \tag{14}$$

Constraints on the input/output domain. Constraints on z^* are introduced in (11) with g of the form:

$$g(c,x) = \beta_l^T p(c,x) + k_l(x), \tag{15}$$

where vector β_l weights the model outputs and $k_l(x)$ is a vector of polynomial user-defined functions in x. Hence, depending on the degree of the components of k_l, upper and lower limits for z^* (zero-order constraints) can be stated, or more complex (higher order) constraints on the feasible output region too. Moreover, using SOS programming and Lemma 1, (11) with (15) can be locally enforced in $x \in \mathcal{X}$ as long as $\mathcal{X}$ is defined by polynomial boundaries.

Constraints on the model derivatives. Model slopes and curvatures w.r.t. x get the following functional form for g in (11):

$$g(c,x) = \alpha_d^T \nabla_x p(c,x) + k_d(x), \tag{16}$$

$$g(c,x) = \mathrm{A}^T \nabla_x^2 p(c,x) \mathrm{A} + B(x), \tag{17}$$

where ∇_x stands for the gradient operator w.r.t. x. ∇_x^2 denotes the Hessian matrix and α_d, $k_d(x)$, $B(x)$ and A are user-defined elements with suitable dimensions. As derivatives of polynomials are also polynomials, (11) with (16) and/or (17) can be locally checked for SOS in $x \in \mathcal{X}$ using the results in Section 3.2.

For example, suppose that *global* convexity is to be ensured in a regression candidate model $p(x_1, x_2) = c_0 + c_1 x_1 + c_2 x_2 + c_3 x_1 x_2^2 + c_4 x_1^2 x_2$. The Hessian matrix for it is:

$$\mathrm{H}(c, x_1, x_2) = \begin{bmatrix} 2c_4 x_2 & 2c_3 x_2 + 2c_4 x_1 \\ 2c_3 x_2 + 2c_4 x_1 & 2c_3 x_1 \end{bmatrix}.$$

The classical approach to ensure convexity in p is forcing the determinant of H to be non-negative. Unfortunately, $-c_3 c_4 x_1 x_2 - c_4^2 x_1^2 - c_3^2 x_2^2 \geq 0$ is nonconvex in c and would transform (10) and (11) into a quadratically constrained regression problem. However, global convexity on p can be easily enforced using SOS programming by just setting (11) to:

$$\begin{bmatrix} 2c_4 x_2 & 2c_3 x_2 + 2c_4 x_1 \\ 2c_3 x_2 + 2c_4 x_1 & 2c_3 x_1 \end{bmatrix} \in \Sigma_{x_1,x_2}^2. \tag{18}$$

Boundary constraints. Boundary conditions (Dirichlet, Neumann, Robin or Cauchy) require equality constraints in (11), enforced over some $x_i = \mathrm{x}_i^* \in \mathcal{X}$. In this case, the general representation for g is:

$$g(c,x) = (\beta_b^T p(c,x) + \alpha_b^T \nabla_x p(c,x) + \kappa^T \nabla_x^2 p(c,x)\kappa + k_b(x))|_{x_i=\mathrm{x}_i^*} \tag{19}$$

and their local enforcement in $x \in \mathcal{X}$ can be proven again by Lemma 1 and SOS programming. Note that $g(c,x) = 0$ is equivalent to check $g(c,x) \in \Sigma_x$ jointly with $-g(c,x) \in \Sigma_x$. Moreover, $g(c,x) \in \Sigma_x$ is equivalent to $g(c,x) - s(x) = 0$ and $s(x) \in \Sigma_x$.

6. Illustrative Examples

Two examples to show the potential benefits of our proposed methodology are presented in this section. The first one is a simple academic example with artificially created data to face the SOS constrained regression against least-squares (LS) polynomial fitting with regularization, a basic approach in the machine-learning literature. The second one is an industrial example of grey-box modeling in an evaporation plant. In particular, the example shows how to build a model for the heat-transfer in a series of exchangers which suffer from fouling due to depositions of organic material.

6.1. SOS Constrained Regression versus Regularization

The purpose of this simple example is to demonstrate the improved features of our physics-based regression approach w.r.t. the "blind" regularization summarized in Section 3.3.

Assume that a dataset of 20 samples is collected from an ill-known SISO process, and that a polynomial model for it is to be sought. For building such model, the data is randomly divided in two sets, $\{X_T, Y_T\}$ with 11 samples for training and $\{X_V, Y_V\}$ with the rest for validation:

$$\begin{aligned}
X_T &= [0.6978, 1.0811, -0.5991, 0.648, -0.3354, 1.3677, 1.3317, -0.9742, 0.4538, 0.329, -1.4],\\
Y_T &= [0.1917, 0.5362, 0.554, 0.1629, 0.1718, 1.2121, 1.4415, 1.3438, 0.2583, -0.0378, 1.5],\\
X_V &= [1.4798, -0.9409, -0.7277, -1.5231, 1.7593, 1.13, -0.0821, 0.5573, 0.1789],\\
Y_V &= [1.64, 1.173, 0.8318, 1.6, 1.706, 0.64, 0.027, 0.2193, 0.1025].
\end{aligned}$$

Looking at the plotted data in Figure 2, one may infer that the "obscure" process could be convex, so fitting with quadratic candidate models would be satisfactory enough. However, this is not the case as we will explain later, and note that this visual inspection would not be possible in high-dimensional systems. Therefore, for the shake of better fitness, the candidate model will be a polynomial of, at most, degree $d = 8$:

$$p(c,x) = c_0 + c_1x + c_2x^2 + c_3x^3 + c_4x^4 + c_5x^5 + c_6x^6 + c_7x^7 + c_8x^8. \tag{20}$$

As expected, using classical unconstrained LS with (20) and just 11 samples for training leads to a totally useless overfitted model (orange curve in Figure 2) with two local minima and drastically falling down around $|x| \geq 1.5$.

6.1.1. Least Squares with Regularization

In order to avoid overfitting, regularization in c is recalled (Section 3.3). In this approach, the user must set the metaparameter Γ a priori and then perform an exploration in γ to find the best fitting for such Γ. This means that the performance in this approach is very tailored to have a good guess for Γ. Unfortunately, the metaparameter cannot be easily related to any physical insight, but only to reduce the influence of some non-preferred monomials, normally the higher-degree ones. Following this idea, two typical alternatives for the metaparameter were tested:

$$\text{[M-1]}\ \Gamma = [0,0,0,1,1,10,10,100,100]^T; \qquad \text{[M-2]}\ \Gamma = [0,0,1,e^2,e^3,e^4,e^5,e^6,e^7]^T.$$

Note that coefficients of the zero-order and linear terms are not penalized in both alternatives (at least the best linear prediction will be found in the worst case). Moreover, the quadratic term is also freed due to such intuition of convexity from data visual inspection, whereas the higher-order monomials are progressively penalized. In M-2, the usual exponential penalty with the monomials degree is set in order to balance fitness to data with model complexity.

After exploration in γ for both setups, the model with less total fitting error (chosen to be training plus validation errors) is found at $\gamma = 0.4$ with the chosen exploration granularity. The best model (coefficients below $c < 10^{-4}$ are disregarded) obtained with the metaparameter choice [M-1] is a polynomial of degree 7 (dashed blue curve in Figure 2), whereas [M-2] is a polynomial of degree 5 (dotted pink curve in Figure 2). Table 1 gives the fitting error for these "best" models, as well as some values resulting from the exploration in the regularization scaling parameter *gamma*.

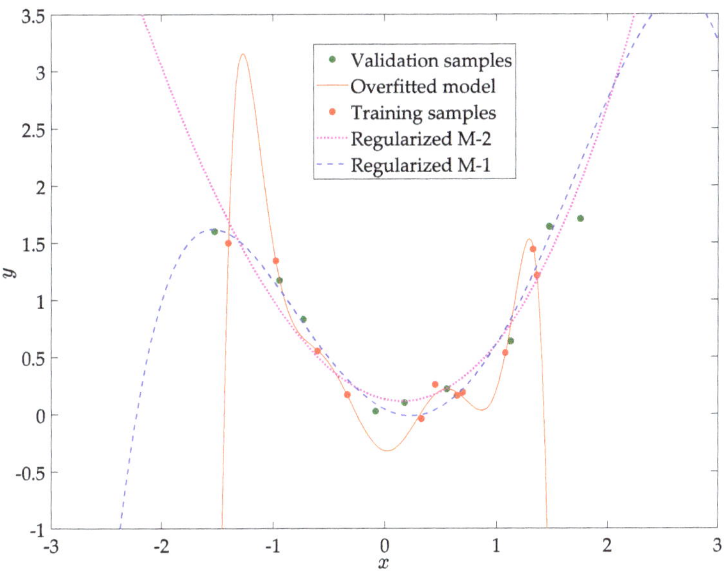

Figure 2. Sampled data and polynomial models fitted by standard LS approaches.

Table 1. Exploration in γ for M-1 and M-2 regularizations.

Γ	γ	Training Error	Validation Error	Total
M-1	0.01	0.1517	1.84	2
	0.1	0.206	0.366	0.572
	0.4	0.218	0.324	0.541
	1	0.23	0.372	0.602
	10	0.34	0.49	0.83
	100	0.416	0.55	0.967
M-2	0.001	0.184	1.021	1.2
	0.01	0.231	0.834	1.065
	0.5	0.405	0.422	0.826
	2	0.63	0.42	1.05
	10	1.671	1.698	3.37

Remark 2. *Looking at Figure 2, the model obtained by the usual exponential regularization [M-2] is preferable to the one obtained by the ad hoc [M-1] because it is quite symmetric and convex (at least in the depicted region), so it would be more "reliable" a priori for extrapolation in* $\mathcal{X} := \{x : 2 < |x| < 3\}$. *However, note that simple*

visual inspection is not available for high-dimensional systems. Thus, without visual information, one would have chosen the model by [M-1], as it is the one which best fit the data.

6.1.2. SOS Constrained Regression

Alternatively to the "blind" regularization, some desired features with physical insight on the model response could have been searched. Thus, as an initial idea, non-negativity and convexity were forced on (20) via SOS constrained regression (LS objective, Section 5) with the following constraints:

$$p(c,x) = c_0 + c_1x + c_2x^2 + c_3x^3 + c_4x^4 + c_5x^5 + c_6x^6 + c_7x^7 + c_8x^8 \quad \in \quad \Sigma_x, \tag{21}$$

$$\frac{\mathrm{d}^2p(c,x)}{\mathrm{d}x^2} = 2c_2 + 6c_3x + 12c_4x^2 + 20c_5x^3 + 30c_6x^4 + 42c_7x^5 + 56c_8x^6 \quad \in \quad \Sigma_x. \tag{22}$$

The convex polynomial found to best fit the training data ($\mathcal{E} = 0.226$) incurs in a high error on the validation data ($\mathcal{E} = 14.46$). By inspecting the modeling error with the data points (visual inspection omitted, as this possibility is hardly available in models with multiple inputs) it was found that the highest deviations appear mainly around the boundaries of the training region. Thence, it might be inferred that the generating process flattens far away from the origin, so it is probably nonconvex (or not strongly convex at least).

A simple way to find a model whose response fits better with this insight of flatness in extrapolation is to set up *local* upper and lower bounds on $p(c,x)$: $\overline{y} - p(c,x) \geq 0 \forall x \in \{x : |x| < 3\}$; $p(c,x) - \underline{y} \geq 0 \forall x \in \{x : 2 < |x| < 3\}$ or better by locally bounding the slope to small values in $\psi := \{x : 2 < |x| < 3\}$. Using Lemma 1, this last condition is enforced by the following SOS constraints:

$$\begin{aligned} p(c,x) - s_1(a_1,x)\cdot(3^2 - x^2) &\in \Sigma_x, \\ 0.3 - \frac{\mathrm{d}p(c,x)}{\mathrm{d}x} - s_2(a_2,x)\cdot(3^2-x^2) - s_3(a_3,x)\cdot(x^2-2) &\in \Sigma_x, \\ \frac{\mathrm{d}p(c,x)}{\mathrm{d}x} + 0.3 - s_4(a_4,x)\cdot(3^2-x^2) - s_5(a_5,x)\cdot(x^2-2) &\in \Sigma_x, \end{aligned} \tag{23}$$

with $s_i(a_i,x) \in \Sigma_x$ being SOS polynomial multipliers whose highest degree is $d \geq 6$, as $p(c,x)$ can be of degree 8. Note that local non-negativity of p on $\mathcal{X} := \{x : |x| < 3\}$ is also enforced, as there is no need to force global positivity outside the region considered for extrapolation, thus reducing conservatism.

The model obtained with this approach is the solid orange curve in Figure 3, labelled as [P-1]. This desired response was obtained with a total regression error (training plus validation) of $\mathcal{E} = 0.41$, beating by 25% the best fit obtained by the regularization approach.

Nonetheless, the response shows several local minima in $\mathcal{X}$. If this surrogate model is to be integrated in a larger grey-box model for real-time optimization purposes, getting a quasi-monotonous model (single global minimum) could be more interesting than achieving the lowest fitting error, in order to reduce the probability of getting stuck in local optima with gradient-based NLP solvers. Several alternative ways are available to handle this issue via SOS constrained regression:

[P-2] Positive curvature in $\mathcal{X}$, tending to zero when $x \in \psi$ (dashed-dotted pink curve in Figure 3):

$$p(c,x) \geq 0, \; \frac{\mathrm{d}^2p(c,x)}{\mathrm{d}x^2} \geq 0, \; \forall x \in \mathcal{X}; \qquad \frac{\mathrm{d}^2p(c,x)}{\mathrm{d}x^2} \leq 0.25 \; \forall x \in \psi,$$

[P-3] Upper bound on p in $\mathcal{X}$ and bounded negative curvature in $x \in \psi$ (dashed green curve):

$$2.5 \geq p(c,x) \geq 0 \; \forall x \in \mathcal{X}; \qquad 0 \geq \frac{\mathrm{d}^2p(c,x)}{\mathrm{d}x^2} > -0.8 \; \forall x \in \psi,$$

[P-4] Symmetrically bounding the slope between two values in $x \in \psi$ (dotted blue curve):

$$p(c,x) \geq 0 \, \forall x \in \mathcal{X}; \qquad 0.1 < \frac{\mathrm{d}p(c,x)}{\mathrm{d}x} < 0.6 \, \forall x \in \{2 \leq x \leq 3\};$$
$$-0.1 \geq \frac{\mathrm{d}p(c,x)}{\mathrm{d}x} \geq -0.6 \, \forall x \in \{-2 \geq x \geq -3\}.$$

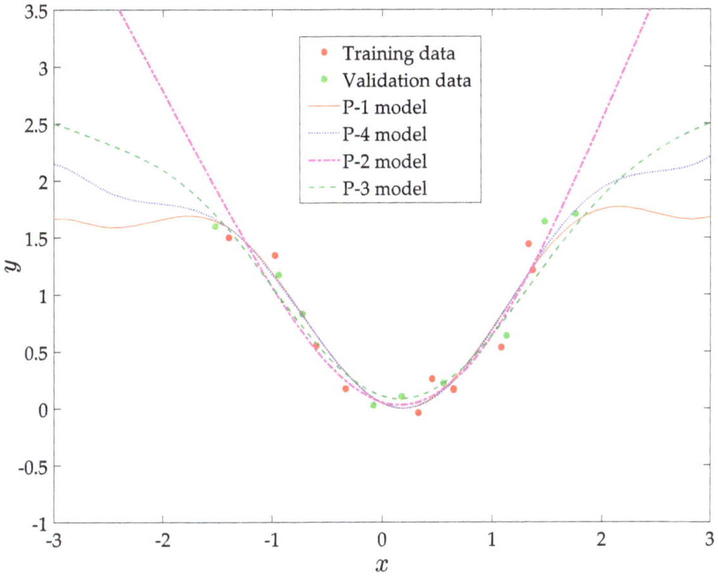

Figure 3. Sampled data and polynomial models fitted by SOS constrained regression.

As can be seen in Figure 3, the three approaches (P-2 to P-4) get *quasi-monotonous* surrogate models which are suitable for optimization purposes. The total regression error is quite similar in all the approaches (Table 2) and the small differences between them could be just a product of sheer luck (fitting the sensor noise). Thus, the choice of one over the other would only depend on the engineer physical intuition.

Table 2. Least-squared errors for the SOS constrained approaches.

Constraint	Training Error	Validation Error	Total
P-1	0.26	0.15	0.41
P-2	0.31	0.364	0.674
P-3	0.372	0.255	0.627
P-4	0.257	0.144	0.4

Remark 3. *Note that the standard LS regularization was not able to find these more feasible models obtained with the SOS approach, at least with the tested values for the metaparameter* Γ. *Anyway, although it may be found, there is no clear and direct relation between* Γ *and the features desired in the model response.*

6.2. Modeling the Heat-Transfer in an Evaporation Plant

In this example, we make use of the proposed methodology in Section 4 to build up a grey-box model for a multiple-effect evaporation plant of a man-made cellulose fiber production factory.

The plant is formed by several evaporation chambers and some heat exchangers in serial connection, a mixing steam condenser and a cooling tower, forming a multiple-effect evaporation system. See Figure 4, where individual equipment have been lumped together for confidentiality reasons and due to the lack of measurements in between. The plant receives a liquid input, mixture of water with chemical components and leftovers of organic material. The goal is to concentrate the liquid by removing a certain amount of water.

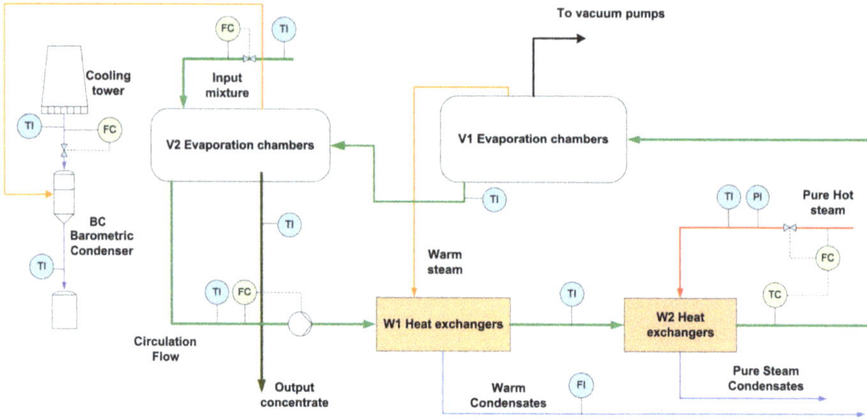

Figure 4. Simplified schema of the evaporation plant with existent instrumentation.

The process operates as follows: the liquid enters the system by chamber V_2 and then goes sequentially through the sets of heat exchangers W_1 and W_2 to increase its temperature up to a desired setpoint. In W_1, the temperature rise is achieved from saturated-steam flows recirculated from the evaporation chambers V_1. Then, the temperature setpoint is reached in W_2 thanks to a fresh steam inlet from boilers. Afterwards, the hot liquid enters sequentially into the low-pressure set of chambers V_1, where a partial evaporation of water takes place. The remaining evaporation is achieved in the last chambers V_2 thanks to the pressure drop in the mixing condenser BC, linked to the cooling tower. Finally, part of the concentrated solution leaves the plant by V_2 and the rest mixes again with the inlet, being recirculated to the heat exchangers.

6.2.1. Stage 1: Estimation

Our modeling approach starts from a nonlinear set of equations of the plant in steady state, obtained from first principles. These equations have been omitted here for brevity, but the reader is referred to the previous works of the authors [45,46] to get a detailed description of both the plant and the physical model equations. Then, in the *estimation* phase (Stage 1 of the proposed methodology), DR is performed to "clean" the process data from incoherent sensor values and to get suitable estimates for the internal-model variables and time-varying parameters, in particular for the average heat-transfer coefficient $UA(t)$ in the lumped heat exchangers. Note that this time-varying parameter depends on the conduction and convection effects plus the exchange surface, values that are not precisely known or complex to model.

Here, the focus is on UA because an accurate modeling of the long-term fouling dynamics in the heat-exchangers pipes is key for a realistic optimization of the operation, and the right scheduling of the maintenance tasks. Indeed, this issue is shared with other industrial systems like furnaces or catalyst deactivation in chemical reactors. All have in common a system-efficiency degradation, which may be palliated or worsened by the way the equipment operates.

Thus, a set of experiments were performed on site, running the plant in different operating conditions (setting different values for the main control variables: the circulation flow and the temperature setpoint). Moreover, in order to get significant information from the actual fouling process, the plant historian for several months of operation (including some stops for cleaning) has been also provided as experimental data for reconciliation. Figure 5 shows the estimated UA for exchangers W_1, provided by the DR (details omitted for brevity).

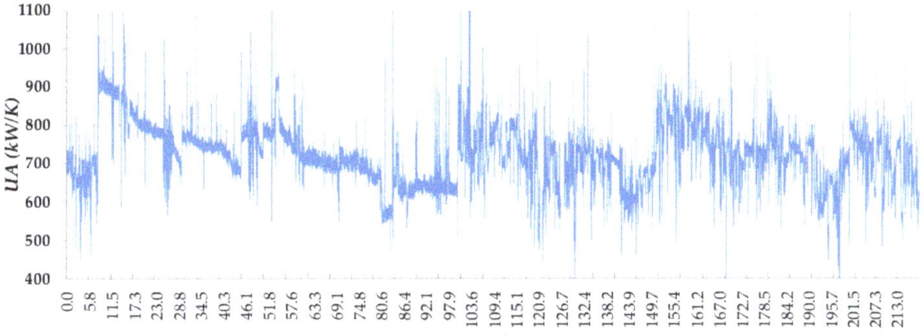

Figure 5. Estimated values of UA over seven months of operation.

6.2.2. Stage 2: Regression

The objective now in the *regression* stage is to build up a polynomial regression model $UA = p(c, F, t)$ to link/predict the heat-transfer coefficient UA with the circulating flow through the exchangers F and with the time t that the plant is in operation since last cleaning.

The first issue arises when selecting the samples for training and validation. Although the recorded dataset of seven months with a sampling time of 5 min may look huge, the quality of data is much more important that the quantity of samples. In addition, in this case, the plant was usually working at high circulating flows, except in the few experiments executed on purpose and in particular situations (product changeovers). Therefore, lots of data for the plant operating in a local region are available, but a significant amount of information of the convection and fouling behaviors at medium/low flows is missing.

Note importantly that, although there is no major computational issue in performing regularized or SOS constrained regression with hundreds of data, if lots of such data are agglomerated around the same operating point, the fitted model might specialize too much in such region, as the model structure for regression will not likely contain the same nonlinearities as the actual plant which generated the data. Hence, the prediction capabilities out of this region can be seriously compromised with such a model. Therefore, the data points must be "triaged" according to their degree of uniqueness (data containing almost-redundant information should get lower weights in the regression problem, or be directly removed from the training set) in order to prevent this possible model bias due to strong non-uniform data densities.

Consequently, after inspecting and analyzing the plant historian, we ended up with a selected subset of 22 samples (UA, F, t) for training plus 20 samples for validation. These data, depicted in Figure 6, contains nearly all the information available in the feasible region of operation:

$$\Omega := \{F, t \in \mathbb{R}^+ : 100 \leq F \leq 200 \text{ m}^3/\text{h},\ t \leq 60 \text{ days}\}. \tag{24}$$

As it can be observed by simple visual inspection, there are many samples covering Ω at high flows, but there is a significant lack of information at lower flows, especially when the plant works after cleaning and when it is in operation for more than 40 days.

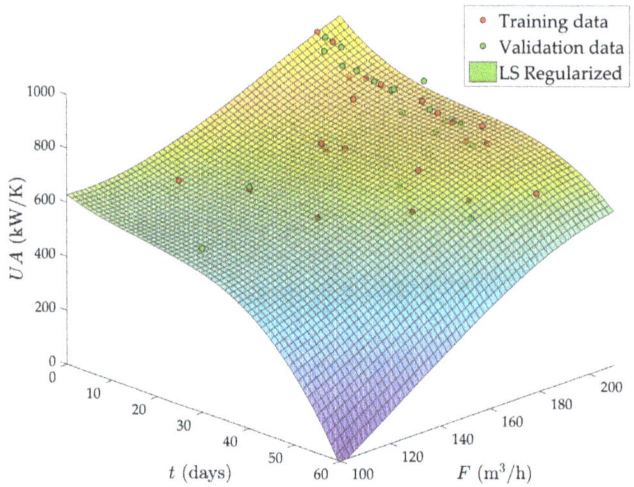

Figure 6. Modeling UA through standard LS regularization.

After normalizing these data to zero mean and $\sigma = 1$, a standard LS identification was initially tested with exponential regularization in the coefficients corresponding to the higher-degree monomials of $p(c, F, t)$, analogous to M-2 in the previous example. Thus, after exploring in γ, the best fitting (lower total error) is achieved with a polynomial of coordinate degree at most $d = 3$ ($\gamma = 0.025$):

$$UA(F,t) = -2.5335 \cdot 10^{-4}F^3 - 7.0692 \cdot 10^{-4}F^2t + 2.0131 \cdot 10^{-3}Ft^2 - 5.5415 \cdot 10^{-3}t^3 + 0.13823F^2 + 0.14058Ft + 0.066824t^2 - 21.0228F - 13.8979t + 1602.0089. \quad (25)$$

Independently of the fitting error (shown in Table 3), there are two aspects in this model which are unacceptable:

- The circulating flow is fixed by a pump in this plant. Therefore, the fouling due to deposition of organic material must tend to a saturation limit with the time. This is because the flow speed increases as the effective pipe area reduces by fouling and, from basic physics, the deposition of particles in the pipes must always decrease with the flow speed. Therefore, the abrupt falling of the UA from day 30 onwards is not possible. Indeed, the predicted UA even reaches zero and negative values after two months of operation with low F.
- With a nearly constant exchange area, UA always decreases as F does, by convective thermodynamics. Hence, the mild increase observed at low F when the evaporator is fully clean (see Figure 7a) is also physically impossible.

Now, SOS constrained regression in Section 5 is recalled to incorporate the above physical requirements in the identification problem. Hence, regularization in the model coefficients is removed but, instead, the following constraints are added:

$$\frac{\mathrm{d}p(c,F,t)}{\mathrm{d}t} < 0, \; \frac{\mathrm{d}p(c,F,t)}{\mathrm{d}F} > 0 \quad \forall \quad F, t \in \Omega, \quad (26)$$

$$\frac{\mathrm{d}p(c,F,t)}{\mathrm{d}t} > -\lambda_t, \; \frac{\mathrm{d}p(c,F,t)}{\mathrm{d}F} < \lambda_F \quad \forall \quad F, t \in \Omega \cap \{t : 30 < t < 60\}, \; \lambda_t, \lambda_F \in \mathbb{R}^+, \quad (27)$$

with λ_t, λ_F, being user-defined bounds on the model slopes that are set up to force the expected nearly-flat response in UA after a month of operation.

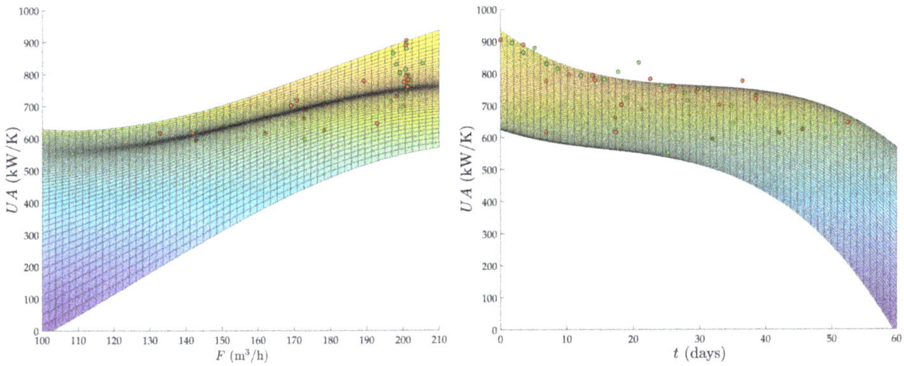

(a) UA versus F **(b)** UA versus t

Figure 7. Partial 2D views of the LS regularized model.

Performing the change of variables in (F, t) corresponding to the normalized data, and casting (26) and (27) to SOS constraints, locally enforced in the corresponding regions via multipliers $s_i(a_i, F, t) \in \Sigma_{F,t}$ with highest degree $d = 2$, the model which best fit the experimental data ($\lambda_t = \lambda_F = 0.6$ are set for regression, as the input data for both F and t are normalized) is found to be a polynomial of $d = 4$:

$$UA(F,t) = 7.0667 \cdot 10^{-8} F^4 + 2.9544 \cdot 10^{-6} F^3 t + 1.6325 \cdot 10^{-6} F^2 t^2 - 2.4195 \cdot 10^{-6} F t^3 + 1.0012 \cdot 10^{-4} t^4 - 1.9868 \cdot 10^{-4} F^3 - 1.5847 \cdot 10^{-3} F^2 t + 5.0898 \cdot 10^{-5} F t^2 - 0.013865 t^3 + 0.088883 F^2 + 0.23223 F t + 0.62707 t^2 - 10.8758 F - 22.7836 t + 1000.2034. \quad (28)$$

Figure 8 shows how the model (28) behaves more coherent with the process physics, and just with an $\sim$1.3% of relative fitness deterioration with respect to the LS-regularized model (25) (see Table 3).

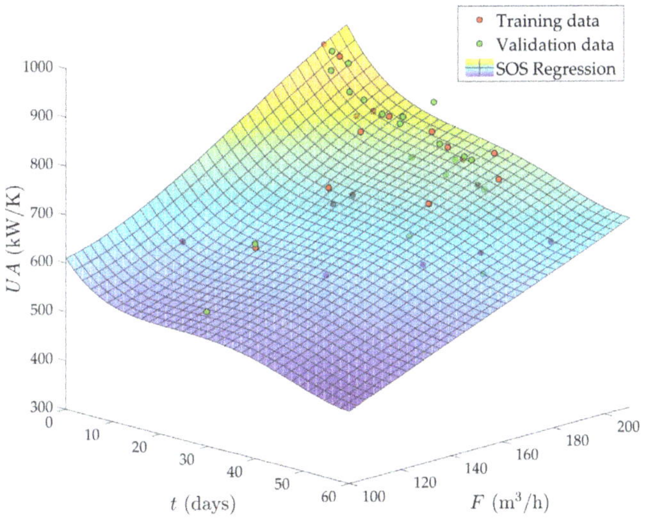

Figure 8. Modeling UA through SOS constrained regression.

6.2.3. Comparison with Previous Works

In the previous work [45], we followed the typical grey-box modeling approach of setting up a functional form for $UA(F,t)$ and then solving an unconstrained LS-optimization problem. This was a quite time-consuming task, though we arrived to satisfactory enough results with the physically-based functional model:

The best fitting of (30) to the selected experimental data ($UA_0 = 1744.5$, $k_1 = 4928.1$, $k_2 = 214.64$ and $\tau = 2096.3$), here with no distinction between training and validation sets as the model structure is fully fixed, gives us a quite desirable response (see Figure 9). However, the fitting to the experimental data degrades in ∼25% w.r.t. model (28).

Furthermore, in [10], we assumed the hypothesis that the increase of specific-steam consumption in the plant due to fouling was linear with the operation time. This was done based on direct measurements and to facilitate the resolution of the maintenance-scheduling problem formulated in [10]. Now, we analyze whether this assumption was reasonably true.

$$\frac{\mathrm{d}^2 p(c,F,t)}{\mathrm{d}t^2} = 0 \; \forall F,t \in \Omega, \tag{29}$$

which can be cast as an SOS equality constraint and checked locally in Ω via Lemma 1. Doing this, effectively the obtained model is affine in t and nonlinear in F, as Figure 10 shows.

$$UA(F,t) = UA_0 - \frac{k_1}{F^{1/3}} - k_2 e^{1-\frac{\tau}{F^2}t}, \quad UA_0, k_1, k_2, \tau \in \mathbb{R}^+. \tag{30}$$

For this aim, the polynomial model $p(c,F,t)$ is forced to be affine in t. This requirement can be easily achieved by adding the following constraint on the curvature to the SOS regression problem:

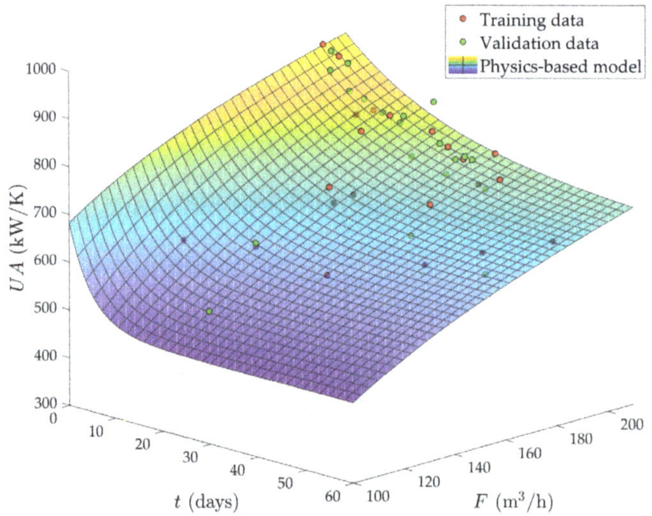

Figure 9. Physics-based model for UA.

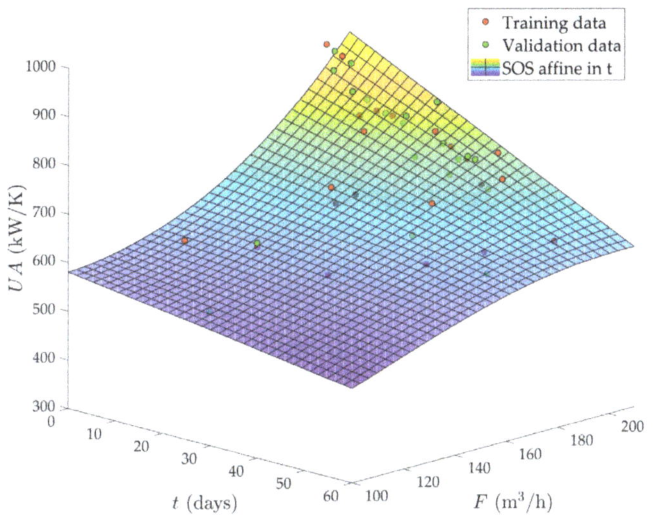

Figure 10. Polynomial affine-in-t model for UA.

Table 3. Actual least-squared errors for the tested approaches.

Method	Training Error	Validation Error	Total	Relative Deterioration
LS regularized	13,448	14,282	27,730	-
SOS constrained	14,751	13,362	28,113	1.36%
SOS affine	20,147	15,131	35,278	21.39%
Physics-based model	-	-	37,361	25.78%

This model also incurs in an $\sim$20% fitting degradation w.r.t. the optimal (28). In this case, what is more relevant than the fitting error is the observed variation of the slope in t at different flows. This indicates that the assumption in [10] was acceptable as long as F remains nearly constant. Indeed, as the plant was normally operating at high flows when the data was collected from the plant historian, we did (could) not realize of this varying behavior with the flow.

7. Discussion

The above examples show how high-degree polynomials with well-behaved output can be obtained via SOS-constrained regression even in the presence of scarce measurements, though there is no major issue in using larger datasets, especially with the economic singular-value decomposition (12). Indeed, the main computational effort with SOS programming is employed in casting the SOS polynomial constraints as SDP ones, not in the number of samples to fit in the objective function.

The average computational time (Intel® i7-4510U machine) required to solve the SOS-constrained regression problems in Section 6.2 is 1.5 s (calling the SDP solver SeDuMi), whereas the nonlinear unconstrained LS regression in Section 6.2.3 needs 0.15 s (using the NLP solver IPOPT with MUMPS as linear solver). Hence, at the price of this increase in the computational demands, the modeler is allowed to naturally include the physical insight into the regression problem via (polynomial) bounds on value, slope and curvature or convexity-related constraints, thus getting coherent model responses.

Note also that the proposed SOS-constrained regression can be an alternative to standard regularization, but both can also be complementary if desired because there is no impediment in weighting the coefficients of the regression model with some metaparameter Γ, whereas, at the same time, some polynomial constraints are to be enforced. In fact, if Γ is chosen according to some information criteria, this combined approach might provide an automatic selection of the suitable model structure by balancing the fitness to data with the model complexity. This possibility will be explored for further works.

The two-stage methodology for grey-box modeling proposed in Section 4, enhanced with the application of SOS-constrained regression (Section 5), has demonstrated its advantages to face a modeling problem for optimization purposes in a real industrial evaporation plant. The main limitation of the approach is that candidate regression submodels are limited to being polynomials, though polynomial basis functions are flexible, used in practice, and the dominant process nonlinearities are taken into account in the first-principles equations. Nevertheless, the possible extension of the SOS-regression approach to include other non-polynomial basis functions will be explored via multimodel and polynomial bounding.

As a final remark, note that, although SOS programming is convex optimization and the proposed methodology is performed offline, its scalability is mainly limited by the number of independent variables in x and by the degree of the involved polynomials d. This fact may pose limitation in applying the proposed ideas to perform regression in complex chemical-reaction problems with many components involved. Nonetheless, the ALAMO approach does not get clear computational benefits in this sense either. Moreover, it is worth noting that the main aim of the SOS approach is not to get complete-plant surrogate models (though they can be sought at relatively small process scales), but just local relationships among a few process variables to complete a grey-box model based on physics.

Author Contributions: Conceptualization, J.L.P. and C.d.P.; methodology, J.L.P. and C.d.P.; software, J.L.P.; validation, J.L.P. and A.S.; formal analysis, A.S.; investigation J.L.P. and A.S.; resources, C.d.P.; writing–original draft preparation, J.L.P.; writing–review and editing, A.S.; visualization, J.L.P.; supervision, A.S. and C.d.P.; project administration, C.d.P.; funding acquisition, C.d.P.

Funding: This research received funding from the European Union, Horizon 2020 research and innovation programme under Grant No. 723575 (CoPro), and from the EU plus the Spanish Ministry of Economy, grant DPI2016-81002-R (AEI/FEDER).

Acknowledgments: The authors especially thank the industrial partner Lenzing AG (Lenzing, Austria) for the data acquisition and experimental tests carried out in the evaporation plant.

Conflicts of Interest: The authors declare no conflict of interest. The funders had no role in the design of the study; in the collection, analysis, or interpretation of data; in the writing of the manuscript, or in the decision to publish the results.

References

1. Davies, R. *Industry 4.0: Digitalization for Productivity and Growth*; Document pe 568.337; European Parliamentary Research Service: Brussels, Belgium, 2015.
2. Krämer, S.; Engell, S. *Resource Efficiency of Processing Plants: Monitoring and Improvement*; John Wiley & Sons: Weinheim, Germany, 2017.
3. Palacín, C.G.; Pitarch, J.L.; Jasch, C.; Méndez, C.A.; de Prada, C. Robust integrated production-maintenance scheduling for an evaporation network. *Comput. Chem. Eng.* **2018**, *110*, 140–151. [CrossRef]
4. Maxeiner, L.S.; Wenzel, S.; Engell, S. Price-based coordination of interconnected systems with access to external markets. *Comput. Aided Chem. Eng.* **2018**, *44*, 877–882.
5. Afram, A.; Janabi-Sharifi, F. Black-box modeling of residential HVAC system and comparison of gray-box and black-box modeling methods. *Energy Build.* **2015**, *94*, 121–149. [CrossRef]
6. Witten, I.H.; Frank, E.; Hall, M.A.; Pal, C.J. *Data Mining: Practical Machine Learning Tools and Techniques*; Morgan Kaufmann: San Francisco, CA, USA, 2016.
7. Olsen, I.; Endrestøl, G.O.; Sira, T. A rigorous and efficient distillation column model for engineering and training simulators. *Comput. Chem. Eng.* **1997**, *21*, S193–S198. [CrossRef]
8. Galan, A.; de Prada, C.; Gutierrez, G.; Sarabia, D.; Gonzalez, R. Predictive Simulation Applied to Refinery Hydrogen Networks for Operators' Decision Support. In Proceedings of the 12th IFAC Symposium on Dynamics and Control of Process Systems, Including Biosystems (DYCOPS), Florianópolis, Brazil, 23–26 April 2019.

9. Kar, A.K. A hybrid group decision support system for supplier selection using analytic hierarchy process, fuzzy set theory and neural network. *J. Comput. Sci.* **2015**, *6*, 23–33. [CrossRef]
10. Kalliski, M.; Pitarch, J.L.; Jasch, C.; de Prada, C. Apoyo a la Toma de Decisión en una Red de Evaporadores Industriales. *Revista Iberoamericana de Automática e Informática Industrial* **2019**, *16*, 26–35. [CrossRef]
11. Zorzetto, L.; Filho, R.; Wolf-Maciel, M. Processing modelling development through artificial neural networks and hybrid models. *Comput. Chem. Eng.* **2000**, *24*, 1355–1360. [CrossRef]
12. Cellier, F.E.; Greifeneder, J. *Continuous System Modeling*; Springer Science & Business Media: New York, NY, USA, 2013.
13. Zou, W.; Li, C.; Zhang, N. A T–S Fuzzy Model Identification Approach Based on a Modified Inter Type-2 FRCM Algorithm. *IEEE Trans. Fuzzy Syst.* **2018**, *26*, 1104–1113. [CrossRef]
14. Neumaier, A. Solving Ill-Conditioned and Singular Linear Systems: A Tutorial on Regularization. *SIAM Rev.* **1998**, *40*, 636–666. [CrossRef]
15. Kim, S.; Koh, K.; Lustig, M.; Boyd, S.; Gorinevsky, D. An Interior-Point Method for Large-Scale ℓ_1-Regularized Least Squares. *IEEE J. Sel. Top. Signal Process.* **2007**, *1*, 606–617. [CrossRef]
16. Llanos, C.E.; Sanchéz, M.C.; Maronna, R.A. Robust Estimators for Data Reconciliation. *Ind. Eng. Chem. Res.* **2015**, *54*, 5096–5105. [CrossRef]
17. de Prada, C.; Hose, D.; Gutierrez, G.; Pitarch, J.L. Developing Grey-box Dynamic Process Models. *IFAC-PapersOnLine* **2018**, *51*, 523–528. [CrossRef]
18. Cozad, A.; Sahinidis, N.V.; Miller, D.C. A combined first-principles and data-driven approach to model building. *Comput. Chem. Eng.* **2015**, *73*, 116–127. [CrossRef]
19. Cozad, A.; Sahinidis, N.V. A global MINLP approach to symbolic regression. *Math. Programm.* **2018**, *170*, 97–119. [CrossRef]
20. Reemtsen, R.; Rückmann, J.J. *Semi-Infinite Programming*; Springer Science & Business Media: New York, NY, USA, 1998; Volume 25.
21. Parrilo, P.A. Semidefinite programming relaxations for semialgebraic problems. *Math. Programm.* **2003**, *96*, 293–320. [CrossRef]
22. Nauta, K.M.; Weiland, S.; Backx, A.C.P.M.; Jokic, A. Approximation of fast dynamics in kinetic networks using non-negative polynomials. In Proceedings of the 2007 IEEE International Conference on Control Applications, Singapore, 1–3 October 2007; pp. 1144–1149.
23. Tan, K.; Li, Y. Grey-box model identification via evolutionary computing. *Control Eng. Pract.* **2002**, *10*, 673–684. [CrossRef]
24. Biegler, L.T. *Nonlinear Programming: Concepts, Algorithms, and Applications to Chemical Processes*; MOS-SIAM Series on Optimization: Philadelphia, PA, USA, 2010; Volume 10.
25. Schuster, A.; Kozek, M.; Voglauer, B.; Voigt, A. Grey-box modelling of a viscose-fibre drying process. *Math. Comput. Model. Dyn. Syst.* **2012**, *18*, 307–325. [CrossRef]
26. Tulleken, H.J. Grey-box modelling and identification using physical knowledge and bayesian techniques. *Automatica* **1993**, *29*, 285–308. [CrossRef]
27. Leibman, M.J.; Edgar, T.F.; Lasdon, L.S. Efficient data reconciliation and estimation for dynamic processes using nonlinear programming techniques. *Comput. Chem. Eng.* **1992**, *16*, 963–986. [CrossRef]
28. Bendig, M. Integration of Organic Rankine Cycles for Waste Heat Recovery in Industrial Processes. Ph.D. Thesis, Institut de Génie Mécanique, École Polytechnique Fédérale de Lausanne, Lausanne, Switzerland, 2015.
29. Lasserre, J.B. Sufficient conditions for a real polynomial to be a sum of squares. *Archiv der Mathematik* **2007**, *89*, 390–398. [CrossRef]
30. Parrilo, P.A. Structured Semidefinite Programs and Semialgebraic Geometry Methods in Robustness and Optimization. Ph.D. Thesis, California Institute of Technology, Pasadena, CA, USA, 2000.
31. Papachristodoulou, A.; Anderson, J.; Valmorbida, G.; Prajna, S.; Seiler, P.; Parrilo, P. SOSTOOLS version 3.00 sum of squares optimization toolbox for MATLAB. *arXiv* **2013**, arXiv:1310.4716, .
32. Hilbert, D. Ueber die Darstellung definiter Formen als Summe von Formenquadraten. *Mathematische Annalen* **1888**, *32*, 342–350. [CrossRef]
33. Scherer, C.W.; Hol, C.W.J. Matrix Sum-of-Squares Relaxations for Robust Semi-Definite Programs. *Math. Programm.* **2006**, *107*, 189–211. [CrossRef]

34. Putinar, M. Positive Polynomials on Compact Semi-algebraic Sets. *Indiana Univ. Math. J.* **1993**, *42*, 969–984. [CrossRef]
35. Pitarch, J.L. Contributions to Fuzzy Polynomial Techniques for Stability Analysis and Control. Ph.D. Thesis, Universitat Politècnica de València, Valencia, Spain, 2013.
36. Scherer, C.; Weiland, S. *Linear Matrix Inequalities in Control*; Lecture Notes; Dutch Institute for Systems and Control: Delft, The Netherlands, 2000; Volume 3, p. 2.
37. Wilson, Z.T.; Sahinidis, N.V. The ALAMO approach to machine learning. *Comput. Chem. Eng.* **2017**, *106*, 785–795. [CrossRef]
38. Hindmarsh, A.C.; Brown, P.N.; Grant, K.E.; Lee, S.L.; Serban, R.; Shumaker, D.E.; Woodward, C.S. SUNDIALS: Suite of nonlinear and differential/algebraic equation solvers. *ACM Trans. Math. Softw. (TOMS)* **2005**, *31*, 363–396. [CrossRef]
39. Gill, P.; Murray, W.; Saunders, M. SNOPT: An SQP Algorithm for Large-Scale Constrained Optimization. *SIAM Rev.* **2005**, *47*, 99–131. [CrossRef]
40. Reynoso-Meza, G.; Sanchis, J.; Blasco, X.; Martínez, M. Design of Continuous Controllers Using a Multiobjective Differential Evolution Algorithm with Spherical Pruning. In *Applications of Evolutionary Computation*; Di Chio, C., Cagnoni, S., Cotta, C., Ebner, M., Ekárt, A., Esparcia-Alcazar, A.I., Goh, C.K., Merelo, J.J., Neri, F., Preuß, M., et al., Eds.; Springer: Berlin/Heidelberg, Germany, 2010; pp. 532–541.
41. Andersson, J.; Åkesson, J.; Diehl, M. CasADi: A Symbolic Package for Automatic Differentiation and Optimal Control. In *Recent Advances in Algorithmic Differentiation*; Forth, S., Hovland, P., Phipps, E., Utke, J., Walther, A., Eds.; Springer: Berlin/Heidelberg, Germany, 2012; pp. 297–307.
42. Hart, W.E.; Laird, C.D.; Watson, J.P.; Woodruff, D.L.; Hackebeil, G.A.; Nicholson, B.L.; Siirola, J.D. *Pyomo—Optimization Modeling in Python*; Springer Science & Business Media: New York, NY, USA, 2017; Volume 67.
43. Wächter, A.; Biegler, L.T. On the implementation of an interior-point filter line-search algorithm for large-scale nonlinear programming. *Math. Programm.* **2006**, *106*, 25–57. [CrossRef]
44. Sahinidis, N.V. BARON: A general purpose global optimization software package. *J. Glob. Optim.* **1996**, *8*, 201–205. [CrossRef]
45. Pitarch, J.L.; Palacín, C.G.; de Prada, C.; Voglauer, B.; Seyfriedsberger, G. Optimisation of the Resource Efficiency in an Industrial Evaporation System. *J. Process Control* **2017**, *56*, 1–12. [CrossRef]
46. Pitarch, J.L.; Palacín, C.G.; Merino, A.; de Prada, C. Optimal Operation of an Evaporation Process. In *Modeling, Simulation and Optimization of Complex Processes HPSC 2015*; Bock, H.G., Phu, H.X., Rannacher, R., Schlöder, J.P., Eds.; Springer International Publishing: Cham, Switzerland, 2017; pp. 189–203.

processes MDPI

Article

Toward a Distinct and Quantitative Validation Method for Predictive Process Modelling—On the Example of Solid-Liquid Extraction Processes of Complex Plant Extracts

Maximilian Sixt, Lukas Uhlenbrock and Jochen Strube *

Institute for Separation and Process Technology, Clausthal University of Technology, 38678 Clausthal-Zellerfeld, Germany; sixt@itv.tu-clausthal.de (M.S.); uhlenbrock@itv.tu-clausthal.de (L.U.)
* Correspondence: strube@itv.tu-clausthal.de

Received: 2 May 2018; Accepted: 23 May 2018; Published: 1 June 2018

Abstract: Physico-chemical modelling and predictive simulation are becoming key for modern process engineering. Rigorous models rely on the separation of different effects (e.g., fluid dynamics, kinetics, mass transfer) by distinct experimental parameter determination on lab-scale. The equations allow the transfer of the lab-scale data to any desired scale, if characteristic numbers like e.g., Reynolds, Péclet, Sherwood, Schmidt remain constant and fluid-dynamics of both scales are known and can be described by the model. A useful model has to be accurate and therefore match the experimental data at different scales and combinations of process and operating parameters. Besides accuracy as one quality attribute for the modelling depth, model precision also has to be evaluated. Model precision is considered as the combination of modelling depth and the influence of experimental errors in model parameter determination on the simulation results. A model is considered appropriate if the deviation of the simulation results is in the same order of magnitude as the reproducibility of the experimental data to be substituted by the simulation. Especially in natural product extraction, the accuracy of the modelling approach can be shown through various studies including different feedstocks and scales, as well as process and operating parameters. Therefore, a statistics-based quantitative method for the assessment of model precision is derived and discussed in detail in this paper to complete the process engineering toolbox. Therefore a systematic workflow including decision criteria is provided.

Keywords: process model validation; partial least square regression; phytochemicals; natural extracts

1. Introduction

Verification and validation present an issue for different kinds of predictive models applied e.g., in economics and banking [1], climate [2], traffic [3], and not least, of course, in process technology [4–7].

Sargent [5] defines model verification as "ensuring that the program of the model and its implementation are correct" and model validation as "substantiation that a model within its domain of applicability possesses a satisfactory range of accuracy consistent with the intended application of the model" based on [4]. A general procedure, including statistical analysis and quantification of whether a model is valid or not, is still missing in process design and development in chemical and pharmaceutical industries. In most cases, the effort to ensure model validity is the major obstacle for decision makers in industry to expand the use of physico chemical–based predictive process modelling instead of or in addition to experimental data from mini- or pilot-plants. This is the main discussion point regarding modelling in process engineering. This is often a point of dissent between academia and industry in many working groups [8–12]. Academia favors rigorous process modelling as the only scientifically sound method for process understanding in combination with laboratory scale experiments for model parameter determination. In contrast, industry avoids the early project efforts in model development due to costs, efforts, and resources. Major obstacle is that no investment decision would be made based on theoretic results without experimental proof. Therefore, any process modelling activity has to prove—distinct and self-explanatory—the model validity a priori.

This study is an attempt to propose a general procedure to assess model validity, based on quantitative decision criteria on the example of an industrial relevant complex component mixture from plants with natural variable feedstock.

There is a need for detailed and precise rigorous process models in chemical engineering because a significant amount of resources in early process development and optimization can be saved. Moreover, predictive models allow evaluation of the process at critical points of operation, e.g., start-up/shut-down, unstable operating points in terms of energy supply and removal and feedstock variations. Especially, feedstock variations are a hot topic in the processing of phyto-pharmaceuticals due to the natural content fluctuations of target molecules in the plants. Besides engineering and economic considerations, especially in the pharmaceutical industry, a modern concept of quality assurance through the whole lifecycle of the product, reaching from early stage of research and development to production, called Quality-by-Design (QbD), gains more and more influence and acceptance and is demanded by authorities [13–15]. The basis of QbD is the evaluation of a so-called design space in which the process is kept while maintaining constant quality attributes of the final product. The establishment of a design space demands multi-parameter optimizations, requiring significant experimental effort. Rigorous process models can contribute significantly by substituting a part of the experiments and therefore, lead from pure empirical process design to a model and data driven process assessment. The process models utilized have to be rigorous (strictly derived from physico-chemistry) in order to be predictive with regard to parameter range and scale. To define a design space, critical quality attributes have to be determined and ranked, first. The principle of the QbD approach is depicted in Figure 1 and discussed in detail in the cited literature [13].

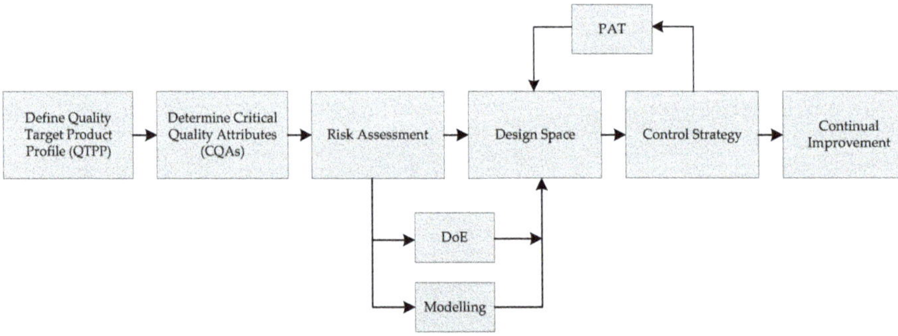

Figure 1. The Quality-by-Design (QbD)-approach to quality assurance [13], PAT process analytical technology, DoE design of experiments.

If the variance of those critical product quality attributes is narrow, a broad and multi-parameter design space is needed, in order to cope appropriately with natural feedstock variability, exemplified by Figure 2.

In contrast, a narrow operation parameter space is sufficient to fulfil the quality requirements, but may cause troubles in manufacturing, if approved in a too small range, due to equipment limitations and breeding success toward high active component content of the used plants. If Quality-by-Design is strictly applied, a reconsideration of the critical quality attributes, the risk assessment, and the derived design space becomes necessary if there is a significant change in the boundary conditions of the process or the production is out of specification for some time and the initial risk assessment did not address these issues in an adequate way. This is explicitly in conflict to a stringent approval procedure, submitting of one single point of operation with narrow parameter ranges to the regulatory agencies, as it is historically common practice in phytoextraction.

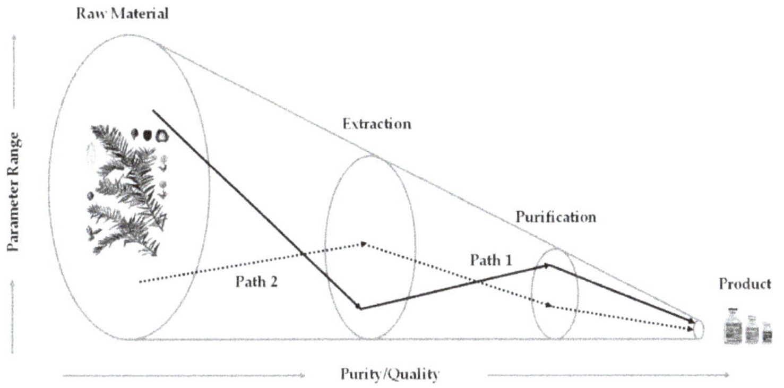

Figure 2. Basic idea of design spaces in phytoextraction [13].

To ensure the proper definition of a design space with the aid of a physico-chemical process model, the model has to fulfil the accuracy and precision criteria. Accuracy is the ability to predict the experimental data correctly within a whole set of parameters. In case the model fails, the model depth has to be increased to reflect the real behavior of the system in higher detail. Precision is the feedback of the errors and uncertainties of the model parameter determination on the simulation results. If the resulting deviation is below the reproducibility of the experimental data that is to be substituted by the model, the model precision is sufficient. If the model fails, the underlying parameter determination concept has to be improved. The difference between accuracy and prediction is depicted in Figure 3.

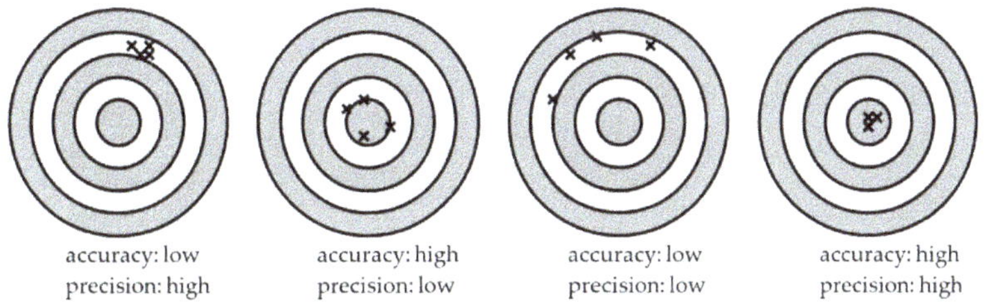

Figure 3. Visualization of accuracy and precision.

Validated and predictive process models for solid-liquid extractions exist for more than two decades. Kassing et al. [16,17] give an overview of different process modelling approaches. These studies, based on pepper and vanilla, predict influences of different particle diameters, solid/liquid ratios, and residence times. Both et al. [18–21] extended the validation on sugar and tea for different process concepts like maceration and percolation, recycling-mode percolation, as well as different residence times and solid/liquid ratios. They even showed the scale independent validity of the model by means of a scale up industrial level study for the extraction of sugar beet [18]. Further-on, salvia, fennel and yew [22,23] were added. Recently, Sixt et al. [24] added annual mugwort, assessing the influence of pressure, temperature, and solvent ratios on the extraction and enhanced the research on yew [25]. As a consequence, various studies exist, which prove that those models are valid to predict experimental data at various scales and different points of operation and feedstocks, thus they are accurate and therefore the model depth is sufficient.

A detailed discussion and quantification of model precision has not yet been provided. Model precision rates the impacts of the errors of experimental model parameter determination on the simulation results. As a consequence, a model is precise, if the feedback of the experimental errors on the simulation is smaller compared to the data gained through field experiments. In that case, the proven accurate model has a sufficient precision to substitute experiments for process optimization, design and control. Kassing et al. [17] did extensive research on the equilibrium determination and the error propagation but did not show the feedback of the parameter determination on the simulation results. To close this gap, a generalized statistics-based comparison of the simulation error and the experimental error is shown to complete the engineering toolbox.

2. Modelling of Solid-Liquid Extraction

Data-driven process design is the key to efficient chemical engineering. Especially for important unit operations like continuous distillation, liquid-liquid extraction, and adsorption, graphical methods for process design as well as detailed models are available. Both models (stage construction for binary mixtures after McCabe-Thiele and ternary mixtures in Gibbs diagrams (mostly liquid-liquid extraction) [26]) are sketched in Figure 4. They rely on equilibrium stages that are reached in the system due to thermodynamics. The real behavior is then taken into account by the stage efficiency. There is an approach to adapt these methods to phytoextraction processes, but a comparable accuracy and wide spread use was not achieved [27]. This is mainly due to botanical and equipment constraints. On the botanical side, the plant tissue is highly compartmented and therefore different mass transport phenomena and limitations have to be taken into account. Moreover, there are target molecules that are not entirely adsorbed to the plant matrix, e.g., essential oils, that are often located in oil seams or trachoma cells in liquid state. Therefore, they can easily be washed out of the plant matrix, which is why an equilibrium stage model fails because no explicit phase equilibrium occurs. A look at the equipment side reveals a variety of different apparatus for phytoextraction [28]. Besides the often used batch equipment, like maceration and hydro-distillation, a number of continuously operated equipment is commercially available with their individual fluid dynamics that have to be taken into account for proper process design. Moreover, the fluid dynamic behavior of these devices is not steady but is subject to cyclic switch times like in carousel extractors.

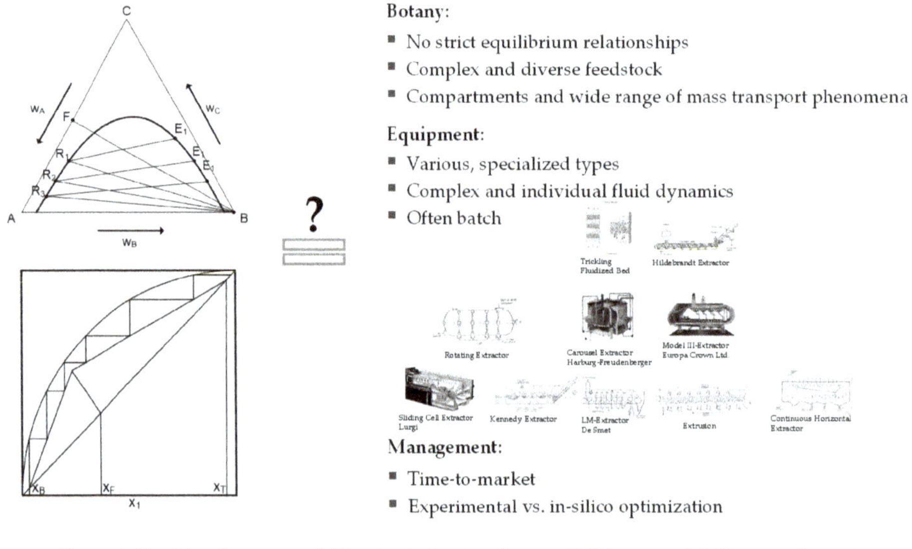

Figure 4. Need for rigorous modelling in phytoextraction, see [16] for more details on equipment.

The model description of solid-liquid extraction has been researched for some time. The equation system can usually be divided into an apparatus-specific part, for example mass balances for flow tubes or stirred vessels, and a matrix-specific part which images the mass transport of individual components of the plant into the solvent. The goals of the modelling are either a reduction of attempts for process optimization or the deepening of the basic understanding. In addition to purely empirical models, which usually can only interpolate measured values, a number of predictive models for the matrix effects have been established [28].

- **Shrinking Core:** In the Shrinking Core model, a solvent front passes through a spherical particle. At the boundary layer between solvent and solid, the mass transfer of the target components take place. The model is based on gas-solid reactions in porous pellets [29].
- **Broken and Intact Cells:** The Broken and Intact Cells model is based on the idea that the target components are found both inside the plant particles, as well as in broken vacuoles or oil channels. This assumption is based on real extraction experiments in which extraction is carried out very rapidly at the beginning (near-surface components or broken vacuoles and oil channels). Subsequently, the extraction rate is greatly retarded (intact cells and oil channels). In the first case, there is no diffusion limitation, but in the second case there is a strong diffusion limitation of the extraction [30,31].
- **Pore Diffusion model:** The Pore Diffusion model originates from chromatography. The solvent diffuses into the porous particle and desorbs the components. Subsequently, the back diffusion and the subsequent removal take place in the core flow. Again, the basic idea of the Broken and Intact Cell model can be implemented by means of radial pore size and active substance distribution [17].

Table 1 gives an overview of various modelling approaches, material systems, solvents and apparatuses.

The extraction can be carried out as a leaching process (percolation) or as an equilibrium process (maceration). In the following, the percolation model is described in more detail.

The percolation process is represented by several sub-models. On the one hand, the distributed-plug-flow model (DPF) is used to represent the macroscopic mass transport within the percolation column, and on the other hand, the diffusion in the porous particle is modeled using a transport equation. In order to map the relationship between the residual load of the respective component in the particle and in the solvent, various equilibrium relationships are used.

The Shrinking Core and the Broken and Intact Cells model are not implemented because a pore diffusion approach gives the highest degree of detail compared to the other two theories. Also, the model is already established and widely used in chromatography.

All sub models are explained in more detail below.

Table 1. Overview of modelling approaches, plug flow (PF), stirred tank reactor (STR), distributed plug flow (DPF), pulsed electrical field (PEF).

Author and Year	Ref.	Fluid	Plant Material	Target Component	Equilibrium	Particle/Shape/Model	Flux
Akgün 2000	[32]	scCO_2	Lavender flower	Essential oil	Constant	Porous particle, Shrinking Core	PF
Al-Jabari 2003	[33]	scCO_2	-	-	Langmuir	-	STR
Bulley 1984	[34]	scCO_2	Rape	Fatty oil	-	-	PF
Cacace 2003	[35]	Ethanol, SO_2 in Water	Berries	Phenols, Anthocyanins	Linear	-	STR
Carrin 2008	[36]	Hexane	Sunflower	Fatty Oil	Linear	Porous particle	DPF, cross-current
Catchpole 1996	[37]	Liquid CO_2	Salvia, celery and coriander seed	Essential and fatty oil	Linear	Sphere, Cylinder, parabolic concentration profile	PF
Chalermachat 2003	[38]	Water	Beetroot	Pigments	-	Porous cylinder	STR, PEF
Chia 2015	[39]	scCO_2 (Soxhlet)	Rice bran oil	Tocopherol	-	Logistic, Simple Single Plate, Diffusion	-
Cocero and Garcia 2001	[40]	scCO_2	Sunflower	Fatty oil	Linear	No internal diffusion	DPF
De Franca 2000	[41]	SCF	Palm oil	Fatty oil, Carotenoids	Constant	-	PF
Del Valle 2000	[42]	scCO_2	Rape oil, basil	Essential and fatty oil	Linear	Sphere	PF
Del Valle 2003	[43]	scCO_2	Chili	Essential oil	Linear	Sphere	PF
Del Valle 2005	[44]	scCO_2	Different Latin American plants	Essential and fatty oil	Linear	Shrinking core	DPF
Del Valle 2006	[45]	scCO_2	Oilseed	Fatty oil	Linear	Shrinking core	DPF
Diankov 2008	[46]	Water	Tabaco	-	-	Plates, Shrinking core	STR
Egorov 2015	[47]	scCO_2	Pumpkinseed	-	-	Shrinking core, particle size distribution	DPF
Espinoza-Perez 2007	[48]	Water	Coffee beans	Caffeine	Linear	Sphere	PF
Esquivel 1999	[49]	scCO_2	Olives bowl	Fatty oil	Linear	Porous particle	PF
Ferreira 2002	[50]	scCO_2	Black pepper	Essential oil	-	Broken and intact cells	PF
Fiori 2007	[51]	scCO_2	Vegetable seed	Fatty oil	Linear	Broken and intact cells	DPF
Fiori 2008	[52]	scCO_2	Grape kernels	Fatty oil	Linear	Broken and intact cells	DPF
Fiori 2009	[53]	scCO_2	Oilseed	Fatty oil	-	Broken and intact cells und shrinking core, particle size distribution	DPF

Table 1. *Cont.*

Author and Year	Ref.	Fluid	Plant Material	Target Component	Equilibrium	Particle/Shape/Model	Flux
Goodarznia and Eikani 1998	[54]	$scCO_2$	Rosemary, basil, caraway, marjoram	Essential oil	Linear	Sphere	DPF
Goto 1990	[55]	$scCO_2$	Wood	Lignin	Linear	Porous particle, parabolic concentration profile	PF
Goto 1993	[56]	$scCO_2$	Peppermint	Essential oil	Linear	Porous particle	PF
Goto 1996	[57]	$scCO_2$	Rape oil	Fatty oil	Constant	Shrinking core	DPF
Guerrero 2008	[58]	Ethanol/Water	Grape pomace	Polyphenols	-	Sphere	PF
Ji 2006	[59]	Water	Gardenia fruit	Geniposide	Langmuir	Shrinking core	STR, ultrasound
Jokic 2015	[60]	$scCO_2$	Soy	Fatty oil	-	Logistic	-
Kim and Hong 2001	[61]	$scCO_2$	Spearmint	Essential oil	Constant	-	PF
Kim and Hong 2002	[62]	$scCO_2$	Spearmint	Essential oil	Constant	Shrinking core	PF
Lee 1986	[63]	$scCO_2$	Rape oil	Fatty oil	Constant	No internal diffusion	PF
López-Padilla 2017	[64]	$scCO_2$	Marigold	Fatty oil	BIC-type	Broken and intact cells	PF
Lucas 2007	[65]	$scCO_2$	Wheat sprouts	Fatty oil	Linear	-	PF
Machmudah 2006	[66]	$scCO_2$	Nutmeg	Fatty oil	BIC-type	Broken and intact cells Shrinking core	PF
Macias-Sanchez 2009	[67]	$scCO_2$ + 5% Ethanol	Micro algae	Carotenoids	Linear	Sphere	PF
Madras 1994	[68]	$scCO_2$	Soil	Organic pollutants	Freundlich	Shrinking core	DPF
Mantell 2002	[69]	Methanol	Grape pomace	Anthocyanins	Linear	Sphere	PF
Marrone 1998	[70]	$scCO_2$	Almond oil	Fatty oil	BIC-type	Broken and intact cells	PF
Martinez 2003	[71]	$scCO_2$	Ginger	Oleoresin	-	Logistic	PF
Nagy 2008	[72]	$scCO_2$	Chili	Essential oil	-	Particle size distribution	PF
Özkal 2005	[73]	$scCO_2$	Apricot kernels	Apricot kernel oil	BIC-type	Broken and intact cells	PF
Peker 1992	[74]	$scCO_2$	Coffee beans	Caffeine	Linear	Sphere	PF
Perrut 1997	[75]	$scCO_2$	Sunflower seed	Fatty oil	BIC-type	Porous particle	PF
Pinelo 2006	[76]	Ethanol	Grape by-products	Antioxidants	-	Sphere	STR
Poletto and Reverchon 1996	[77]	$scCO_2$	Vegetable	Essential and fatty oil	Linear	-	PF
Reis-Vasco 2000	[78]	$scCO_2$	Pennyroyal	Essential oil	Linear	Broken and intact cells	DPF
Reverchon 1996	[79]	$scCO_2$	Salvia oil	Essential oil	Linear	Sphere, cylinder, rod	PF
Reverchon and Marrone 1997	[80]	$scCO_2$	Cloves	Essential oil	Linear	No internal diffusion	DPF
Reverchon 1999	[81]	$scCO_2$	Fennel	Essential and fatty oil	BIC-type	Broken and intact cells	PF

Table 1. *Cont.*

Author and Year	Ref.	Fluid	Plant Material	Target Component	Equilibrium	Particle/Shape/Model	Flux
Reverchon 2000	[82]	$scCO_2$	Rosehip oil	Fatty oil	BIC-type	Broken and intact cells	PF
Reverchon and Marrone 2001	[83]	$scCO_2$	Vegetable oil	Fatty oil	BIC-type	Broken and intact cells	PF
Rosa 2016	[84]	$scCO_2$	Green coffee beans	Cafestole, Kahweole	Linear	No internal diffusion	DPF
Roy 1996	[85]	$scCO_2$	Ginger oil	Essential oil	Constant	Shrinking core	DPF
Salamatin 2017	[86]	$scCO_2$	Pumpkin seed	-	-	Shrinking core und Broken and intact cells	-
Seikova 2003	[87]	Water	Belladonna	Alkaloids	-	Sphere, cylinder, rod	STR
Seikova 2004	[88]	Water pH 9 (NaOH)	Tomato seed	Proteins	-	Sphere, cylinder, rod	STR
Simeonov 1999	[89]	Water	Tabaco leaves, oak bark	-	Linear	Sphere, cylinder, rod	STR
Simeonov 2003	[90]	Methanol, Petrol ether	Indigo, coriander	Essential oil, Fatty oil, Isoflavonoids	-	Sphere, cylinder, rod	STR
Simeonov 2008	[91]	70/30 v/v Ethanol/Water	Root of bloody geranium	-	-	Sphere, cylinder, rod	STR
Skerget 2001	[92]	$scCO_2$	Milk thistle, pepper, chili, cacao	-	Linear	Porous particle, parabolic concentration profile	STR
Sovova 1994	[30]	$scCO_2$	Vegetable	Fatty oil	Constant	Broken and intact cells	PF
Sovova 1994	[93]	$scCO_2$	Caraway	Essential oil	Linear	Broken and intact cells	PF
Sovova 2005	[31]	$scCO_2$	-	-	BIC-type	Broken and intact cells	PF
Stamenic 2008	[94]	$scCO_2$	Thyme, celery, valerian root	Essential oil	-	Broken and intact cells, trichoma cells	DPF
Stastova 1996	[95]	$scCO_2$	Sea buckthorn	Fatty oil	Constant	Broken and intact cells	PF
Strube 2008	[96]	20% (w/w) Water/Ethanol	Brazilian amargo	Terpenoids	Langmuir	Porous particle, parabolic concentration profile	DPF
Strube 2012	[17]	Ethanol, Ethyl acetate	Pepper, vanilla	Piperine, Vanillin	Langmuir	Porous particle	DPF
Strube 2017	[25]	Water (PHWE)	Yew	10-deacetylbaccatin III	Constant	Porous particle	DPF with degradation kinetics
Teixera de Souza 2008	[97]	$scCO_2$	Candeia tree	Essential oil	-	-	PF
Veloso 2008	[98]	Hexane, Water, Alcohols	Oil seed	Fatty oil	Linear	Porous particle, no internal diffusion	DPF, cross-current
Winitsorn 2008	[99]	Ethanol	Tamarind, green tea	-	-	Porous particle	STR
Wu and Hou 2001	[100]	$scCO_2$	Egg yolk	Fatty oil	BIC-type	No internal diffusion	PF
Zizovic 2005	[101]	$scCO_2$	Basil, rosemary, marjoram, pennyroyal	Essential oil	-	Trichoma cells	DPF
Zizovic 2007	[102]	$scCO_2$	Marigold, chamomile	Essential oil	-	Sphere with channels, no internal diffusion	DPF

2.1. Distributed-Plug-Flow (DPF) Model

The DPF model describes macroscopic mass transport in the liquid phase for each component and is given in Equation (1).

$$\frac{\partial c_L(z,t)}{\partial t} = D_{ax} \cdot \frac{\partial^2 c_L(z,t)}{\partial z^2} - \frac{u_z}{\varepsilon} \cdot \frac{\partial c_L(z,t)}{\partial z} - \frac{1-\varepsilon}{\varepsilon} \cdot k_f \cdot a_P \cdot [c_L(z,t) - c_P(r=R,z,t)] \quad (1)$$

The model equation consists of several terms. The first term describes the accumulation, i.e., the time-related enrichment of the target and minor components in the solvent. The second term represents the so-called axial dispersion. D_{ax} is the axial dispersion coefficient that has to be adapted to the real flow conditions. If this term is neglected, only the flow profile of the ideal flow tube (PFR) is displayed. However, if D_{ax} is greater than zero, an expansion of the residence time distribution up to the behavior of the ideal stirred tank can be modeled. In the present case, the axial dispersion is determined by a correlation over the Reynolds and Péclet numbers [17]. First, the Reynolds number Re must be calculated. This sets the inertial forces in relation to the viscous forces in fluids. Here, u_z is the empty tube velocity of the fluid, $d_{P,mean}$ is the mean particle diameter, ρ_L is the density of the fluid, η is the dynamic viscosity of the fluid and ε is the void fraction of the medium through which it flows.

$$Re = \frac{u_z \cdot d_{P,mean} \cdot \rho_L}{\eta \cdot \varepsilon} \quad (2)$$

Based on the Reynolds number, the Péclet number can be determined. It forms the relationship between convective and dispersive mass transfer. The correlation proposed by Chung applies only to Reynolds numbers between 10^{-3} and 10^3 [103].

$$Pe = \frac{0.2}{\varepsilon} + \frac{0.011}{\varepsilon} (\varepsilon \cdot Re)^{0.48} \qquad 10^{-3} < Re < 10^3 \quad (3)$$

From the Péclet number in turn follows the axial dispersion coefficient D_{ax} according to the Equation (4). In addition, the value of D_{ax} can be determined by tracer experiments, which are common practice in chromatography [104–106].

$$D_{ax} = \frac{d_{P,mean} \cdot u_z}{\varepsilon \cdot Pe} \quad (4)$$

The third term in Equation (1) describes the convection that results from the pumping of the solvent. The empty tube velocity is represented by u_z and the void level of the fixed bed by ε. The differential $\partial c_L(z,t)/\partial z$ is the local concentration profile of the respective component in the axial direction.

The last term describes the mass transfer from the particle into the fluid of each component. The mass transfer coefficient k_f and the specific surface area of the particles a_P represent the model parameters. If spherical particles are assumed, the following relationship arises between the specific surface area and the particle radius:

$$a_P = \frac{6}{d_{P,mean}} \quad (5)$$

The mass transport coefficient k_f is also determined by a correlation of the Schmidt (6), the Sherwood (7), and the Reynolds number (2) for the particles [16].

$$Sc = \frac{\eta}{\rho_L \cdot D_{12}} \quad (6)$$

$$Sh = \frac{k_f \cdot d_{P,mean}}{D_{12}} \quad (7)$$

D_{12} is the binary diffusion coefficient between the respective component and the solvent. This model parameter must be determined from an actual extraction. By means of the correlation below, the mass transfer coefficient k_f is then calculated [16].

$$Sh = 2 + 1.1 \cdot Sc^{0.33} \cdot Re^{0.6} \tag{8}$$

The DPF model is a partial differential equation of the second order. The solution therefore requires two boundary conditions and an initial condition. At the beginning of the process, the solvent is unloaded, so the concentration of extracted components is zero.

$$c_L(z,t) = 0 \qquad \begin{matrix} t = 0 \\ 0 \leq z \leq L \end{matrix} \tag{9}$$

The boundary condition for the axial dispersion represents a material balance around the inlet zone of the flow tube [107]. The fluid is first conveyed by convection to percolation. If the fluid enters, it is transported away inside the tube by convection and diffusion, which is represented by Equation (10).

$$u_z \cdot c_L(z,t) = D_{ax} \cdot \left.\frac{\partial c_L(z,t)}{\partial z}\right|_{z=0} \qquad \begin{matrix} t > 0 \\ z = 0 \end{matrix} \tag{10}$$

At the outlet of the flow tube, the axial concentration change is negligible. The derivative of the concentration in axial direction is therefore zero.

$$\left.\frac{\partial c_L(z,t)}{\partial z}\right|_{z=L} = 0 \qquad \begin{matrix} t > 0 \\ z = L \end{matrix} \tag{11}$$

2.2. Pore Diffusion (PD) Model

The mass transport of the components from the pores of the plant material into the fluid is represented by a pore diffusion model. The model assumes that the solvent enters the particle by diffusion, where it dissolves the respective component and diffuses back into the liquid core, as shown in Figure 5.

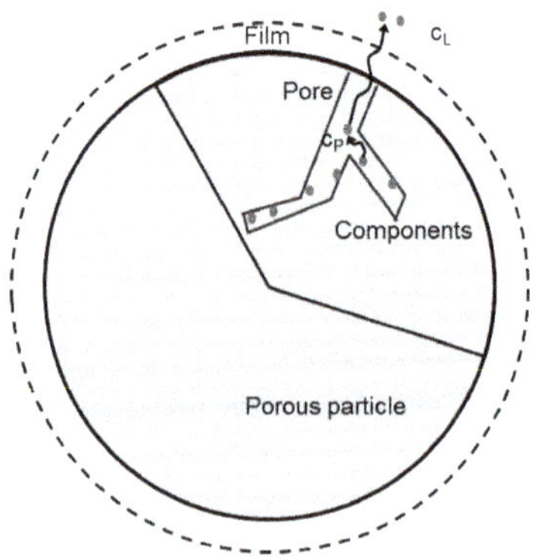

Figure 5. Principle of the pore diffusion model.

In the case of spherical particles, Equation (12) results for the pore diffusion model from Fick's second law.

$$\frac{\partial q(z,r,t)}{\partial t} = D_{eff}(r) \cdot \left(\frac{\partial^2 c_P(z,r)}{\partial r^2} + \frac{2}{r} \cdot \frac{\partial c_P(z,r)}{\partial r}\right) + \frac{\partial D_{eff}(r)}{\partial r} \cdot \frac{\partial c_P(z,r)}{\partial r} \tag{12}$$

In the equation, $\partial q(z, r, t)/\partial t$ represents the time-dependent loading of the solid with the respective component. These will be discussed in more detail in the following chapter. The concentration of the target component is expressed by c_P, which is radially resolved. D_{eff} represents the effective diffusion coefficient, which is calculated starting from the binary diffusion coefficient D_{12}, by means of the porosity of the plant material ε_P, the tortuosity τ and the constrictivity factor δ [17].

$$D_{eff} = \frac{D_{12} \cdot \varepsilon_P \cdot \delta}{\tau} \tag{13}$$

For the parameters ε_P, τ, and δ, corresponding correlations exist [17]. In addition, they can be determined by suitable methods, such as the mercury penetration method. In this work, they are summarized to a sum parameter and determined by an extraction experiment.

Since the pore diffusion model is also a partial differential equation of the second order, two boundary conditions and an initial condition are needed. At the beginning of the extraction, the extraction material is maximally loaded at each location in the axial and in the radial direction, which is taken into account by Equation (14).

$$q(z, r, t) = q_m(r) \qquad \begin{matrix} t = 0 \\ 0 < z \leq L \\ 0 < r \leq R \end{matrix} \tag{14}$$

During the extraction, a concentration profile is formed in the particle. Based on the assumption that the solvent penetrates evenly into the spherical particle from all sides and does not diffuse beyond the center of the particle, the local derivative of the radial concentration profile must be zero in the particle center.

$$\frac{\partial c_P(z, r, t)}{\partial r} = 0 \qquad \begin{matrix} t > 0 \\ 0 < z \leq L \\ r = 0 \end{matrix} \tag{15}$$

At the outer particle edge, a mass balance serves as a boundary condition. The extracted component first passes by diffusion to the particle edge and from there into the fluid. The mass transfer coefficient k_f links the pore diffusion model with the DPF model.

$$D_{eff}(r) \cdot \frac{\partial c_P(z, r, t)}{\partial r} = k_f \cdot [c_L(z, t) - c_P(r, z, t)] \qquad r = R \tag{16}$$

2.3. Equilibrium

For the design and modelling of phytoextraction, the plant particles are considered to be porous spheres in which the components are adsorbed. The solvent must diffuse into the pores, there dissolve the components and move back into the core of the fluid. Within the pores is an adsorption/desorption equilibrium that can be described by equilibrium lines. The approaches available in the literature are described in more detail below [28,108]. The loading q represents the linking of the equilibrium lines to the pore diffusion model.

2.3.1. Henry

The simplest form of equilibrium is Henry's linear approach. K_H is the Henry coefficient, which is accessible experimentally and represents the proportionality factor between the concentration c and the loading q.

$$q = K_H \cdot c \tag{17}$$

2.3.2. Freundlich

The equilibrium relationship according to Freundlich represents an exponential approach. The Henry's approach is extended by the exponent n, which can also be determined experimentally.

$$q = K_F \cdot c^n \tag{18}$$

2.3.3. Langmuir

The Langmuir equilibrium relationship results from first order adsorption and desorption kinetics. The coefficient K_L is the quotient of the rate constants of adsorption and desorption. It is also determinable by equilibrium experiments. In contrast to the two above-mentioned equilibrium approaches, the Langmuir shape converges to a limit value, i.e., to a maximum loading of the solid q_{max}.

$$q = q_{max} \cdot \frac{K_L \cdot c}{1 + K_L \cdot c} \tag{19}$$

2.3.4. Modified-Langmuir

Especially in phytoextraction, there are no pure adsorption or desorption equilibria. A high proportion of the valuable substance in dissolved e.g., in the vacuole or in cell spaces. This corresponds to the model concept "Broken and Intact Cells" by Sovová [30]. Kassing et al. [16,17] have adopted this concept in the pore diffusion model and implemented it using a radial target component distribution. If the target component distribution measurable by methods such as Raman mapping [109], this can be implemented. If no corresponding data is available, the introduction of the capacity factor a as a macroscopic parameter reflects the basic idea of the Broken and Intact Cells model, which is based on non-steady-differentiable equilibrium lines [31,75]. If the solvent is able to dissolve and transport away a large amount of the target substances, or if it is accessible, only a small fraction, which is actually adsorbed in the cell or between the cells, remains. The capacity factor a leads then to a modified version of the Langmuir equilibrium.

$$q = q_{max} \cdot a \cdot \frac{K_L \cdot c}{1 + K_L \cdot c} \tag{20}$$

3. Model Parameter Determination

Figure 6 explains the procedure of stepwise determination of model parameters with separated independent effects. Altenhöner [104] proposed this procedure of equation assembly at first for chromatography process modelling, which was transferred successfully to solid liquid extractions [17].

At first, the laboratory scale equipment is characterized once by tracer experiments, allowing a direct transfer to other scales, if their fluid dynamics is characterized as well [18]. This fills the first boxed term of the DPF-model equation; all other terms are not needed and neglected. Therefore, no interference of model parameter occurs, and the effects and their characteristic parameters are discriminated distinctively. Afterward, thermodynamic equilibrium phase behavior is implemented by i.e., a Langmuir-type equation, where K is representative for the slope of the equilibrium curve at infinite dilution and q_{max} is the overall amount of the regarded component within the plant material by nature. Finally, the mass transfer coefficients are determined for the DPF model by i.e., correlations, using the Péclet and Reynolds number, as shown before. The effective diffusion coefficient needed for the pore diffusion model is measured by means of an extraction experiment. Therefore, such an equation setup is assembled stepwise. As the index *i* in Figure 6 indicates, this procedure is only limited by the number of distinguishable und quantifiable components. In most cases, the parameter determination is carried out in a standardized apparatus [16,17] consuming about 150 g of plant material and a few liters of solvent.

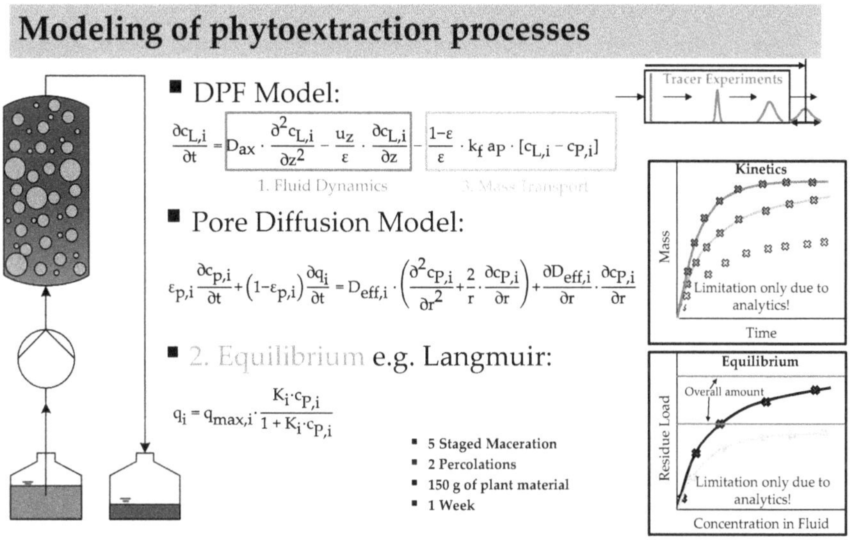

Figure 6. Stepwise equation assembly in phytoextraction (solid liquid extraction).

3.1. *Overall Amount*

The overall amount of each considered component is determined by means of an exhaustive percolation and Equation (21). Where q_{max} is the loading of the plant material with the respective component, c_{Comp} the concentration in the extract, $V_{Extract}$ the extract volume and m_{Plant} the amount of plant material used. RM is the residual moisture of the plant material and $\rho_{Extract}$ the extract's density.

$$q_{max} = \frac{c_{Comp} \cdot m_{Extract}}{m_{Plant} \cdot (1 - RM) \cdot \rho_{Extract}} \tag{21}$$

The extraction yield is referred to the overall amount. The error propagation is given in Equation (22).

$$\begin{aligned} \Delta q_{max} = & \left| \frac{m_{Extract}}{m_{Plant} \cdot (1-RM) \cdot \rho_{Extract}} \right| \cdot \Delta c_{Comp} + \left| \frac{c_{Comp}}{m_{Plant} \cdot (1-RM) \cdot \rho_{Extract}} \right| \\ & \cdot \Delta m_{Extract} + \left| \frac{c_{Comp} \cdot m_{Extract}}{(m_{Plant} \cdot (1-RM) \cdot \rho_{Extract})^2} \right| \cdot (m_{Plant} \cdot (1-RM)) \\ & \cdot \Delta \rho_{Extract} + \left| \frac{c_{Comp} \cdot m_{Extract}}{(m_{Plant} \cdot (1-RM) \cdot \rho_{Extract})^2} \right| \cdot (\rho_{Extract} \cdot (1-RM)) \\ & \cdot \Delta m_{Plant} + \left| \frac{c_{Comp} \cdot m_{Extract}}{(m_{Plant} \cdot (1-RM) \cdot \rho_{Extract})^2} \right| \cdot m_{Plant} \cdot \rho_{Extract} \cdot \Delta RM \end{aligned} \tag{22}$$

The error calculation for a representative example is depicted below. The data is:

- mass of plant material 20 g;
- mass of solvent 5000 g;
- density of solvent 791 g/L;
- concentration 0.03 g/L;
- and the residual moisture is 8%.

The resulting overall amount is 1.03% ± 0.0257% referred to dry mass. The relative deviation therefore is ±2.57%. The resulting individual errors in each step are given in Table 2.

Table 2. Error propagation for the overall amount measurement.

Parameter	Deviation	Origin	Error
Δc_{Comp}	±2% = ±0.0006 g/L@0.03 g/L	HPLC-analytics	±0.0206%
Δm_{Plant}	±0.01 g = ±0.05%@20 g	Last digit of balance	±0.0005%
$\Delta m_{Extract}$	±0.01 g = ±0.0002%@5000 g	Last digit of balance	±0.000002%
$\Delta \rho_{Extract}$	±0.1 g/L = 0.012%@791 g/L	Last digit of digital density meter	±0.00013%
ΔRM	±0.5%	Reproducibility	±0.0045%
	Overall		±0.0257%

3.2. Equilibrium

The equilibrium between the respectively considered component in the solvent and in the solid is determined by means of multi-stage maceration. For this purpose, a defined amount of plant material and solvent is extracted for 24 h in a maceration vessel. This is followed by the removal of a sample before additional solvent is added and then a new equilibrium point is established. This process can be repeated as often and with any ratio of plant to solvent, as long as the mixture is still stirrable. The residual charge results per step as the ratio of the respective component in the extract to the total amount in the plant by means of Equation (23).

$$q_{Comp} = \frac{c_{Comp} \cdot m_{Extract}}{q_{max} \cdot m_{Plant} \cdot (1 - RM) \cdot \rho_{Extract}} \tag{23}$$

The error propagation is given in Equation (24).

$$\begin{aligned} \Delta q_{Comp} = & \left| \frac{m_{Extract}}{q_{max} \cdot m_{Plant} \cdot (1-RM) \cdot \rho_{Extract}} \right| \cdot \Delta c_{Comp} \\ & + \left| \frac{c_{Comp.}}{q_{max} \cdot m_{Plant} \cdot (1-RM) \cdot \rho_{Extract}} \right| \cdot \Delta m_{Extract} \\ & + \left| \frac{c_{Comp.} \cdot m_{Extract}}{(q_{max} \cdot m_{Plant} \cdot (1-RM) \cdot \rho_{Extract})^2} \right| \cdot (q_{max} \cdot m_{Plant} \cdot (1 - RM)) \\ & \cdot \Delta \rho_{Extract} + \left| \frac{c_{Comp} \cdot m_{Extract}}{(q_{max} \cdot m_{Plant} \cdot (1-RM) \cdot \rho_{Extract})^2} \right| \\ & \cdot (q_{max} \cdot (1 - RM) \cdot \rho_{Extract}) \cdot \Delta m_{Plant} \\ & + \left| \frac{c_{Comp} \cdot m_{Extract}}{(q_{max} \cdot m_{Plant} \cdot (1-RM) \cdot \rho_{Extract})^2} \right| \cdot q_{max} \cdot m_{Plant} \cdot \rho_{Extract} \\ & \cdot \Delta RM + \left| \frac{c_{Comp} \cdot m_{Extract}}{(q_{max} \cdot m_{Plant} \cdot (1-RM) \cdot \rho_{Extract})^2} \right| \cdot m_{Plant} \cdot (1 - RM) \\ & \cdot \rho_{Extract} \cdot \Delta q_{max} \end{aligned} \tag{24}$$

The error calculation for a representative example is depicted below. The data is

- mass of plant material 20 g;
- mass of solvent 300 g;
- density of solvent 791 g/L;
- overall amount 1%;
- concentration 0.3 g/L;
- and the residual moisture is 8%.

The residual load is 0.618% ± 0.0314%. The relative deviation therefore is ±5.08%. The details can be seen in Table 3.

Table 3. Error propagation for the equilibrium measurement.

Parameter	Deviation	Origin	Error
$\Delta c_{Comp.}$	$\pm 2\% = \pm 0.006$ g/L@0.3 g/L	HPLC-analytics	$\pm 0.0124\%$
Δm_{Plant}	± 0.01 g = $\pm 0.05\%$@20 g	Last digit of balance	$\pm 0.0003\%$
$\Delta m_{Extract}$	± 0.01 g = $\pm 0.0002\%$@300 g	Last digit of balance	$\pm 0.00002\%$
$\Delta \rho_{Extract}$	± 0.1 g/L = 0.012%@791 g/L	Last digit of digital density meter	$\pm 0.00008\%$
ΔRM	$\pm 0.5\%$	Reproducibility	$\pm 0.0027\%$
Δq_{max}	$\pm 2.5\%$	Error Propagation	$\pm 0.0159\%$
	Overall		$0.618\% \pm 0.0314\%$

4. Model Validation

The aim of this study is to provide a method for the analysis of the model precision in order to complete the toolbox towards proved model validation. Accuracy was shown through numerous successful applications of the model with different feedstocks and scales as listed above.

4.1. Sensitivity Analysis

The depicted experimental errors in the model parameter determination doubtlessly have a feedback on the simulation results and therefore on the model precision. The scattering of the simulation results due to minimum and maximum values of the model parameters for a representative extraction experiment are assessed in the following. The mean value as well as the minimum and the maximum of the individual parameter are listed in Table 4.

Table 4. Parameters for the sensitivity analysis.

Parameter	Min.	Mean	Max.	Deviation	Origin
q	0.0010237	0.00105	0.001076	±2.5%	Error propagation
a	0.117	0.13	0.143	±10%	Error propagation and reproducibility
V	0.98 mL/min	1 mL/min	1.02 mL/min	±2%	Data sheet
m	18.26 g	18.35 g	18.44 g	±0.5%	Error propagation
K_L	67.5	75	82.5	±10%	Error propagation and reproducibility
d	800 μm	900 μm	1000 μm	±100 μm	Mesh space of sieves

The simulation results as well as experimental values are given in Figure 7A. For each simulation, only one parameter has been changed, the remaining parameters remain at their mean value. The experimentally determined extracted mass has a relative deviation of ±4.26% including reproducibility and error propagation (Figure 7B). The greatest influences on the extracted mass have the initial overall amount and the mean particle diameter. The first parameter leads to a higher or respectively lower loading of the plant material. The particle diameter leads to a faster or respectively slower extraction due to the changing accessibility of the target component. An intermediate effect has the capacity factor and the initial plant mass. Hardly any deviation in the simulation results is referred to the Langmuir-coefficient and the volume flow. None of these parameter intervals leads to a deviation that is more than the experimental error, thus the experimental parameter determination is precise enough in that case and therefore the model is adequate.

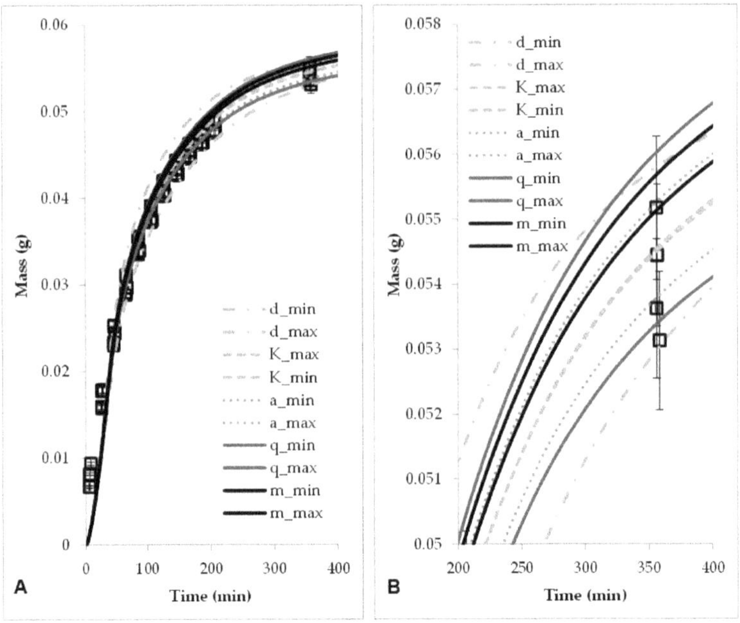

Figure 7. Sensitivity analysis one-parameter-at-a-time. (**A**) whole extraction; (**B**) zoom to the last data point (highest mass of extracted components).

4.2. Statistical Evaluation

The previous one-parameter-at-a-time study showed that the simulation results vary below the experimental error. To evaluate the resulting precision of the simulation results, due to random parameter combinations, the parameters were mixed by the rules of DoE, resulting in 64 individual simulation runs around one point of operation. The results are given in Figure 8.

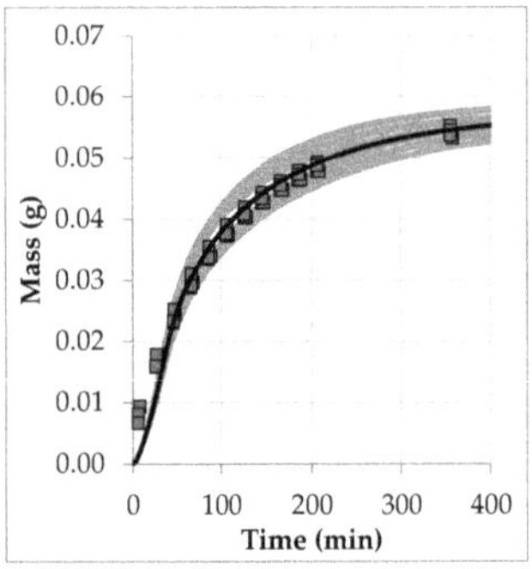

Figure 8. Sensitivity analysis, DoE.

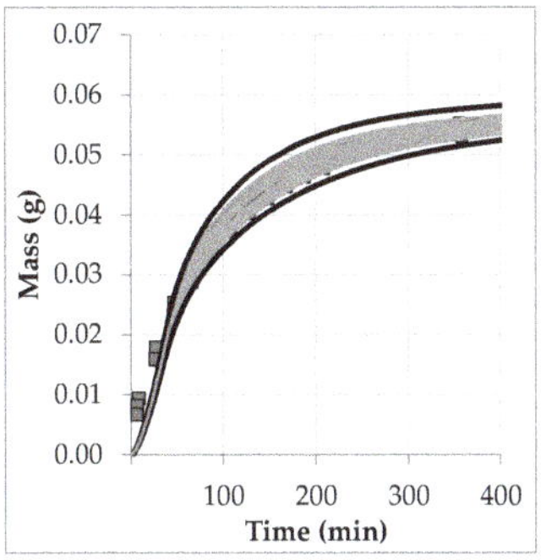

Figure 9. Monte Carlo–based sensitivity analysis; extremes drawn in black from DoE (Figure 8).

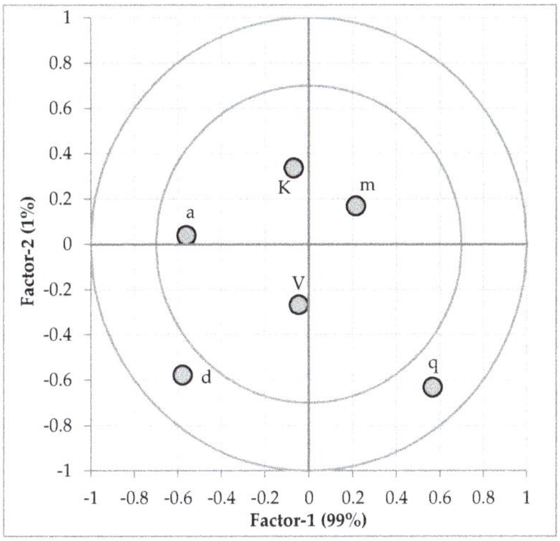

Figure 10. PLS correlations loadings plot.

Result: The deviation is ±4.5% and therefore only about 5% above the experimental error. The error of the modelling and simulation is definitely in the same order of magnitude as the experimental extraction data. This shows that the concept of physico-chemical modelling and experimental model parameter determination is valid and precise enough for the prediction of experimental extraction curves. The DoE plan consists of the mean value and the individual minima and maxima of the respective parameter. Highly non-linear influences are therefore not observed. To fill this gap, either a statistical plan including more factor levels or a Monte Carlo simulation with a random distribution of the investigated parameters serves for closing this gap. Both are equally applicable. In that specific case, the effects of the parameters on the simulation results is linear and an additional Monte Carlo simulation does not result in higher or lower envelopes as the initial DoE-plan did, as shown in Figure 9. The application of Monte Carlo simulations has already been shown in [110].

For further statistical analysis, a PLS regression was calculated. The correlation loadings plot is shown in Figure 10. Only one PLS factor is capable of describing 99% of the data set taken from the DoE plan. The mean particle diameter d and the loading q behave reciprocal. Both, a small mean particle diameter and a high loading lead to a high mass of extracted component at a fixed time. These two parameters contribute significantly to the first principal component of the PLS model.

The two equilibrium parameters K and a lead to a high mass of extracted components if they have their smallest values, hence the equilibrium limitation in the system is small.

The plant mass m is positively correlated to the loading q. Consequently, a high mass leads to a high amount of extracted components. The flux V has nearly no influence on the first PLS factor and can be regarded as not significant. *Result:* The correlations loadings plot shows that the implemented equations, i.e., the modelling depth of a verified model, behave in a physically consistent way. It also indicates that the measurement precisions of the overall amount and the mean particle diameter have the largest impacts on the simulation results. That supports the results shown in Figure 7.

The same result is obtained by the evaluation of the statistical plan. The resulting main effects are shown in Figure 11. The statistical evaluation shows the same results as the PLS model. The loading q and the mass m are positively correlated to the extracted mass. The equilibrium parameters a and K are negatively correlated to the mass, as well as the mean particle diameter d. The flux V has nearly no influence on the extraction, in that case.

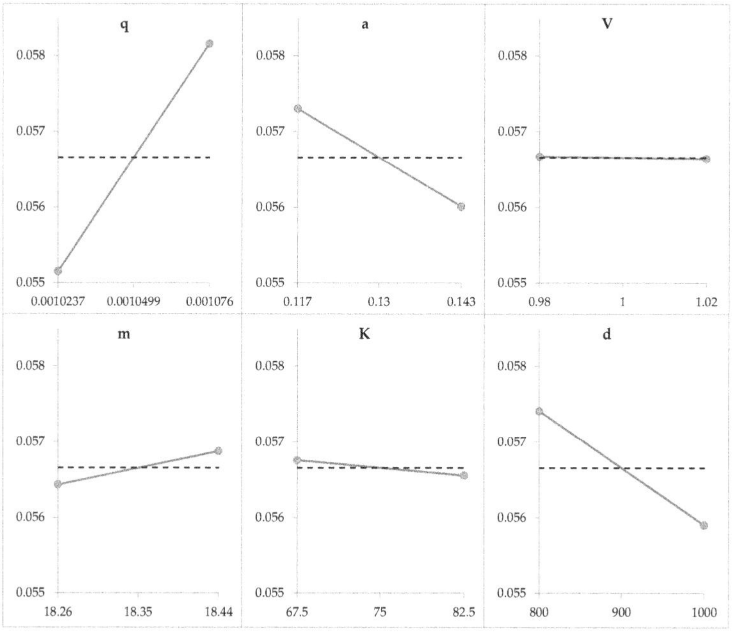

Figure 11. PLS correlations loadings plot.

5. Conclusions and Discussion

In order to evaluate the proposed approach as a consistent method with clear quantitative decision criteria based on distinctly defined work plan steps, efforts and benefits have to be discussed.

5.1. *Effort Analysis*

Figure 12 shows a schedule for a complete parameter determination, which is to be handled by one single person with the appropriate equipment. The actual working time is colored in black, the times during which the experiment does not need to be supervised are dark gray, while light gray is the time in which the analysis takes place.

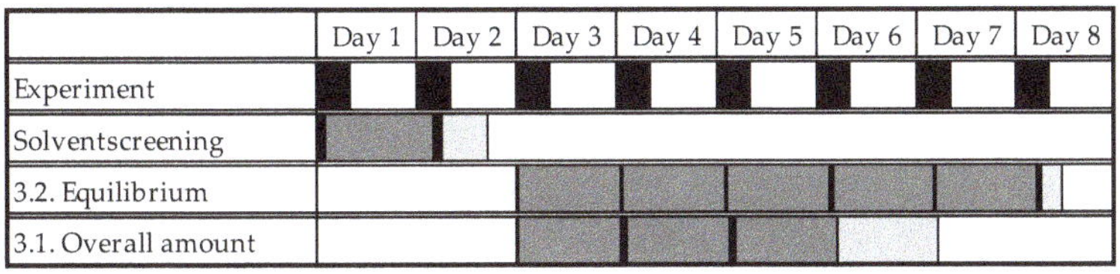

Figure 12. Schedule of the parameter determination process.

- A complete parameter determination with each experiment being run three times, takes about eight working days;
- To determine all parameters, only about 150 g of plant material and about 8 L of solvent are consumed.

If a statistical test plan is used instead of model parameter determination and simulation, one percolation can be carried out daily with the same equipment. Even with only three parameters and one center point as a triplicate, a full-factorial experimental design results in 11 experiments, which together have a much lower information density than the shown path of model parameter determination and proven accurate and precise modelling. The model-based approach is therefore to be preferred in any case especially since with increasing experience the results can be very quickly checked for plausibility or can be transferred to other material systems.

5.2. Modelling in Modern Process Engineering

The modern approach for process development and quality assessment Quality-by-Design (QbD) is increasingly widespread, as demanded by regulatory authorities [15]. The idea is to define a design space of operating parameters in which the product always fulfils given quality attributes. This leads to multi-parameter optimizations and a significant experimental effort. Small-scale models or rather physico chemical–process models are capable of filling this gap in terms of model-assisted process design [13], as indicated in Figure 13.

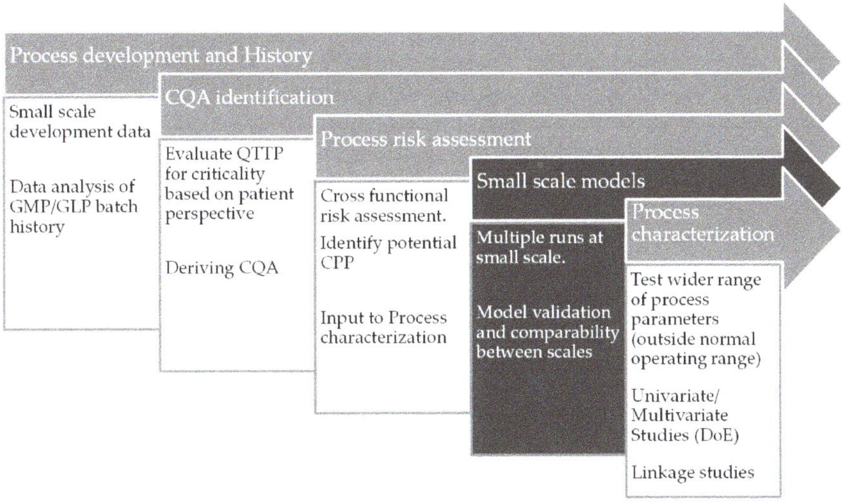

Figure 13. Missing link in process development [13].

As explained before, process models are more efficient than experimental scale-down models, but they are only of any use if they are quantitatively distinctly proven to be valid, accurate, and precise, as proposed before, to base any relevant approval or investment decisions on.

For small- and medium-sized enterprises (SMEs) or start-up companies that have never operated on a pilot or manufacturing scale, it is impossible to establish valid experimental scale-down models due to the lack of knowledge, data, and experience how they should correctly scale down in their laboratory. This would partly be overcome by the process modelling approach and enable the activities of such innovative companies towards regulatory approval and fully integrated manufacturing.

5.3. *Workflow*

In order to become a routine method even in small and medium companies, a clear and general workflow of model setup, implementation, verification, and validation with regard to accuracy, e.g., modelling depth and precision, is needed. Such a workflow is proposed as a final conclusion in Figure 14, including relevant tools and decision criteria for every task and evaluation to be performed.

At first, a model task has to be defined. Afterwards, a conceptual model is derived and implemented. The first decision criterion serves for model verification. Here, magnitude of characteristic numbers Reynold, Péclet, Sherwood, Schmidt, have to be in the same magnitude as literature data. The simulation of simplified case studies with known data and outcome serves for further verification. The second decision criterion proves accuracy, whether identical effects are similar significant or not between simulation and reality.

Afterwards, the third decision criterion for model precision is approved, whether the magnitudes of the error bars of modelling errors are smaller than the experimental ones. A DoE-based approach is used here to ensure the maximum degree of information whilst maintaining low efforts and time. To take into account highly non-linear behavior of different parameters, additional Monte Carlo simulations with equally distributed parameters can be done. Twenty to thirty simulations serve to ensure about 95% probability and 200–300 simulations for 99% probability [111]. The DoE-based approach gains identical results at fewer efforts, utilizing only the upper and lower limit of the parameter range. Nevertheless, Monte-Carlo simulations taking into account a distinct parameter distribution, do deliver an additional probability distribution of the results. This is especially useful to rate the chance for outliers and failures during operation. Therefore, the advice generated is to apply DoE at first with only extreme values, than rise the factor level and finally enlarge the simulation runs if needed to Monte-Carlo studies.

The final decision criteria is based on the comparison of field experiments, that are independent and not any part of the experimental model parameter determination setup, which needs at first to be proven to consistency by data reconciliation methods. Such consistent data sets are analyzed due to targets values (yield, purity etc.), parameter range and their sensitivity with regards to analogues simulation data setups. If yield, purity, space-time-yield, specific auxiliary/energy amount, and the slope of parameter interactions on them as well as identical quadrant position of parameter at identical magnitude of correlation coefficients in the PLS regression are coincident, then the model is distinctly quantitative proven to be valid for its at first defined task and application.

It is shown that a model-based process design can be successfully implemented. A corresponding workflow has been designed for the case of liquid-solid extraction. Opportunity is available for expansion to other unit operations and combinations thereof.

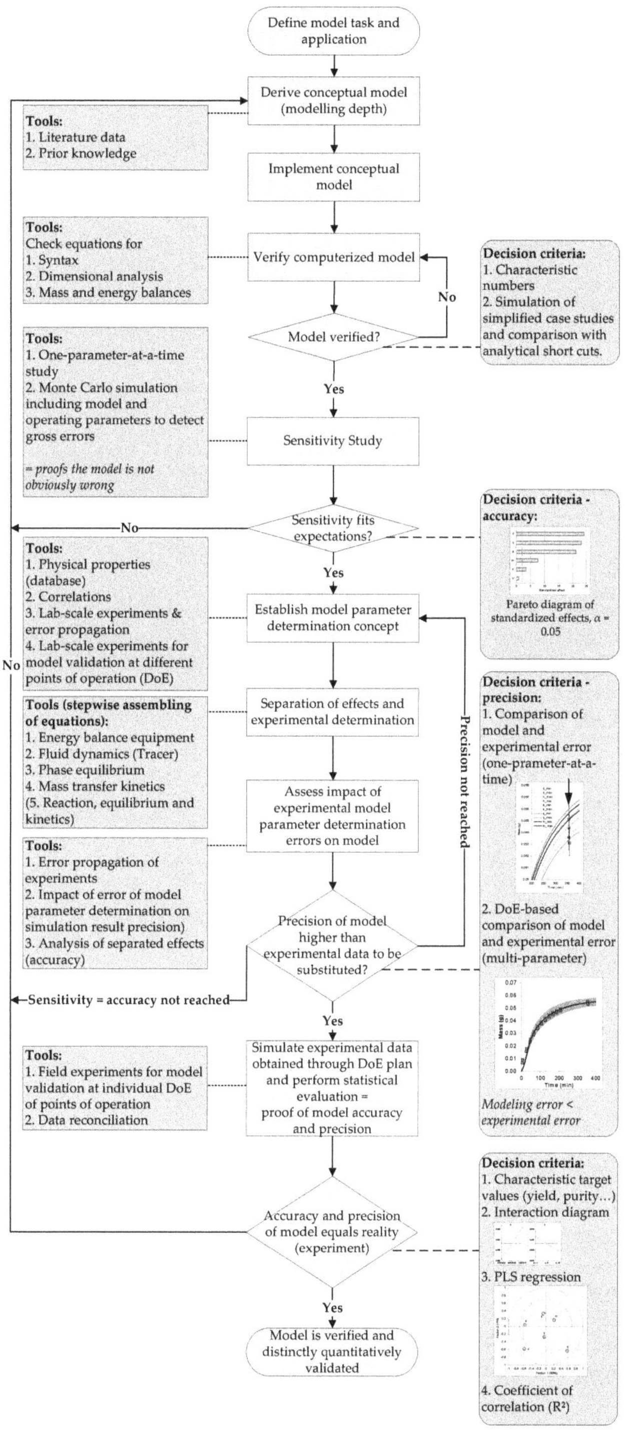

Figure 14. Workflow of process modelling and simulation.

Author Contributions: M.S. developed the process model and wrote the paper. L.U. contributed to statistical evaluation and visualization. J.S. was responsible for conception and supervision.

Acknowledgments: The authors would like to thank Reinhard Ditz/formerly Merck KGaA, Darmstadt for paper revision and fruitful discussions as well as the experienced ITVP laboratory team. Especially, the lecturers of FAH Bonn and PDA Berlin organized trainings and education courses annually in Clausthal are acknowledged.

Conflicts of Interest: The authors declare no conflict of interest.

Symbols and Abbreviations

a_P	Specific surface area, 1/m
c_L	Concentration in the liquid phase, kg/m^3
c_P	Concentration in the porous particle, kg/m^3
D_{ax}	Axial dispersion coefficient, m/s^2
D_{eff}	Effective diffusion coefficient, m^2/s
DPF	Distributed plug flow
K_L	Equilibrium constant, m^3/kg
k_f	Mass transport coefficient, m/s
Pe	Péclet number
PEF	Pulsed electrical field
PF	Plug flow
PLS	Partial Least Square Regression
q	Loading, kg/m^3
q_{max}	Maximum Loading, kg/m^3
Re	Reynolds number
r	Radius, m
Sc	Schmidt number
Sh	Sherwood number
SME	Small and medium-sized enterprise
STR	Stirred tank reactor
t	Time, s
u_z	Superficial velocity, m/s
V	Volume flow, m^3/s
z	Coordinate in axial direction, m
ε	Voids fraction, -
ρ	Density, kg/m^3

References

1. Kennedy, R.C.; Xiang, X.; Madey, G.R.; Cosimano, T.F. *Verification and Validation of Scientific and Economic Models*; PROC Agent: Chicago, IL, USA, 2005.
2. Oreskes, N.; Shrader-Frechette, K.; Belitz, K. Verification, validation, and confirmation of numerical models in the Earth sciences. *Science* **1994**, *263*, 641–646. [CrossRef] [PubMed]
3. Ni, D.; Leonard, J.; Guin, A.; Williams, B. Systematic Approach for Validating Traffic Simulation Models. *Transp. Res. Rec.* **2004**, *1876*, 20–31. [CrossRef]
4. Schleisinger, S.; Crosbie, R.E.; Gagné, R.E. Terminology for model credibility. *Simulation* **1979**, *32*, 103–104. [CrossRef]
5. Jain, S. Modeling and Analysis for Semiconductor Manufacturing. In Proceedings of the 2011 Winter Simulation Conference, Phoenix, AZ, USA, 11–14 December 2011.
6. Schuler, H. *Prozeßsimulation*; Vancouver Coastal Health: Weinheim, Germany, 1995.
7. Strube, J. Prädiktive Modellierung von Trennverfahren. *Chem. Ing. Tech.* **2012**, *84*, 867. [CrossRef]
8. Dechema. *Roadmap Chemical Reaction Engineering: An Initiative of the ProcessNet Subject Division Chemical Reaction Engineering*, 2nd ed.; Gesellschaft für Chemische Technik und Biotechnologie: Frankfurt, Germany, 2017.

9. Ditz, R.; Gerard, D.; Hagels, H.; Igl, N.; Schäffler, M.; Schulz, H.; Stürtz, M.; Tegtmeier, M.; Treutwein, J.; Strube, J.; et al. *Phytoextracts. Proposal towards a New Comprehensive Research Focus*; Dechema: Frankfurt, Germany, 2017.
10. Dechema. *Empfehlungen für Grundständige Studiengänge Biotechnologie mit Naturwissenschaftlichem und Mit Verfahrenstechnischem Schwerpunkt*; Dechema: Frankfurt, Germany, 2017.
11. Dechema. *Prozessintensivierung: Eine Standortbestimmung*; Dechema: Frankfurt, Germany, 2008.
12. Kreysa, G.; Marquardt, R. *Biotechnologie 2020*; Dechema: Frankfurt, Germany, 2005.
13. Uhlenbrock, L.; Sixt, M.; Strube, J. Quality-by-Design (QbD) process evaluation for phytopharmaceuticals on the example of 10-deacetylbaccatin III from yew. *Resource* **2017**, *3*, 137–143. [CrossRef]
14. Food and Drug Administration. *Guidance for Industry: PAT—A Framework for Innovative Pharmaceutical Development, Manufacturing, and Quality Assurance*; Food and Drug Administration: Silver Spring, MD, USA, 2004.
15. Food and Drug Administration. *Guidance for Industry: Q9 Quality Risk Management*; Food and Drug Administration: Silver Spring, MD, USA, 2006.
16. Kaßing, M. *Process Development for Plant-Based Extract Production*; Shaker: Aachen, Germany, 2012.
17. Kaßing, M.; Jenelten, U.; Schenk, J.; Hänsch, R.; Strube, J. Combination of Rigorous and Statistical Modeling for Process Development of Plant-Based Extractions Based on Mass Balances and Biological Aspects. *Chem. Eng. Technol.* **2012**, *35*, 109–132. [CrossRef]
18. Both, S.; Eggersglüß, J.; Lehnberger, A.; Schulz, T.; Schulze, T.; Strube, J. Optimizing Established Processes like Sugar Extraction from Sugar Beets—Design of Experiments versus Physicochemical Modeling. *Chem. Eng. Technol.* **2013**, *36*, 2125–2136. [CrossRef]
19. Both, S. *Systematische Verfahrensentwicklung für Pflanzlich Basierte Produkte im Regulatorischen Umfeld*; Shaker: Aachen, Germany, 2015.
20. Both, S.; Chemat, F.; Strube, J. Extraction of polyphenols from black tea—Conventional and ultrasound assisted extraction. *Ultrason. Sonochem.* **2014**, *21*, 1030–1034. [CrossRef] [PubMed]
21. Both, S.; Koudous, I.; Jenelten, U.; Strube, J. Model-based equipment-design for plant-based extraction processes—Considering botanic and thermodynamic aspects. *C. R. Chim.* **2014**, *17*, 187–196. [CrossRef]
22. Koudous, I. *Stoffdatenbasierte Verfahrensentwicklung zur Isolierung von Wertstoffen aus Pflanzenextrakten*; Shaker: Herzogenrath, Germany, 2017.
23. Sixt, M.; Koudous, I.; Strube, J. Process design for integration of extraction, purification and formulation with alternative solvent concepts. *C. R. Chim.* **2016**, *19*, 733–748. [CrossRef]
24. Sixt, M.; Strube, J. Systematic and Model-Assisted Evaluation of Solvent Based- or Pressurized Hot Water Extraction for the Extraction of Artemisinin from *Artemisia annua* L. *Processes* **2017**, *5*, 86. [CrossRef]
25. Sixt, M.; Strube, J. Pressurized hot water extraction of 10-deacetylbaccatin III from yew for industrial application. *Resource* **2017**, *3*, 177–186. [CrossRef]
26. Sattler, K. *Thermische Trennverfahren: Grundlagen, Auslegung, Apparate*; Wiley-VCH: Weinheim, Germany, 2001.
27. Chémat, F.; Strube, J. *Green Extraction of Natural Products. Theory and Practice*; Wiley-VCH Verlag: Weinheim, Germany, 2015.
28. Kaßing, M.; Jenelten, U.; Schenk, J.; Strube, J. A New Approach for Process Development of Plant-Based Extraction Processes. *Chem. Eng. Technol.* **2010**, *33*, 377–387. [CrossRef]
29. Levenspiel, O. *Chemical Reaction Engineering*, 3rd ed.; Wiley: New York, NY, USA, 1999.
30. Sovová, H. Rate of the vegetable oil extraction with supercritical CO_2-I. Modelling of extraction curves. *Chem. Eng. Sci.* **1994**, *49*, 409–414. [CrossRef]
31. Sovová, H. Mathematical model for supercritical fluid extraction of natural products and extraction curve evaluation. *J. Supercrit. Fluids* **2005**, *33*, 35–52. [CrossRef]
32. Akgün, M.; Akgün, N.A.; Dinçer, S. Extraction and Modeling of Lavender Flower Essential Oil Using Supercritical Carbon Dioxide. *Ind. Eng. Chem. Res.* **2000**, *39*, 473–477. [CrossRef]
33. Al-Jabari, M. Modeling analytical tests of supercritical fluid extraction from solids with langmuir kinetics. *Chem. Eng. Commun.* **2003**, *190*, 1620–1640. [CrossRef]
34. Bulley, N.R.; Fattori, M.; Meisen, A.; Moyls, L. Supercritical fluid extraction of vegetable oil seeds. *J. Am. Oil Chem. Soc.* **1984**, *61*, 1362–1365. [CrossRef]
35. Cacace, J.E.; Mazza, G. Mass transfer process during extraction of phenolic compounds from milled berries. *J. Food Eng.* **2003**, *59*, 379–389. [CrossRef]

36. Carrín, M.E.; Crapiste, G.H. Mathematical modeling of vegetable oil-solvent extraction in a multistage horizontal extractor. *J. Food Eng.* **2008**, *85*, 418–425. [CrossRef]
37. Catchpole, O.J.; Grey, J.B.; Smallfield, B.M. Near-critical extraction of sage, celery, and coriander seed. *J. Supercrit. Fluids* **1996**, *9*, 273–279. [CrossRef]
38. Chalermchat, Y.; Fincan, M.; Dejmek, P. Pulsed electric field treatment for solid-liquid extraction of red beetroot pigment: Mathematical modelling of mass transfer. *J. Food Eng.* **2004**, *64*, 229–236. [CrossRef]
39. Chia, S.L.; Sulaiman, R.; Boo, H.C.; Muhammad, K.; Umanan, F.; Chong, G.H. Modeling of Rice Bran Oil Yield and Bioactive Compounds Obtained Using Subcritical Carbon Dioxide Soxhlet Extraction (SCDS). *Ind. Eng. Chem. Res.* **2015**, *54*, 8546–8553. [CrossRef]
40. Cocero, M.J.; García, J. Mathematical model of supercritical extraction applied to oil seed extraction by CO_2 + saturated alcohol—I. Desorption model. *J. Supercrit. Fluids* **2001**, *20*, 229–243. [CrossRef]
41. De França, L.F.; Meireles, M.A.A. Modeling the extraction of carotene and lipids from pressed palm oil (*Elaes guineensis*) fibers using supercritical CO_2. *J. Supercrit. Fluids* **2000**, *18*, 35–47. [CrossRef]
42. Del Valle, J.M.; Napolitano, P.; Fuentes, N. Estimation of Relevant Mass Transfer Parameters for the Extraction of Packed Substrate Beds Using Supercritical Fluids. *Ind. Eng. Chem. Res.* **2000**, *39*, 4720–4728. [CrossRef]
43. Del Valle, J.M.; Jiménez, M.; Napolitano, P.; Zetzl, C.; Brunner, G. Supercritical carbon dioxide extraction of pelletized *Jalapeño peppers*. *J. Sci. Food Agric.* **2003**, *83*, 550–556. [CrossRef]
44. Del Valle, J.M.; de la Fuente Juan, C.; Cardarelli, D.A. Contributions to supercritical extraction of vegetable substrates in Latin America. *J. Food Eng.* **2005**, *67*, 35–57. [CrossRef]
45. Del Valle, J.M.; de la Fuente Juan, C. Supercritical CO_2 extraction of oilseeds: Review of kinetic and equilibrium models. *Crit. Rev. Food. Sci. Nutr.* **2006**, *46*, 131–160. [CrossRef] [PubMed]
46. Diankov, S.; Simeonov, E.; Tomova, K. Modelling of Multistage Extraction Kinetics for *Nicotiana tabacum* L.—Water System. *J. Univ. Chem. Technol. Metall.* **2008**, *43*, 119–124.
47. Egorov, A.G.; Salamatin, A.A. Bidisperse Shrinking Core Model for Supercritical Fluid Extraction. *Chem. Eng. Technol.* **2015**, *38*, 1203–1211. [CrossRef]
48. Espinoza-Pérez, J.D.; Vargas, A.; Robles-Olvera, V.J.; Rodríguez-Jimenes, G.C.; Garcia-Alvarado, M.A. Mathematical modeling of caffeine kinetic during solid-liquid extraction of coffee beans. *J. Food Eng.* **2007**, *81*, 72–78. [CrossRef]
49. Esquível, M.M.; Bernardo-Gil, M.G.; King, M.B. Mathematical models for supercritical extraction of olive husk oil. *J. Supercrit. Fluids* **1999**, *16*, 43–58. [CrossRef]
50. Ferreira, S.R.S.; Meireles, M.A.A. Modeling the supercritical fluid extraction of black pepper (*Piper nigrum* L.) essential oil. *J. Food Eng.* **2002**, *54*, 263–269. [CrossRef]
51. Fiori, L.; Calcagno, D.; Costa, P. Sensitivity analysis and operative conditions of a supercritical fluid extractor. *J. Supercrit. Fluids* **2007**, *41*, 31–42. [CrossRef]
52. Fiori, L.; Basso, D.; Costa, P. Seed oil supercritical extraction: Particle size distribution of the milled seeds and modeling. *J. Supercrit. Fluids* **2008**, *47*, 174–181. [CrossRef]
53. Fiori, L.; Basso, D.; Costa, P. Supercritical extraction kinetics of seed oil: A new model bridging the 'broken and intact cells' and the 'shrinking-core' models. *J. Supercrit. Fluids* **2009**, *48*, 131–138. [CrossRef]
54. Goodarznia, I.; Eikani, M.H. Supercritical carbon dioxide extraction of essential oils. *Chem. Eng. Sci.* **1998**, *53*, 1387–1395. [CrossRef]
55. Goto, M.; Smith, J.M.; McCoy, B.J. Kinetics and mass transfer for supercritical fluid extraction of wood. *Ind. Eng. Chem. Res.* **1990**, *29*, 282–289. [CrossRef]
56. Goto, M.; Sato, M.; Hirose, T. Extraction of Peppermint Oil by Supercritical Carbon Dioxide. *J. Chem. Eng. Jpn.* **1993**, *26*, 401–407. [CrossRef]
57. Goto, M.; Roy, B.C.; Hirose, T. Shrinking-core leaching model for supercritical-fluid extraction. *J. Supercrit. Fluids* **1996**, *9*, 128–133. [CrossRef]
58. Guerrero, M.S.; Torres, J.S.; Nunez, M.J. Extraction of polyphenols from white distilled grape pomace: Optimization and modelling. *Bioresour. Technol.* **2008**, *99*, 1311–1318. [CrossRef] [PubMed]
59. Ji, J.-B.; Lu, X.-H.; Cai, M.-Q.; Xu, Z.-C. Improvement of leaching process of Geniposide with ultrasound. *Ultrason. Sonochem.* **2006**, *13*, 455–462. [CrossRef] [PubMed]
60. Jokic, S.; Svilovic, S.; Vidovic, S. Modelling the supercritical CO_2 extraction kinetics of soybean oil. *Croat. J. Food Sci. Technol.* **2015**, *7*, 52–57. [CrossRef]

61. Kim, K.H.; Hong, J. Desorption kinetic model for supercritical fluid extraction of spearmint leaf oil. *Sep. Sci. Technol.* **2001**, *36*, 1437–1450. [CrossRef]
62. Kim, K.H.; Hong, J. A mass transfer model for super- and near-critical CO_2 extraction of spearmint leaf oil. *Sep. Sci. Technol.* **2002**, *37*, 2271–2288. [CrossRef]
63. Lee, A.K.K.; Bulley, N.R.; Fattori, M.; Meisen, A. Modelling of supercritical carbon dioxide extraction of Canola oilseed in fixed beds. *J. Am. Oil Chem. Soc.* **1986**, *63*, 921–925. [CrossRef]
64. López-Padilla, A.; Ruiz-Rodriguez, A.; Reglero, G.; Fornari, T. Supercritical carbon dioxide extraction of *Calendula officinalis*: Kinetic modeling and scaling up study. *J. Supercrit. Fluids* **2017**, *130*, 292–300. [CrossRef]
65. Lucas, S.; Calvo, M.P.; García-Serna, J.; Palencia, C.; Cocero, M.J. Two-parameter model for mass transfer processes between solid matrixes and supercritical fluids: Analytical solution. *J. Supercrit. Fluids* **2007**, *41*, 257–266. [CrossRef]
66. Machmudah, S.; Sulaswatty, A.; Sasaki, M.; Goto, M.; Hirose, T. Supercritical CO_2 extraction of nutmeg oil: Experiments and modeling. *J. Supercrit. Fluids* **2006**, *39*, 30–39. [CrossRef]
67. Macías-Sánchez, M.D.; Serrano, C.M.; Rodríguez, M.R.; Martínez de la Ossa, E. Kinetics of the supercritical fluid extraction of carotenoids from microalgae with CO_2 and ethanol as cosolvent. *Chem. Eng. J.* **2009**, *150*, 104–113. [CrossRef]
68. Madras, G.; Thibaud, C.; Erkey, C.; Akgerman, A. Modeling of supercritical extraction of organics from solid matrices. *AIChE J.* **1994**, *40*, 777–785. [CrossRef]
69. Mantell, C.; Rodríguez, M.; Martínez de la Ossa, E. Semi-batch extraction of anthocyanins from red grape pomace in packed beds: Experimental results and process modelling. *Chem. Eng. Sci.* **2002**, *57*, 3831–3838. [CrossRef]
70. Marrone, C.; Poletto, M.; Reverchon, E.; Stassi, A. Almond oil extraction by supercritical CO_2: Experiments and modelling. *Chem. Eng. Sci.* **1998**, *53*, 3711–3718. [CrossRef]
71. Martínez, J.; Monteiro, A.R.; Rosa, P.T.V.; Marques, M.O.M.; Meireles, M.A.A. Multicomponent Model to Describe Extraction of Ginger Oleoresin with Supercritical Carbon Dioxide. *Ind. Eng. Chem. Res.* **2003**, *42*, 1057–1063. [CrossRef]
72. Nagy, B.; Simándi, B. Effects of particle size distribution, moisture content, and initial oil content on the supercritical fluid extraction of paprika. *J. Supercrit. Fluids* **2008**, *46*, 293–298. [CrossRef]
73. Özkal, S.G.; Yener, M.E.; Bayındırlı, L. Mass transfer modeling of apricot kernel oil extraction with supercritical carbon dioxide. *J. Supercrit. Fluids* **2005**, *35*, 119–127. [CrossRef]
74. Peker, H.; Srinivasan, M.P.; Smith, J.M.; McCoy, B.J. Caffeine extraction rates from coffee beans with supercritical carbon dioxide. *AIChE J.* **1992**, *38*, 761–770. [CrossRef]
75. Perrut, M.; Clavier, J.Y.; Poletto, M.; Reverchon, E. Mathematical Modeling of Sunflower Seed Extraction by Supercritical CO_2. *Ind. Eng. Chem. Res.* **1997**, *36*, 430–435. [CrossRef]
76. Pinelo, M.; Sineiro, J.; Núñez, M.J. Mass transfer during continuous solid-liquid extraction of antioxidants from grape byproducts. *J. Food Eng.* **2006**, *77*, 57–63. [CrossRef]
77. Poletto, M.; Reverchon, E. Comparison of Models for Supercritical Fluid Extraction of Seed and Essential Oils in Relation to the Mass-Transfer Rate. *Ind. Eng. Chem. Res.* **1996**, *35*, 3680–3686. [CrossRef]
78. Reis-Vasco, E.M.C.; Coelho, J.A.P.; Palavra, A.M.F.; Marrone, C.; Reverchon, E. Mathematical modelling and simulation of pennyroyal essential oil supercritical extraction. *Chem. Eng. Sci.* **2000**, *55*, 2917–2922. [CrossRef]
79. Reverchon, E. Mathematical modeling of supercritical extraction of sage oil. *AIChE J.* **1996**, *42*, 1765–1771. [CrossRef]
80. Reverchon, E.; Marrone, C. Supercritical extraction of clove bud essential oil: Isolation and mathematical modeling. *Chem. Eng. Sci.* **1997**, *52*, 3421–3428. [CrossRef]
81. Reverchon, E.; Daghero, J.; Marrone, C.; Mattea, M.; Poletto, M. Supercritical Fractional Extraction of Fennel Seed Oil and Essential Oil: Experiments and Mathematical Modeling. *Ind. Eng. Chem. Res.* **1999**, *38*, 3069–3075. [CrossRef]
82. Reverchon, E.; Kaziunas, A.; Marrone, C. Supercritical CO_2 extraction of hiprose seed oil: Experiments and mathematical modelling. *Chem. Eng. Sci.* **2000**, *55*, 2195–2201. [CrossRef]
83. Reverchon, E.; Marrone, C. Modeling and simulation of the supercritical CO_2 extraction of vegetable oils. *J. Supercrit. Fluids* **2001**, *19*, 161–175. [CrossRef]

84. Rosa, R.H.; von Atzingen, G.V.; Belandria, V.; Oliveira, A.L.; Bostyn, S.; Rabi, J.A. Lattice Boltzmann simulation of cafestol and kahweol extraction from green coffee beans in high-pressure system. *J. Food Eng.* **2016**, *176*, 88–96. [CrossRef]
85. Roy, B.C.; Goto, M.; Hirose, T. Extraction of Ginger Oil with Supercritical Carbon Dioxide: Experiments and Modeling. *Ind. Eng. Chem. Res.* **1996**, *35*, 607–612. [CrossRef]
86. Salamatin, A.A. Detection of Microscale Mass-Transport Regimes in Supercritical Fluid Extraction. *Chem. Eng. Technol.* **2017**, *40*, 829–837. [CrossRef]
87. Seikova, I.; Simeonov, E. Determination of Solid Deformation Effect on the Effective Diffusivity during Extraction from Plants. *Sep. Sci. Technol.* **2003**, *38*, 3713–3729. [CrossRef]
88. Seikova, I.; Simeonov, E.; Ivanova, E. Protein leaching from tomato seed—Experimental kinetics and prediction of effective diffusivity. *J. Food Eng.* **2004**, *61*, 165–171. [CrossRef]
89. Simeonov, E.; Tsibranska, I.; Minchev, A. Solid-liquid extraction from plants—Experimental kinetics and modelling. *Chem. Eng. J.* **1999**, *73*, 255–259. [CrossRef]
90. Simeonov, E.; Seikova, I.; Pentchev, I.; Mintchev, A. Modeling of a Screw Solid-Liquid Extractor through Concentration Evolution Experiments. *Ind. Eng. Chem. Res.* **2003**, *42*, 1433–1438. [CrossRef]
91. Simeonov, E.; Koleva, V. Solid-Liquid Extraction from Roots of *Geranium sanquineum* L. *J. Univ. Chem. Technol. Metall.* **2008**, *43*, 409–412.
92. Škerget, M.; Knez, Ž. Modelling high pressure extraction processes. *Comput. Chem. Eng.* **2001**, *25*, 879–886. [CrossRef]
93. Sovová, H.; Komers, R.; Kučera, J.; Jež, J. Supercritical carbon dioxide extraction of caraway essential oil. *Chem. Eng. Sci.* **1994**, *49*, 2499–2505. [CrossRef]
94. Stamenić, M.; Zizovic, I.; Orlović, A.; Skala, D. Mathematical modelling of essential oil SFE on the micro-scale—Classification of plant material. *J. Supercrit. Fluids* **2008**, *46*, 285–292. [CrossRef]
95. Šťastová, J.; Jež, J.; Bártlová, M.; Sovová, H. Rate of the vegetable oil extraction with supercritical CO_2—III. Extraction from sea buckthorn. *Chem. Eng. Sci.* **1996**, *51*, 4347–4352. [CrossRef]
96. Ndocko Ndocko, E.; Bäcker, W.; Strube, J. Process Design Method for Manufacturing of Natural Compounds and Related Molecules. *Sep. Sci. Technol.* **2008**, *43*, 642–670. [CrossRef]
97. DeSouza, A.T.; Benazzi, T.L.; Grings, M.B.; Cabral, V.; Antônio da Silva, E.; Cardozo-Filho, L.; Ceva Antunes, O.A. Supercritical extraction process and phase equilibrium of Candeia (*Eremanthus erythropappus*) oil using supercritical carbon dioxide. *J. Supercrit. Fluids* **2008**, *47*, 182–187. [CrossRef]
98. Veloso, G.O.; Thomas, G.C.; Krioukov, V.G. A mathematical model of extraction in countercurrent crossed flows. *Chem. Eng. Process.* **2008**, *47*, 1470–1477. [CrossRef]
99. Winitsorn, A.; Douglas, P.L.; Douglas, S.; Pongampai, S.; Teppaitoon, W. Modeling the extraction of valuable substances from natural plants using solid-liquid extraction. *Chem. Eng. Commun.* **2008**, *195*, 1457–1464. [CrossRef]
100. Wu, W.; Hou, Y. Mathematical modeling of extraction of egg yolk oil with supercritical CO_2. *J. Supercrit. Fluids* **2001**, *19*, 149–159. [CrossRef]
101. Zizovic, I.; Stamenić, M.; Orlović, A.; Skala, D. Supercritical carbon dioxide essential oil extraction of Lamiaceae family species: Mathematical modelling on the micro-scale and process optimization. *Chem. Eng. Sci.* **2005**, *60*, 6747–6756. [CrossRef]
102. Zizovic, I.; Stamenić, M.; Orlović, A.; Skala, D. Supercritical carbon dioxide extraction of essential oils from plants with secretory ducts: Mathematical modelling on the micro-scale. *J. Supercrit. Fluids* **2007**, *39*, 338–346. [CrossRef]
103. Chung, S.F.; Wen, C.Y. Longitudinal dispersion of liquid flowing through fixed and fluidized beds. *AIChE J.* **1968**, *14*, 857–866. [CrossRef]
104. Altenhöner, U.; Meurer, M.; Strube, J.; Schmidt-Traub, H. Parameter estimation for the simulation of liquid chromatography. *J. Chromatogr. A* **1997**, *769*, 59–69. [CrossRef]
105. Strube, J.; Altenhöner, U.; Meurer, M.; Schmidt-Traub, H.; Schulte, M. Dynamic simulation of simulated moving-bed chromatographic processes for the optimization of chiral separations. *J. Chromatogr. A* **1997**, *769*, 81–92. [CrossRef]
106. Schmidt-Traub, H. *Preparative Chromatography*, 2nd ed.; Wiley-VCH Verlag: Weinheim, Germany, 2013.
107. Ndocko Ndocko, E.; Ditz, R.; Josch, J.-P.; Strube, J. New Material Design Strategy for Chromatographic Separation Steps in Bio-Recovery and Downstream Processing. *Chem. Ing. Tech.* **2011**, *83*, 113–129. [CrossRef]

108. Baerns, M. *Technische Chemie*; Wiley-VCH-Verlag: Weinheim, Germany, 2006.
109. Gudi, G.; Krähmer, A.; Koudous, I.; Strube, J.; Schulz, H. Infrared and Raman spectroscopic methods for characterization of *Taxus baccata* L.—Improved taxane isolation by accelerated quality control and process surveillance. *Talanta* **2015**, *143*, 42–49. [CrossRef] [PubMed]
110. Subramanian, G. *Biopharmaceutical Production Technology*; Wiley-VCH-Verlag: Weinheim, Germany, 2012.
111. Subramani, H.J.; Hidajat, K.; Ray, A.K. Optimization of reactive SMB and Varicol systems. *Comput. Chem. Eng.* **2003**, *27*, 1883–1901. [CrossRef]

processes

MDPI

Article

GEKKO Optimization Suite

Logan D. R. Beal, Daniel C. Hill, R. Abraham Martin and John D. Hedengren *

Department of Chemical Engineering, Brigham Young University, Provo, UT 84602, USA; beal.logan@gmail.com (L.D.R.B.); dhill2522@gmail.com (D.C.H.); abemart@gmail.com (R.A.M.)

* Correspondence: john_hedengren@byu.edu; Tel.: +1-801-477-7341

Received: 1 July 2018; Accepted: 23 July 2018; Published: 31 July 2018

Abstract: This paper introduces GEKKO as an optimization suite for Python. GEKKO specializes in dynamic optimization problems for mixed-integer, nonlinear, and differential algebraic equations (DAE) problems. By blending the approaches of typical algebraic modeling languages (AML) and optimal control packages, GEKKO greatly facilitates the development and application of tools such as nonlinear model predicative control (NMPC), real-time optimization (RTO), moving horizon estimation (MHE), and dynamic simulation. GEKKO is an object-oriented Python library that offers model construction, analysis tools, and visualization of simulation and optimization. In a single package, GEKKO provides model reduction, an object-oriented library for data reconciliation/model predictive control, and integrated problem construction/solution/visualization. This paper introduces the GEKKO Optimization Suite, presents GEKKO's approach and unique place among AMLs and optimal control packages, and cites several examples of problems that are enabled by the GEKKO library.

Keywords: algebraic modeling language; dynamic optimization; model predictive control; moving horizon estimation

---

1. Introduction

Computational power has increased dramatically in recent decades. In addition, there are new architectures for specialized tasks and distributed computing for parallelization. Computational power and architectures have expanded the capabilities of technology to new levels of automation and intelligence with rapidly expanding artificial intelligence capabilities and computer-assisted decision processing. These advancements in technology have been accompanied by a growth in the types of mathematical problems that applications solve. Lately, machine learning (ML) has become the must-have technology across all industries, largely inspired by the recent public successes of new artificial neural network (ANN) applications. Another valuable area that is useful in a variety of applications is dynamic optimization. Applications include chemical production planning [1], energy storage systems [2,3], polymer grade transitions [4], integrated scheduling and control for chemical manufacturing [5,6], cryogenic air separation [7], and dynamic process model parameter estimation in the chemical industry [8]. With a broad and expanding pool of applications using dynamic optimization, the need for a simple and flexible interface to pose problems is increasingly valuable. GEKKO is not only an algebraic modeling language (AML) for posing optimization problems in simple object-oriented equation-based models to interface with powerful built-in optimization solvers but is also a package with the built-in ability to run model predictive control, dynamic parameter estimation, real-time optimization, and parameter update for dynamic models on real-time applications. The purpose of this article is to introduce the unique capabilities in GEKKO and to place this development in context of other packages.

2. Role of a Modeling Language

Algebraic modeling languages (AML) facilitate the interface between advanced solvers and human users. High-end, off-the-shelf gradient-based solvers require extensive information about the problem, including variable bounds, constraint functions and bounds, objective functions, and first and second derivatives of the functions, all in consistent array format. AMLs simplify the process by allowing the model to be written in a simple, intuitive format. The modeling language accepts a model (constraints) and objective to optimize. The AML handles bindings to the solver binary, maintains the required formatting of the solvers, and exposes the necessary functions. The necessary function calls include constraint residuals, objective function values, and derivatives. Most modern modeling languages leverage automatic differentiation (AD) [9] to facilitate exact gradients without explicit derivative definition by the user.

In general, an AML is designed to solve a problem in the form of Equation (1).

$$\min_{u,x} \quad J(x,u) \tag{1a}$$

$$0 = f(x,u) \tag{1b}$$

$$0 \leq g(x,u) \tag{1c}$$

The objective function in Equation (1) is minimized by adjusting the state variables x and inputs u. The inputs u may include variables such as measured disturbances, unmeasured disturbances, control actions, feed-forward values, and parameters that are determined by the solver to minimize the objective function J. The state variables x may be solved with differential or algebraic equations. Equations include equality constraints (f) and inequality constraints (g).

3. Dynamic Optimization

Dynamic optimization is a unique subset of optimization algorithms that pertain to systems with time-based differential equations. Dynamic optimization problems extend algebraic problems of the form in Equation (1) to include the possible addition of the differentials $\frac{dx}{dt}$ in the objective function and constraints, as shown in Equation (2).

$$\min_{u,x} \quad J\left(\frac{\mathrm{d}x}{\mathrm{d}t}, x, u\right) \tag{2a}$$

$$0 = f\left(\frac{\mathrm{d}x}{\mathrm{d}t}, x, u\right) \tag{2b}$$

$$0 \leq g\left(\frac{\mathrm{d}x}{\mathrm{d}t}, x, u\right) \tag{2c}$$

Differential algebraic equation (DAE) systems are solved by discretizing the differential equations to a system of algebraic equations to achieve a numerical solution. Some modeling languages are capable of natively handling DAEs by providing built-in discretization schemes. The DAEs are typically solved numerically and there are a number of available discretization approaches. Historically, these problems were first solved with a direct shooting method [10]. Direct shooting methods are still used and are best suited for stable systems with few degrees of freedom. Direct shooting methods eventually led to the development of multiple shooting, which provided benefits such as parallelization and stability [11]. For very large problems with multiples degrees of freedom, "direct transcription" (also known as "orthogonal collocation on finite elements") is the state-of-the-art method [12]. Some fields have developed other unique approaches, such as pseudospectral optimal control methods [13].

Dynamic optimization problems introduce an additional set of challenges. Many of these challenges are consistent with those of other forms of ordinary differential equation (ODE) and partial differential equation (PDE) systems; only some challenges are unique to discretization in time. These challenges include handling stiff systems, unstable systems, numerical versus analytical solution mismatch, scaling issues (the problems get large very quickly with increased discretization), the number and location in the horizon of discretization points, and the optimal horizon length. Some of these challenges, such as handling stiff systems, can be addressed with the appropriate discretization scheme. Other challenges, such as the necessary precision of the solution and the location of discretizations of state variables, are better handled by a knowledgeable practitioner to avoid excessive computation.

Popular practical implementations of dynamic optimization include model predictive control (MPC) [14] (along with its nonlinear variation NMPC [15] and the economic objective alternative EMPC [16]), moving horizon estimation (MHE) [17] and dynamic real-time optimization (DRTO) [18]. Each of these problems is a special case of Equation (2) with a specific objective function. For example, in MPC, the objective is to minimize the difference between the controlled variable set point and model predictions, as shown in Equation (3).

$$\min_{u,x} \quad \left\| x - x_{sp} \right\| \tag{3}$$

where x is a state variable and x_{sp} is the desired set point or target condition for that state. The objective is typically a 1-norm, 2-norm, or squared error. EMPC modifies MPC by maximizing profit rather than minimizing error to a set point, but uses the same dynamic process model, as shown in Equation (4).

$$\max_{u,x} \quad \text{Profit} \tag{4}$$

MHE adjusts model parameters to minimize the difference between measured variable values (x_{meas}) and model predictions (x), as shown in Equation (5).

$$\min_{u,x} \quad \| x - x_{meas} \| \tag{5}$$

4. Previous Work

There are many software packages and modeling languages currently available for optimization and optimal control. This section, while not a comprehensive comparison, attempts to summarize some of the distinguishing features of each package.

Pyomo [19] is a Python package for modeling and optimization. It supports automatic differentiation and discretization of DAE systems using orthogonal collocation or finite-differencing. The resulting nonlinear programming (NLP) problem can be solved using any of several dozen AMPL Solver Library (ASL) supported solvers.

JuMP [20] is a modeling language for optimization in the Julia language. It supports solution of linear, nonlinear, and mixed-integer problems through a variety of solvers. Automatic differentiation is supplied, but, as of writing, JuMP does not include built-in support for differential equations.

Casadi [21] is a framework that provides a symbolic modeling language and efficient automatic differentiation. It is not a dynamic optimization package itself, but it does provides building blocks for solving dynamic optimization problems and interfacing with various solvers. Interfaces are available in MATLAB, Python, and C++.

GAMS [22] is a package for large-scale linear and nonlinear modeling and optimization with a large and established user base. It connects to a variety of commercial and open-source solvers, and programming interfaces are available for it in Excel, MATLAB, and R. Automatic differentiation is available.

AMPL [23] is a modeling system that integrates a modeling language, a command language, and a scripting language. It incorporates a large and extensible solver library, as well as fast automatic differentiation. AMPL is not designed to handle differential equations. Interfaces are available in C++, C#, Java, MATLAB, and Python.

The gPROMS package [24] is an advanced process modeling and flow-sheet environment with optimization capabilities. An extensive materials property library is included. Dynamic optimization is implemented through single and multiple shooting methods. The software is used through a proprietary interface designed primarily for the process industries.

JModelica [25] is an open-source modeling and optimization package based on the Modelica modeling language. The platform brings together a number of open-source packages, providing ODE integration through Sundials, automatic differentiation through Casadi, and NLP solutions through IPOPT. Dynamic systems are discretized using both local and pseudospectral collocation methods. The platform is accessed through a Python interface.

ACADO [26] is a self-contained toolbox for optimal control. It provides a symbolic modeling language, automatic differentiation, and optimization of differential equations through multiple shooting using the built in QP solver. Automatic C++ code generation is available for online predictive control applications, though support is limited to small to medium-sized problems. Interfaces are available in MATLAB and C++.

DIDO [27] is an object-oriented MATLAB toolbox for dynamic optimization and optimal control. Models are formulated in MATLAB using DIDO expressions, and differential equations are handled using a pseudospectral collocation approach. At this time, automatic differentiation is not supported.

GPOPS II [28] is a MATLAB-based optimal control package. Dynamic models are discretized using hp-adaptive collocation, and automatic differentiation is supported using the ADiGator package. Solution of the resulting NLP problem is performed using either the IPOPT or SNOPT solvers.

PROPT [29] is an optimal control package built on top of the TOMLAB MATLAB optimization environment. Differential equations are discretized using Gauss and Chebyshev collocation, and solutions of the resulting NLP are found using the SNOPT solver. Derivatives are provided through source transformation using TOMLAB's symbolic differentiation capabilities. Automatic scaling and integer states are also supported. Access is provided through a MATLAB interface.

PSOPT [30] is an open-source C++ package for optimal control. Dynamic systems are discretized using both local and global pseudospectral collocation methods. Automatic differentiation is available by means of the ADOL-C library. Solution of NLPs is performed using either IPOPT or SNOPT.

In addition to those listed above, many other software libraries are available for modeling and optimization, including AIMMS [31], CVX [32], CVXOPT [33], YALMIP [34], PuLP [35], POAMS, OpenOpt, NLPy, and PyIpopt.

5. GEKKO Overview

GEKKO fills the role of a typical AML, but extends its capabilities to specialize in dynamic optimization applications. As an AML, GEKKO provides a user-friendly, object-oriented Python interface to develop models and optimization solutions. Python is a free and open-source language that is flexible, popular, and powerful. IEEE Spectrum ranked Python the #1 programming language in 2017. Being a Python package allows GEKKO to easily interact with other popular scientific and numerical packages. Further, this enables GEKKO to connect to any real system that can be accessed through Python.

Since Python is designed for readability and ease rather than speed, the Python GEKKO model is converted to a low-level representation in the Fortran back-end for speed in function calls. Automatic differentiation provides the necessary gradients, accurate to machine precision, without extra work from the user. GEKKO then interacts with the built-in open-source, commercial, and custom large-scale solvers for linear, quadratic, nonlinear, and mixed integer programming (LP, QP, NLP, MILP, and MINLP) in the back-end. Optimization results are loaded back to Python for easy access and further analysis or manipulation.

Other modeling and optimization platforms focus on ultimate flexibility. While GEKKO is capable of flexibility, it is best suited for large-scale systems of differential and algebraic equations with continuous or mixed integer variables for dynamic optimization applications. GEKKO has a graphical user interface (GUI) and several built-in objects that support rapid prototyping and deployment of advanced control applications for estimation and control. It is a full-featured platform with a core that has been successfully deployed on many industrial applications.

As a Dynamic Optimization package, GEKKO accommodates DAE systems with built-in discretization schemes and facilitates popular applications with built-in modes of operation and tuning parameters. For differential and algebraic equation systems, both simultaneous and sequential methods are built in to GEKKO. Modes of operation include data reconciliation, real-time optimization, dynamic simulation, moving horizon estimation, and nonlinear predictive control. The back-end compiles the model to an efficient low-level format and performs model reduction based on analysis of the sparsity structure (incidence of variables in equations or objective function) of the model.

Sequential methods separate the problem in Equation (2) into the standard algebraic optimization routine Equation (1) and a separate differential equation solver, where each problem is solved sequentially. This method is popular in fields where the solution of differential equations is extremely difficult. By separating the problems, the simulator can be fine-tuned, or wrapped in a "black box". Since the sequential approach is less reliable in unstable or ill-conditioned problems, it is often adapted to a "multiple-shooting" approach to improve performance. One benefit of the sequential approach is a guaranteed feasible solution of the differential equations, even if the optimizer fails to find an optimum. Since GEKKO does not allow connecting to black box simulators or the multiple-shooting approach, this feasibility of differential equation simulations is the main benefit of sequential approaches.

The simultaneous approach, or direct transcription, minimizes the objective function and resolves all constraints (including the discretized differential equations) simultaneously. Thus, if the solver terminates without reaching optimality, it is likely that the equations are not satisfied and the dynamics of the infeasible solution are incorrect—yielding a worthless rather than just suboptimal solution. However, since simultaneous approaches do not waste time accurately simulating dynamics that are thrown away in intermediary iterations, this approach tends to be faster for large problems with many degrees of freedom [36]. A common discretization scheme for this approach, which GEKKO uses, is orthogonal collocation on finite elements. Orthogonal collocation represents the state and control variables with polynomials inside each finite element. This is a form of implicit Runga–Kutta methods, and thus it inherits the benefits associated with these methods, such as stability. Simultaneous methods require efficient large-scale NLP solvers and accurate problem information, such as exact second derivatives, to perform well. GEKKO is designed to provide such information and take advantage of the simultaneous approach's benefits in sparsity and decomposition opportunities. Therefore, the simultaneous approach is usually recommended in GEKKO.

GEKKO is an open-source Python library with an MIT license. The back-end Fortran routines are not open-source, but are free to use for academic and commercial applications. It was developed by the PRISM Lab at Brigham Young University and is in version 0.1 at the time of writing. Documentation on the GEKKO Python syntax is available in the online documentation, currently hosted on Read the Docs. The remainder of this text explores the paradigm of GEKKO and presents a few example problems, rather than explaining syntax.

5.1. Novel Aspects

GEKKO combines the model development, solution, and graphical interface for problems described by Equation (2). In this environment, differential equations with time derivatives are automatically discretized and transformed to algebraic form (see Equation (6)) for solution by large-scale and sparse solvers.

$$\min_{u,z} \quad \sum_{i}^{n} J\left(z_i, u_i\right) \tag{6a}$$

$$0 = f\left(z_i, u_i\right) \;\forall\, i \in n \tag{6b}$$

$$0 \leq g(z_i, u_i)\ \forall\, i \in n \tag{6c}$$

$$\text{Collocation Equations} \tag{6d}$$

where n is the number of time points in the discretized time horizon, $z = \left[\frac{\mathrm{d}x}{\mathrm{d}t}, x\right]$ is the combined state vector, and the collocation equations are added to relate differential terms to the state values. The collocation equations and derivation are detailed in [37]. GEKKO provides the following to an NLP solver in sparse form:

- Variables with initial values and bounds
- Evaluation of equation residuals and objective function
- Jacobian (first derivatives) with gradients of the equations and objective function
- Hessian of the Lagrangian (second derivatives) with second derivatives of the equations and objective
- Sparsity structure of first and second derivatives

Once the solution is complete, the results are loaded back into Python variables. GEKKO has several modes of operation. The two main categories are steady-state solutions and dynamic solutions. Both sequential and simultaneous modes are available for dynamic solutions. The core of all modes is the nonlinear model, which does not change between selection of the different modes. Each mode interacts with the nonlinear model to receive or provide information, depending on whether there is a request for simulation, estimation, or control. Thus, once a GEKKO model is created, it can be implemented in model parameter update (MPU), real-time optimization (RTO) [38], moving horizon estimation (MHE), model predictive control (MPC), or steady-state or dynamic simulation modes by setting a single option. The nine modes of operation are listed in Table 1.

Table 1. Modes of operation.

	Non-Dynamic	**Simultaneous Dynamic**	**Sequential Dynamic**
Simulation	Steady-state simulation	Simultaneous dynamic simulation	Sequential dynamic simulation
Estimation	Parameter regression, Model parameter update (MPU)	Moving horizon estimation (MHE)	Sequential dynamic estimation
Control	Real-time optimization (RTO)	Optimal control, Nonlinear control (MPC)	Sequential dynamic optimization

There are several modeling language and analysis capabilities enabled with GEKKO. Some of the most substantial advancements are automatic (structural analysis) or manual (user specified) model reduction, and object support for a suite of commonly used modeling or optimization constructs. The object support, in particular, allows the GEKKO modeling language to facilitate new application areas as model libraries are developed.

5.2. Built-In Solvers

GEKKO has multiple high-end solvers pre-compiled and bundled with the executable program instead of split out as a separate programs. The bundling allows out-of-the-box optimization, without the need of compiling and linking solvers by the user. The integration provides efficient communication between the solver and model that GEKKO creates as a human readable text file. The model text file is then compiled to efficient byte code for tight integration between the solver and the model. Interaction between the equation compiler and solver is used to monitor and modify the equations for initialization [39], model reduction, and decomposition strategies. The efficient compiled byte-code model includes forward-mode automatic differentiation (AD) for sparse first and second derivatives of the objective function and equations.

The popular open-source interior-point solver IPOPT [40] is the default solver. The custom interior-point solver BPOPT and the MINLP active-set solver APOPT [41] are also included. Additional solvers such as MINOS and SNOPT [42] are also integrated, but are only available with the associated requisite licensing.

6. GEKKO Framework

Each GEKKO model is an object. Multiple models can be built and optimized within the same Python script by creating multiple instances of the GEKKO model class. Each variable type is also an object with property and tuning attributes. Model constraints are defined with Python variables and Python equation syntax.

GEKKO has eight types of variables, four of which have extra properties. Constants, Parameters, Variables, and Intermediates are the base types. Constants and Parameters are fixed by the user, while Variables are calculated by the solver and Intermediates are updated with every iteration by the modeling language. Fixed Variables (FV), Manipulated Variables (MV), State Variables (SV), and Controlled Variables (CV) expand Parameters and Variables with extra attributes and features to facilitate dynamic optimization problem formulation and robustness for real-time application use. All variable declarations return references to a new object.

6.1. *User-Defined Models*

This section introduces the standard aspects of AMLs within the GEKKO paradigm. Optimization problems are created as collections of variables, equations, and objectives.

6.1.1. Variable Types

The basic set of variable types includes Constants, Parameters, and Variables. Constants exist for programing style and consistency. There is no functional difference between using a GEKKO Constant, a Python variable, or a floating point number in equations. Parameters serve as constant values, but unlike Constants, they can be (and usually are) arrays. Variables are calculated by the solver to meet constraints and minimize the objective. Variables can be constrained to strict boundaries or required to be integer values. Restricting Variables to integer form then requires the use of a specialized solver (such as APOPT) that iterates with a branch-and-bound method to find a solution.

6.1.2. Equations

Equations are all solved together implicitly by the built-in optimizer. In dynamic modes, equations are discretized across the whole time horizon, and all time points are solved simultaneously. Common unary operators are available with their respective automatic differentiation routines such as absolute value, exponentiation, logarithms, square root, trigonometric functions, hyperbolic functions, and error functions. The GEKKO operands are used in model construction instead of Python Math or NumPy functions. The GEKKO operands are required so that first and second derivatives are calculated with automatic differentiation.

A differential term is expressed in an equation with $x.dt()$, where x is a Variable. Differential terms can be on either the right or the left side of the equations, with equality or inequality constraints, and in objective functions. Some software packages require index-1 or index-0 differential and algebraic equation form for solution. There is no DAE index limit in GEKKO or need for consistent initial conditions. Built-in discretization is only available in one dimension (time). Discretization of the the differential equations is set by the user using the GEKKO model attribute *time*. The time attribute is set as an array which defines the boundaries of the finite elements in the orthogonal collocation scheme. The number of nodes used to calculate the internal polynomial of each finite element is set with a global option. However, these internal nodes only serve to increase solution accuracy since only the values at the boundary time points are returned to the user.

Partial differential equations (PDEs) are allowed within the GEKKO environment through manual discretization in space with vectorized notation. For example, Listing 1 shows the numerical solution to the wave equation shown in Equation (7) through manual discretization in space and built-in discretization in time. The simulation results are shown in Figure 1.

$$\frac{\partial^2 u(x,t)}{\partial^2 t} = c^2 \frac{\partial^2 u(x,t)}{\partial^2 x} \tag{7}$$

Listing 1. PDE Example GEKKO Code with Manual Discretization in Space.

```
from gekko import GEKKO
import numpy as np

# Initialize model
m = GEKKO()

# Discretizations (time and space)
m.time = np.linspace(0,1,100)
npx = 100
xpos = np.linspace(0,2*np.pi,npx)
dx = xpos[1]-xpos[0]

# Define Variables
c = m.Const(value = 10)
u = [m.Var(value = np.cos(xpos[i])) for i in range(npx)]
v = [m.Var(value = np.sin(2*xpos[i])) for i in range(npx)]

m.Equations([u[i].dt()==v[i] for i in range(npx)])
# Manual discretization in space (central difference)
m.Equation(v[0].dt()==c**2 * (u[1] - 2.0*u[0] + u[npx-1])/dx**2 )
m.Equations([v[i+1].dt()== c**2*(u[i+2]-2.0*u[i+1]+u[i])/dx**2 for i in range(npx-2)])
m.Equation(v[npx-1].dt()== c**2 * (u[npx-2] - 2.0*u[npx-1] + u[0])/dx**2 )
# set options
m.options.imode = 4
m.options.solver = 1
m.options.nodes = 3

m.solve()
```

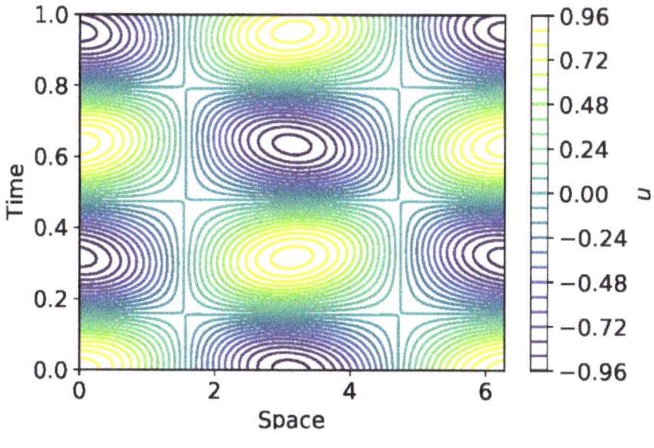

Figure 1. Results of wave equation PDE simulation.

6.1.3. Objectives

All types of GEKKO quantities may be included in the objective function expression, including Constants, Parameters, Variables, Intermediates, FVs, MVs, SVs, and CVs. In some modes, GEKKO models automatically build objectives. MVs and CVs also contain objective function contributions that are added or removed with configuration options. For example, in MPC mode, a CV with a set point automatically receives an objective that minimizes error between model prediction and the set point trajectory of a given norm. There may be multiple objective function expressions within a single GEKKO model. This is often required to express multiple competing objectives in an estimation or control problem. Although there are multiple objective expressions, all objectives terms are summed into a single optimization expression to produce an optimal solution.

6.2. *Special Variable Types*

GEKKO features special variable types that facilitate the tuning of common industrial dynamic optimization problems with numerically robust options that are efficient and easily accessible. These special variable types are designed to improve model efficiency and simplify configuration for common problem scenarios.

6.2.1. Intermediates

Most modeling languages only include standard variables and constraints, where all algebraic constraints and their associated variables are solved implicitly through iterations of the optimizer. GEKKO has a new class of variables termed Intermediates. Intermediates, and their associated equations, are similar to variables except that they are defined and solved explicitly and successively substituted at every solver function call. Intermediate values, first derivatives, and second derivatives are substituted into other successive Intermediates or into the implicit equations. This is done outside of the solver in order to reduce the number of algebraic variables while maintaining the readability of the model. The intermediate equation must be an explicit equality. Each intermediate equation is solved in order of declaration. All variable values used in the explicit equation come from either the previous iteration or as an Intermediate declared previously.

In very large-scale problems, removing a portion of variables from the matrix math of implicit solutions can reduce matrix size, keeping problems within the efficient bounds of hardware limitations. This is especially the case with simultaneous dynamic optimization methods, where a set of model equations are multiplied over all of the collocation nodes. For each variable reduced in the base model, that variable is also eliminated from every collocation node. Intermediates are formulated to be highly memory-efficient during function calls in the back-end with the use of sparse model reduction. Intermediate variables essentially blend the benefits of sequential solver approaches into simultaneous methods.

6.2.2. Fixed Variable

Fixed Variables (FV) inherit Parameters, but potentially add a degree of freedom and are always fixed throughout the horizon (i.e., they are not discretized in dynamic modes). Estimated parameters, measured disturbances, unmeasured disturbances, and feed-forward variables are all examples of what would typically fit into the FV classification.

6.2.3. Manipulated Variable

Manipulated Variables (MV) inherit FVs but are discretized throughout the horizon and have time-dependent attributes. In addition to absolute bounds, relative bounds such as movement ($dmax$), upper movement ($dmax_{hi}$), and lower movement ($dmax_{lo}$) guide the selection by the optimizer. Hard constraints on movement of the value are sometimes replaced with a move suppression factor ($dcost$) to penalize movement of the MV. The move suppression factor is a soft constraint because it discourages movement with use of an objective function factor. *cost* is a penalty to minimize u (or maximize u with a negative sign). The MV object is given in Equation (8) for an ℓ_1-norm objective. The MV internal nodes for each horizon step are also calculated with supplementary equations based on whether it is a first-order or zero-order hold.

$$\min_{\Delta u_-, \Delta u_+} \quad dcost\,(\Delta u_- + \Delta u_+) + cost\; u_n \tag{8a}$$

$$\Delta u = u_n - u_1 \tag{8b}$$

$$0 = dudt\,\Delta t - \Delta u \tag{8c}$$

$$\Delta u_+ - \Delta u \geq 0 \tag{8d}$$

$$\Delta u_- + \Delta u \geq 0 \tag{8e}$$

$$dmax_{lo} \leq \Delta u \leq dmax_{hi} \tag{8f}$$

$$\Delta u_-, \Delta u_+ \geq 0 \tag{8g}$$

where n is the number of nodes in each time interval, Δu is the change of u, Δu_- is the negative change, Δu_+ is the positive change and *dudt* is the slope. The MV object equations and objective are different for a squared error formulation as shown in Equation (9). The additional linear inequality constraints for Δu_+ and Δu_- are not needed and the penalty on Δu is squared as a move suppression factor that is compatible in a trade-off with the squared controlled variable objective.

$$\min_{\Delta u} \quad dcost\,\Delta u^2 + cost\; u_n \tag{9a}$$

$$\Delta u = u_n - u_1 \tag{9b}$$

$$0 = dudt\,\Delta t - \Delta u \tag{9c}$$

$$\Delta u \leq dmax \tag{9d}$$

Consistent with the vocabulary of the process control community, MVs are intended to be the variables that are directly manipulated by controllers. The practical implementations are often simplified, implemented by a distributed control system (DCS) and communicated to lower-level controllers (such as proportional-integral-derivative controllers [PIDs]). Thus, the internal nodes of MVs are calculated to reflect their eventual implementation. Rather than enabling each internal node of an MV as a degree of freedom, if there are internal points ($nodes \geq 3$) then there is an option (*mv_type*) that controls whether the internal nodes are equal to the starting value as a zero-order hold ($mv\_type = 0$) or as a first-order linear interpolation between the beginning and end values of that interval ($mv\_type = 1$). The zero-order hold is common in discrete control where the MVs only change at specified time intervals. A linear interpolation is desirable to avoid sudden increments in MV values or when it is desirable to have a continuous profile. This helps match model predictions with actual implementation. For additional accuracy or less-frequent MV changes, the MV_STEP_HOR option can keep an MV fixed for a given number of consecutive finite elements. There are several other options that configure the behavior of the MV. The MV is either determined by the user with $status = 0$ or the optimizer at the step end points with $status = 1$.

6.2.4. State Variable

State Variables (SV) inherit Variables with a couple of extra attributes that control bounds and measurements.

6.2.5. Controlled Variable

Controlled Variables (CV) inherit SVs but potentially add an objective. The CV object depends on the current mode of operation. In estimation problems, the CV object is constructed to reconcile measured and model-predicted values for steady-state or dynamic data. In control modes, the CV provides a setpoint that the optimizer will try to match with the model prediction values. CV model predictions are determined by equations, not as user inputs or solver degrees of freedom.

The CV object is given in Equation (10) for an ℓ_1-norm objective for estimation. In this case, parameter values (p) such as FVs or MVs with $status = 1$ are adjusted to minimize the difference between model (y_{model}) and measured values (y_{meas}) with weight w_{meas}. The states as well as the parameters are simultaneously adjusted by the solver to minimize the objective and satisfy the equations. There is a deadband with width $meas_{gap}$ around the measured values to avoid fitting noise and discourage unnecessary parameter movement. Unnecessary parameter movement is also avoided by penalizing with weight w_{model} the change (c_{hi}, c_{lo}) away from prior model predictions.

$$\min_p \quad (w_{meas}\, e_{hi} + w_{meas}\, e_{lo} + w_{model}\, c_{hi} + w_{model}\, c_{lo})\, fstatus \tag{10a}$$

$$e_{hi} \geq y_{model} - y_{meas} - \frac{meas_{gap}}{2} \tag{10b}$$

$$e_{lo} \geq -y_{model} + y_{meas} - \frac{meas_{gap}}{2} \tag{10c}$$

$$c_{hi} \geq y_{model} - \hat{y}_{model} \tag{10d}$$

$$c_{lo} \geq -y_{model} + \hat{y}_{model} \tag{10e}$$

$$e_{hi}, e_{lo}, c_{hi}, c_{lo} \geq 0 \tag{10f}$$

where $fstatus$ is the feedback status that is 0 when the measurement is not used and 1 when the measurement reconciliation is included in the overall objective (intermediate values between 0 and 1 are allowed to weight the impact of measurements). A measurement may be discarded for a variety of reasons, including gross error detection and user specified filtering. For measurements from real systems, it is critical that bad measurements are blocked from influencing the solution. If bad measurements do enter, the ℓ_1-norm solution has been shown to be less sensitive to outliers, noise, and drift [43].

The CV object is different for a squared-error formulation, as shown in Equation (11). The desired norm is easily selected through a model option.

$$\min_p \left(w_{meas} \left(y_{model} - y_{meas}\right)^2 + w_{model} \left(y_{model} - \hat{y}_{model}\right)^2 \right) fstatus \tag{11}$$

Important CV options are $fstatus$ for estimation and $status$ for control. These options determine whether the CV objective terms contribute to the overall objective (1) or not (0).

In MPC, the CVs have several options for the adjusting the performance such as speed of reaching a new set point, following a predetermined trajectory, maximization, minimization, or staying within a specified deadband. The CV equations and variables are configured for fast solution by gradient-based solvers, as shown in Equations (12)–(14). In these equations, tr_{hi} and tr_{lo} are the upper and lower reference trajectories, respectively. The wsp_{hi} and wsp_{lo} are the weighting factors on upper or lower errors and $cost$ is a factor that either minimizes (−) or maximizes (+) within the set point deadband between the set point range sp_{hi} and sp_{lo}.

$$\min_p \quad (wsp_{hi}\, e_{hi} + wsp_{lo}\, e_{lo}) + cost\, y_{model} \tag{12a}$$

$$\tau \frac{\mathrm{d}tr_{hi}}{\mathrm{d}t} = tr_{hi} - sp_{hi} \tag{12b}$$

$$\tau \frac{\mathrm{d}tr_{lo}}{\mathrm{d}t} = tr_{lo} - sp_{lo} \tag{12c}$$

$$e_{hi} \geq y_{model} - tr_{hi} \tag{12d}$$

$$e_{lo} \geq -y_{model} + tr_{hi} \tag{12e}$$

$$e_{hi}, e_{lo} \geq 0 \tag{12f}$$

An alternative to Equation (12b–e) is to pose the reference trajectories as inequality constraints and the error expressions as equality constraints, as shown in Equation (13). This is available in GEKKO to best handle systems with dead-time in the model without overly aggressive MV moves to meet a first-order trajectory.

$$\min_{p} \quad \left(wsp_{hi}\, e_{hi} + wsp_{lo}\, e_{lo}\right) + cost\, y_{model} \tag{13a}$$

$$\tau \frac{\mathrm{d}y_{model}}{\mathrm{d}t} \leq -tr_{hi} + sp_{hi} \tag{13b}$$

$$\tau \frac{\mathrm{d}y_{model}}{\mathrm{d}t} \geq -tr_{lo} + sp_{lo} \tag{13c}$$

$$e_{hi} = y_{model} - tr_{hi} \tag{13d}$$

$$e_{lo} = -y_{model} + tr_{hi} \tag{13e}$$

$$e_{hi}, e_{lo} \geq 0 \tag{13f}$$

While the ℓ_1-norm objective is default for control, there is also a squared error formulation, as shown in Equation (14). The squared error introduces additional quadratic terms but also eliminates the need for slack variables e_{hi} and e_{lo} as the objective is guided along a single trajectory (tr) to a set point (sp) with priority weight (wsp).

$$\min_{p} \quad wsp\; e^2 + cost\; y_{model} \tag{14a}$$

$$\tau \frac{\mathrm{d}tr}{\mathrm{d}t} = tr - sp \tag{14b}$$

It is important to avoid certain optimization formulations to preserve continuous first and second derivatives. GEKKO includes both MV and CV tuning with a wide range of options that are commonly used in advanced control packages. There are also novel options that improve controller and estimator responses for multi-objective optimization. One of these novel options is the ability to specify a tier for MVs and CVs. The tier option is a multi-level optimization where different combinations of MVs and CVs are progressively turned on. Once a certain level of MV is optimized, it is turned off and fixed at the optimized values while the next rounds of MVs are optimized. This is particularly useful to decouple the multivariate problem where only certain MVs should be used to optimize certain CVs although there is a mathematical relationship between the decoupled variables. Both MV and CV tuning can be employed to "tune" an application. A common trade-off for control is the speed of CV response to set point changes versus excessive MV movement. GEKKO offers a full suite of tuning options that are built into the CV object for control and estimation.

6.3. Extensions

Two additional extensions in GEKKO modeling are the use of Connections to link variables and object types (such as process flow streams). As an object-oriented modeling environment, there is a library of pre-built objects that individually consist of variables, equations, objective functions, or are collections of other objects.

6.3.1. Connections

All GEKKO variables (with the exception of FVs) and equations are discretized uniformly across the model time horizon. This approach simplifies the standard formulation of popular dynamic optimization problems. To add flexibility to this approach, GEKKO Connections allow custom relationships between variables across time points and internal nodes. Connections are processed after the parameters and variables are parsed but before the initialization of the values. Connections are the merging of two variables or connecting specific nodes of a discretized variable or setting just one unique point fixed to a given value.

6.3.2. Pre-Built Model Objects

The GEKKO modeling language encourages a disciplined approach to optimization. Part of this disciplined approach is to pose well-formed optimization problems that have continuous first and second derivatives for large-scale gradient-based solvers. An example is the use of the absolute value function, which has a discontinuous derivative at $x = 0$. GEKKO features a number of unique model objects that cannot be easily implemented through continuous equation restrictions. By implementing these models in the Fortran back-end, the unique gradients can be hard-coded for efficiency. The objects include an absolute value formulation, cubic splines, and discrete-time state space models.

Cubic splines are appropriate in cases where data points are available without a clear or simple mathematical relationship. When a high-fidelity simulator is too complex to be integrated into the model directly, a set of points from the simulator can act as an approximation of the simulator's relationships. When the user provides a set of input and output values, the GEKKO back-end builds a cubic spline interpolation function. Subsequent evaluation of the output variable in the equations triggers a back-end routine to identify the associated cubic function and evaluate its value, first derivatives, and second derivatives. The cubic spline results in smooth, continuous functions suitable for gradient-based optimization.

6.4. Model Reduction, Sensitivity and Stability

Model reduction condenses the state vector x into a minimal realization that is required to solve the dynamic optimization problem. There are two primary methods of model reduction that are included with GEKKO, namely model construction (manual) and structural analysis (automatic). Manual model reduction uses Intermediate variable types, which reduce the size and complexity of the iterative solve through explicit solution and efficient memory management. Automatic model reduction, on the other hand, is a pre-solve strategy that analyzes the problem structure to explicitly solve simple equations. The equations are eliminated by direct substitution to condense the overall problem size. Two examples of equation eliminations are expressions such as $x = 2$ and $y = 2x$. Both equations can be eliminated by fixing the values $x = 2$ and $y = 4$. The pre-solve analysis also identifies infeasible constraints such as if y were defined with an upper bound of 3. The equation is identified as violating a constraint before handing the problem to an NLP solver. Automatic model reduction is controlled with the option *reduce*, which is zero by default. If *reduce* is set to a non-zero integer value, it scans the model that many times to find linear equations and variables that can be eliminated.

Sensitivity analysis is performed in one of two ways. The first method is to specify an option in GEKKO (*sensitivity*) to generate a local sensitivity at the solution. This is performed by inverting the sparse Jacobian at the solution [44]. The second method is to perform a finite difference evaluation of the solution after the initial optimization problem is complete. This involves resolving the optimization problem multiple times and calculating the resultant change in output with small perturbations in the inputs. For dynamic problems, the automatic time-shift is turned off for sensitivity calculation to prevent advancement of the initial conditions when the problem is solved repeatedly.

Stability analysis is a well-known method for linear dynamic systems. A linear version of the GEKKO model is available from the sparse Jacobian that is available when $diaglevel \geq 1$. A linear dynamic model is placed into a continuous, sparse state space form, as shown in Equation (15).

$$\dot{x} = Ax + Bu \tag{15a}$$

$$y = Cx + Du \tag{15b}$$

If the model can be placed in this form, the open-loop stability of the model is determined by the sign of the eigenvalues of matrix A. Stability analysis can also be performed with the use of a step response for nonlinear systems or with Lyapunov stability criteria that can be implemented as GEKKO Equations.

6.5. Online Application Options

GEKKO has additional options that are tailored to online control and estimation applications. These include *meas*, *bias*, and $time\_shift$. The *meas* attribute facilitates loading in new measurements in the appropriate place in the time horizon, based on the application type.

Gross error detection is critical for automation solutions that use data from physical sensors. Sensors produce data that may be corrupted during collection or transmission, which can lead to drift, noise, or outliers. For FV, MV, SV, and CV classifications, measured values are validated with absolute validity limits and rate-of-change validity limits. If a validity limit is exceeded, there are several configurable options such as "hold at the last good measured value" and "limit the rate of change toward the potentially bad measured value". Many industrial control systems also send a measurement status (*pstatus*) that can signal when a measured value is bad. Bad measurements are ignored in GEKKO and either the last measured value is used or else no measurement is used and the application reverts to a model predicted value.

The value of *bias* is updated from *meas* and the unbiased model prediction ($model_u$). The *bias* is added to each point in the horizon, and the controller objective function drives the biased model ($model_b$) to the requested set point range. This is shown in Equation (16).

$$bias = meas - model_u \tag{16a}$$

$$model_b = model_u + bias \tag{16b}$$

The $time\_shift$ parameter shifts all values through time with each subsequent resolve. This provides both accurate initial conditions for differential equations and efficient initialization of all variables (including values of derivatives and internal nodes) through the horizon by leaning on previous solutions.

6.6. Limitations

The main limitation of GEKKO is the requirement of fitting the problem within the modeling language framework. Most notably, user-defined functions in external libraries or other such connections to "black boxes" are not currently enabled. Logical conditions and discontinuous functions are not allowed but can be reformulated with binary variables or Mathematical Programming with Complementarity Constraints (MPCCs) [45] so they can be used in GEKKO. IF-THEN statements are purposely not allowed in GEKKO to prevent discontinuities. Set-based operations such as unions, exclusive OR, and others are also not supported. Regarding differential equations, only one discretization scheme is available and it only applies in one dimension (*time*). Further discretizations must be performed by the user.

The back-end Fortran routines are only compiled for Windows and Linux at this time. The routines come bundled with the package for these operating systems to enable local solutions. MacOS and ARM processors must use the remote solve options to offload their problems to the main server.

Finally, it must be remembered that the actual optimization occurs in the bundled solvers. While these solvers are state-of-the-art, they are not infallible. GEKKO back-end adjustments, such as scaling, assist the solver, but it falls to the user to pose feasible problems and formulate them to promote convergence. Knowledge of the solver algorithms allows users to pose better problems and get better results.

7. Graphical User Interface

AML development history has moved from low-level models or text-based models to high-level implementations (e.g., Pyomo, JuMP, and Casadi) to facilitate rapid development. The next phase of accelerating the development process involves visual representation of the results. This is especially important in online control and estimation applications so the operator can easily visualize and track the application's progress and intentions. In some modeling languages, simply loading the optimization results into a scripting language for further processing and analysis can be difficult. GEKKO includes a built-in graphical interface to facilitate visualizing results.

The GEKKO GUI uses Vue.js and Plotly to display optimization results quickly and easily. It also tracks past results for MHE and MPC problems, allowing time-dependent solutions to be displayed locally and in real-time as the iterative solution progresses. The GUI itself is implemented through a Python webserver that retrieves and stores optimization results and a Vue.js client that queries the Python webserver over HTTP. Polling between the client and webserver allows for live updating and seamless communication between the client and the webserver. The GUI allows plots to be created and deleted on demand, supporting individual visualization for variables on different scales. Model- and variable-specific details are displayed in tables to the left of the plots (see Figure 2).

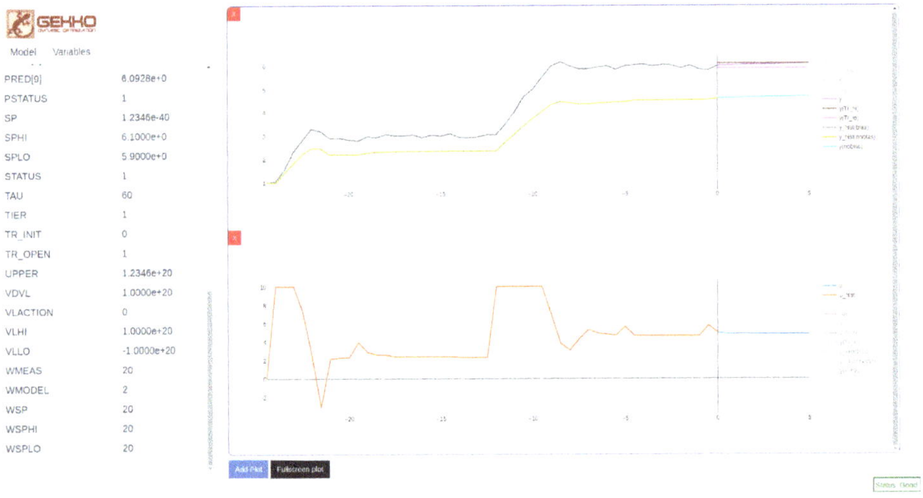

Figure 2. Sample GEKKO GUI screen for rapidly visualizing solutions.

8. Examples

This section presents a set of example GEKKO models in complete Python syntax to demonstrate the syntax and available features. Solutions of each problem are also presented. Additional example problems are shown in the back matter, with an example of an artificial neural network in Appendix A and several dynamic optimization benchmark problems shown in Appendix B. Since the GEKKO Fortran backend is the successor to APMonitor [37], the many applications of APMonitor are also possible within this framework, including recent applications in combined scheduling and control [46], industrial dynamic estimation [43], drilling automation [47,48], combined design and control [49], hybrid energy storage [50], batch distillation [51], systems biology [44], carbon capture [52], flexible printed circuit boards [53], and steam distillation of essential oils [54].

8.1. Nonlinear Programming Optimization

First, problem 71 from the well-known Hock Schitkowski Benchmark set is included to facilitate syntax comparison with other AMLs. The Python code for this problem using the GEKKO optimization suite is shown in Listing 2.

$$\begin{aligned} &\min x_1 x_4(x_1 + x_2 + x_3) + x_3 \\ &\text{subject to} \quad x_1 x_2 x_3 x_4 \geq 25 \\ &\quad x_1^2 + x_2^2 + x_3^2 + x_4^2 = 40 \\ &\quad 1 \leq x_1, x_2, x_3, x_4 \leq 5 \\ &\quad x_0 = (1, 5, 5, 1) \end{aligned}$$

Listing 2. HS71 Example GEKKO Code.

```
from gekko import GEKKO
m = GEKKO() # Initialize gekko
# Initialize variables
x1 = m.Var(1,lb=1,ub=5)
x2 = m.Var(5,lb=1,ub=5)
x3 = m.Var(5,lb=1,ub=5)
x4 = m.Var(1,lb=1,ub=5)
# Equations
m.Equation(x1*x2*x3*x4>=25)
m.Equation(x1**2+x2**2+x3**2+x4**2==40)
m.Obj(x1*x4*(x1+x2+x3)+x3) # Objective
m.options.IMODE = 3 # Steady state optimization
m.solve() # Solve
print('Results')
print('x1: ' + str(x1.value))
print('x2: ' + str(x2.value))
print('x3: ' + str(x3.value))
print('x4: ' + str(x4.value))
```

The output of this code is $x_1 = 1.0$, $x_2 = 4.743$, $x_3 = 3.82115$, and $x_4 = 1.379408$. This is the optimal solution that is also confirmed by other solver solutions.

8.2. Closed-Loop Model Predictive Control

The following example demonstrates GEKKO's online MPC capabilities, including measurements, timeshifting, and MPC tuning. The MPC model is a generic first-order dynamic system, as shown in Equation (17). There exists plant–model mismatch (different parameters from the "process_simulator" function) and noisy measurements to more closely resemble a real system. The code is shown in Listing 3 and the results are shown in Figure 3, including the CV measurements and set points and the implemented MV moves.

$$\tau \frac{\mathrm{d}y}{\mathrm{d}t} = -y + uK \tag{17}$$

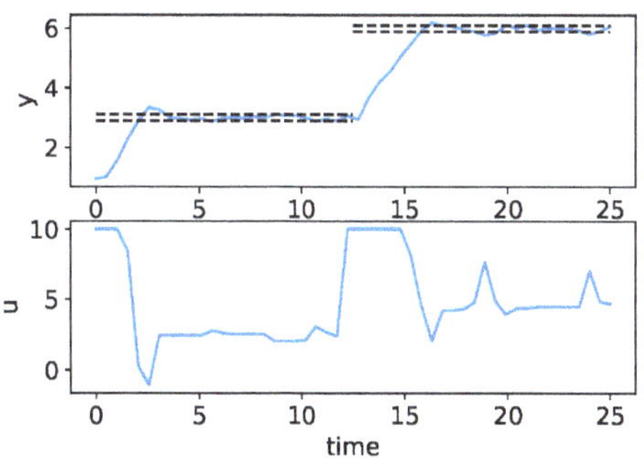

Figure 3. Results of an online MPC example.

Listing 3. Closed-Loop MPC GEKKO Code.

```
from gekko import GEKKO
import numpy as np
import matplotlib.pyplot as plt

#%% MPC Model
c = GEKKO()
c.time = np.linspace(0,5,11) #horizon to 5 with discretization of 0.5

#Parameters
u = c.MV(lb=-10,ub=10) #input
K = c.Param(value=1) #gain
tau = c.Param(value=10) #time constant
#Variables
y = c.CV(1)
#Equations
c.Equation(tau * y.dt() == -y + u * K)
#Options
c.options.IMODE = 6     #MPC
c.options.CV_TYPE = 1   #l1 norm
c.options.NODES = 3

y.STATUS = 1                #write MPC objective
y.FSTATUS = 1               #receive measurements
y.SPHI = 3.1
y.SPLO = 2.9

u.STATUS = 1                #enable optimization of MV
u.FSTATUS = 0               #no feedback
u.DCOST = 0.05              #discourage unnecessary movement

#%% Time loop
cycles = 50
time = np.linspace(0,cycles*.5,cycles)
y_meas = np.empty(cycles)
u_cont = np.empty(cycles)

for i in range(cycles):
    #process
    y_meas[i] = process_simulator(u.NEWVAL)

    #controller
    if i == 24: ##change set point half way through
        y.SPHI = 6.1
        y.SPLO = 5.9
    y.MEAS = y_meas[i]
    c.solve(disp=False)
    u_cont[i] = u.NEWVAL

#%% Plot results
plt.figure()
plt.subplot(2,1,1)
plt.plot(time,y_meas)
plt.ylabel('y')
plt.subplot(2,1,2)
plt.plot(time,u_cont)
plt.ylabel('u')
plt.xlabel('time')
```

8.3. Combined Scheduling and Control

The final example demonstrates an approach to combining the scheduling and control optimization of a continuous, multi-product chemical reactor. Details regarding the model and objectives of this problem are available in [46]. This problem demonstrates GEKKO's ability to efficiently solve large-scale problems, the ease of using the built-in discretization for differential equations, the applicability of special variables and their built-in tuning to various problems, and the flexibility provided by connections and custom objective functions. The code is shown in Listing 4 and the optimized horizons of the process concentrations and temperatures are shown in Figure 4.

Listing 4. Combined Scheduling and Control Example GEKKO Code.

```
from gekko import GEKKO
import numpy as np

tf = 48.    # horizon length, hours
dis = 200   # number of points in time discretization
o = np.ones(dis)

#Define products
num_prod = 3
pCas = [0.35,0.12,0.25]
pdemands = [1920, 2880, 2880]
prices = [2.4,2.7,2.1]
tol = .005

energy_cost = 50                        #USD/MWh
energy_price = energy_cost * tf/dis

#%% Initialize model
m = GEKKO()
m.time = np.linspace(0, tf, dis)

#%% CSTR Control Model
#MVs
Q_cool = m.MV(value=3, lb=0, ub=10)            #kJ/s
q = m.MV(value=120, lb=100, ub=120)            #m^3/s
#Constants
V = m.Param(value=400)                         #m^3
rho = m.Param(value=1000)                      #kg/m^3
Cp = m.Param(value=0.000000239)                #kJ/m^3K
mdelH = m.Param(value=0.05)                    #kJ/mol
ER = m.Param(value=8750)                       #K
k0 = m.Param(value=1.8*10**10)                 #1/s
UA = m.Param(value=0.05)                       #kJ/sK
Ca0 = m.Param(value=1)                         #mol/m^3
T0 = m.Param(value=350)                        #K
rho_cool = m.Param(value=1000)                 #kg/m^3
Cp_cool = m.Param(value=0.000000239)           #kJ/m^3K
V_jacket = m.Param(value=20)                   #m^3
q_cool = m.Param(value=200)                    #m^3/s
#Variables
Ca = m.Var(value=.36, ub=1, lb=0)              #mol/m^3
T = m.Var(value=378, lb=250, ub=500)           #K
Tc_in = m.Var(value=o*215, lb=30, ub=500)      #K
Tc = m.Var(value=o*280, lb=200, ub=500)        #K
#Initialize variables
Ca.value = np.linspace(0.35,0.12,dis)
T.value = np.linspace(360,370,dis)

#Equations
m.Equation(V* Ca.dt() == q*(Ca0-Ca)-V*(k0*m.exp(-ER/T)*Ca))
m.Equation(rho*Cp*V* T.dt() == q*rho*Cp*(T0-T) + V*mdelH*(k0*m.exp(-ER/T)*Ca) + UA*(Tc-T))
m.Equation(Tc.dt() == q_cool/V_jacket*(Tc_in-Tc) + UA/(V_jacket*rho*Cp)*(T-Tc))
m.Equation(Q_cool == -rho_cool*Cp_cool*q_cool*(Tc_in-Tc))
m.Equation(Q_cool <= 4)

#%% Scheduling Model
#scheduling variables
prod = [m.Var(value=0,lb=0,ub=1) for i in range(num_prod)]  #instantaneous production
amt = [m.Var(value=0) for i in range(num_prod)]             #cumulative production
final_amt = [m.FV() for _ in range(num_prod)]               #total production
for i in range(num_prod):
    final_amt[i].STATUS = 1 #calculated values
    m.Connection(final_amt[i],amt[i],pos2='end')

m.Equations([amt[i].dt() == prod[i] * q for i in range(num_prod)])
m.Equations([final_amt[i] <= pdemands[i] for i in range(num_prod)]) #maximum demand of each product

#%% Linking Function - Product to Concentration
m.Equations([prod[i]*100*( (Ca - pCas[i])**2 - tol**2) <= 0 for i in range(num_prod)])

#%% Custom Objective
```

Listing 4. *Cont.*

```
m.Obj(-sum(q*(prod[p])*prices[p] for p in range(num_prod))/(tf))

#%% Options
#Global options
m.options.IMODE = 6
m.options.NODES = 2
m.options.MV_TYPE = 0
m.options.MV_DCOST_SLOPE = 0
#MV tuning
Q_cool.STATUS = 1
Q_cool.DMAX = 0.36
Q_cool.DCOST = 0.003
Q_cool.COST = energy_price/tf
q.STATUS=1
q.DCOST = 0.0001

#%% Solve
m.solve(GUI=True)
```

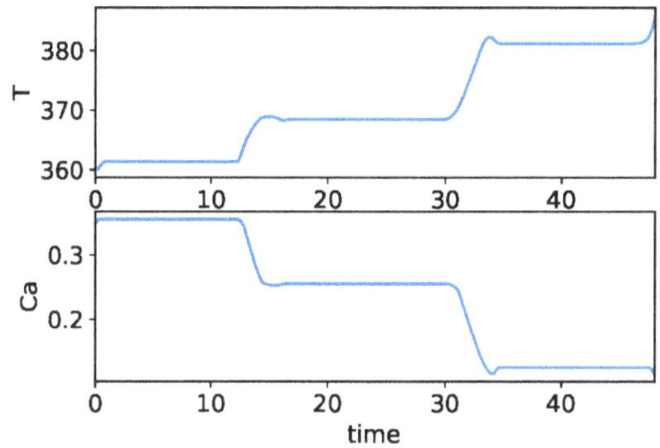

Figure 4. Combined scheduling and control problem results.

9. Conclusions

GEKKO is presented as a fully-featured AML in Python for LP, QP, NLP, MILP, and MINLP applications. Features such as AD and automatic ODE discretization using orthogonal collocation on finite elements and bundled large-scale solvers make GEKKO efficient for large problems. Further, GEKKO's specialization in dynamic optimization problems is explored. Special variable types, built-in tuning, pre-built objects, result visualization, and model-reduction techniques are addressed to highlight the unique strengths of GEKKO. A few examples are presented in Python GEKKO syntax for comparison to other packages and to demonstrate the simplicity of GEKKO, the flexibility of GEKKO-created models, and the ease of accessing the built-in special variables types and tuning options.

Author Contributions: L.D.R.B. developed the Python code; D.C.H. developed the GUI; J.D.H. developed the Fortran backend; and R.A.M. provided assistance in all roles. All authors contributed in writing the paper.

Funding: This research was funded by National Science Foundation grant number 1547110.

Acknowledgments: Contributions made by Damon Peterson and Nathaniel Gates are gratefully acknowledged.

Conflicts of Interest: The authors are the principle developers of GEKKO, an open-source Python package. The Fortran backend belongs to Advanced Process Solutions, LLC which is associated with J.D.H. The founding sponsors had no role in the development of GEKKO; in the writing of the manuscript, and in the decision to publish the results.

Abbreviations

The following abbreviations are used in this manuscript:

AML	Algebraic Modeling Language
DAE	Differential and Algebraic Equations
NMPC	Nonlinear Model Predictive Control
RTO	Real-Time Optimization
MHE	Moving Horizon Estimation
ML	Machine Learning
ANN	Artificial Neural Networks
AD	Automatic (or Algorithmic) Differentiation
ODE	Ordinary Differential Equations
PDE	Partial Differential Equations
MPC	Model Predictive Control
EMPC	Economic Model Predictive Control
DRTO	Dynamic Real-Time Optimization
ASL	AMPL Solver Library
LP	Linear Programming
QP	Quadratic Programming
NLP	Non-Linear Programming
MILP	Mixed-Integer Linear Programming
MINLP	Mixed-Integer Non-Linear Programming
MPU	Model Parameter Update
FV	Fixed Variable
MV	Manipulated Variable
SV	State Variable
CV	Controlled Variable
DCS	Distributed Control System
GUI	Graphical User Interface
MPCC	Mathematical Programming with Complementarity Constraints

Appendix A. Machine Learning with Artificial Neural Network

Machine learning has several areas of application, including regression and classification. This example problem is a simple case study that demonstrates GEKKO's ability to create an artificial neural network, solved with a gradient based optimizer (IPOPT). In this case, the function $y = sin(x)$ is used to generate 20 equally spaced sample points between 0 and $2\,\pi$. These data are used to train the neural network with one input, a linear layer with two nodes, a nonlinear layer of three nodes with hyperbolic tangent activation functions, a linear layer with two nodes, and one output node. An overview of the neural network is shown in Figure A1, with a sample GEKKO implementation in Listing A1 and results in Figure A2.

Listing A1. ANN Sample GEKKO Code.

```
from gekko import GEKKO
import numpy as np
import matplotlib.pyplot as plt

# generate training data
x = np.linspace(0.0,2*np.pi,20)
y = np.sin(x)

# neural network structure
n1 = 2   # hidden layer 1 (linear)
n2 = 3   # hidden layer 2 (nonlinear)
n3 = 2   # hidden layer 3 (linear)

# Initialize gekko
```

Listing A1. *Cont.*

```
m = GEKKO()

# input(s)
m.inpt = m.Param(x)

# layer 1
m.w1 = m.Array(m.FV, (1,n1), value=1)
m.l1 = [m.Intermediate(m.w1[0,i]*m.inpt) for i in range(n1)]

# layer 2
m.w2a = m.Array(m.FV, (n1,n2), value=1)
m.w2b = m.Array(m.FV, (n1,n2), value=0.5)
m.l2 = [m.Intermediate(sum([m.tanh(m.w2a[j,i]+m.w2b[j,i]*m.l1[j]) \
                        for j in range(n1)])) for i in range(n2)]

# layer 3
m.w3 = m.Array(m.FV, (n2,n3), value=1)
m.l3 = [m.Intermediate(sum([m.w3[j,i]*m.l2[j] for j in range(n2)])) for i in range(n3)]

# output(s)
m.outpt = m.CV(y)
m.Equation(m.outpt==sum([m.l3[i] for i in range(n3)]))

# flatten matrices
m.w1 = m.w1.flatten()
m.w2a = m.w2a.flatten()
m.w2b = m.w2b.flatten()
m.w3 = m.w3.flatten()

# Fit parameter weights
m.outpt.FSTATUS = 1
for i in range(len(m.w1)):
    m.w1[i].STATUS=1
for i in range(len(m.w2a)):
    m.w2a[i].STATUS=1
    m.w2b[i].STATUS=1
for i in range(len(m.w3)):
    m.w3[i].STATUS=1
m.options.IMODE = 2
m.options.EV_TYPE = 2
m.solve(disp=False)

# Test sample points
for i in range(len(m.w1)):
    m.w1[i].STATUS=0
for i in range(len(m.w2a)):
    m.w2a[i].STATUS=0
    m.w2b[i].STATUS=0
for i in range(len(m.w3)):
    m.w3[i].STATUS=0

m.inpt.value=np.linspace(-2*np.pi,4*np.pi,100)
m.options.IMODE = 2
m.solve(disp=False)

#Plot
plt.figure()
plt.plot(x,y,'bo',label='data')
plt.plot(m.inpt.value,m.outpt.value, 'r-',label='ANN fit')
plt.ylabel('Output (y)')
plt.xlabel('Input (x)')
plt.legend()
```

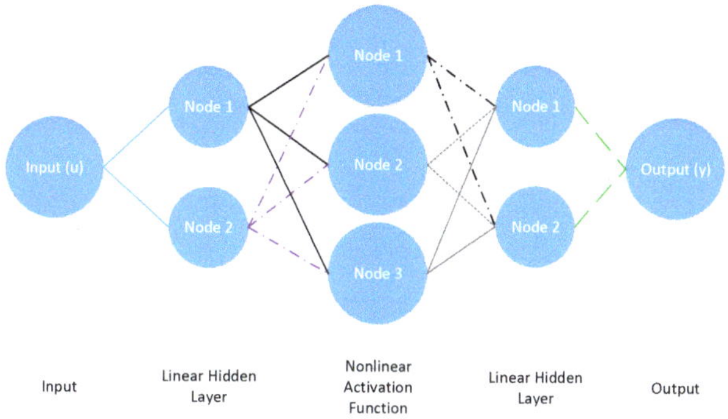

Figure A1. Neural network structure.

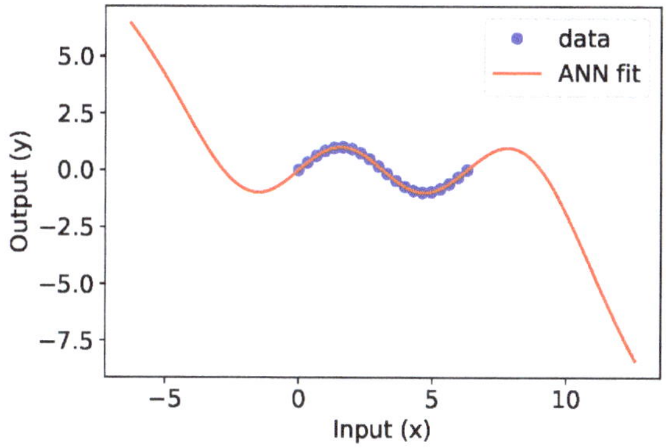

Figure A2. Artificial neural network approximates $y = sin(x)$ function.

Appendix B. Dynamic Optimization Example Problems

The following three problems are examples of GEKKO used in solving classic dynamic optimization problems that are frequently used as benchmarks. The first example problem is a basic problem with a single differential equation, integral objective function, and specified initial condition, as shown in Listing A2. The second example problem is an example of a dynamic optimization problem that uses an economic objective function, similar to EMPC but without the iterative refinement as time progresses, as shown in Listing A3. The third example is a dynamic optimization problem that minimizes final time with fixed endpoint conditions, as shown in Listing A4.

Listing A2. Luus Problem: Integral Objective.

Original Form

$\min_u \frac{1}{2}\int_0^2 x_1^2(t)\,dt$

subject to

$\frac{dx_1}{dt} = u$

$x_1(0) = 1$

$-1 \leq u(t) \leq 1$

Equivalent Form for GEKKO

$\min_u x_2\left(t_f\right)$

subject to

$\frac{dx_1}{dt} = u$

$\frac{dx_2}{dt} = \frac{1}{2}x_1^2(t)$

$x_1(0) = 1$

$x_2(0) = 0$

$t_f = 2$

$-1 \leq u(t) \leq 1$

```
import numpy as np
import matplotlib.pyplot as plt
from gekko import GEKKO
m = GEKKO() # initialize gekko
nt = 101
m.time = np.linspace(0,2,nt)
# Variables
x1 = m.Var(value=1)
x2 = m.Var(value=0)
u = m.Var(value=0,lb=-1,ub=1)
p = np.zeros(nt) # mark final time point
p[-1] = 1.0
final = m.Param(value=p)
# Equations
m.Equation(x1.dt()==u)
m.Equation(x2.dt()==0.5*x1**2)
m.Obj(x2*final) # Objective function
m.options.IMODE = 6 # optimal control mode
m.solve() # solve
plt.figure(1) # plot results
plt.plot(m.time,x1.value,'k-',label=r'$x_1$')
plt.plot(m.time,x2.value,'b-',label=r'$x_2$')
plt.plot(m.time,u.value,'r--',label=r'$u$')
plt.legend(loc='best')
plt.xlabel('Time')
plt.ylabel('Value')
plt.show()
```

Listing A3. Commercial Fishing Dynamic Optimization.

Original Form

$\max_{u(t)} \int_0^{10}\left(E - \frac{c}{x}\right)u\,U_{max}\,dt$

subject to

$\frac{dx}{dt} = r\,x(t)\left(1 - \frac{x(t)}{k}\right) - u\,U_{max}$

$x(0) = 70$

$0 \leq u(t) \leq 1$

$E = 1,\ c = 17.5,\ r = 0.71$

$k = 80.5,\ U_{max} = 20$

Equivalent Form for GEKKO

$\min_{u(t)} -J\left(t_f\right)$

subject to

$\frac{dx}{dt} = r\,x(t)\left(1 - \frac{x(t)}{k}\right) - u\,U_{max}$

$\frac{dJ}{dt} = \left(E - \frac{c}{x}\right)u\,U_{max}$

$x(0) = 70$

$J(0) = 0$

$0 \leq u(t) \leq 1$

$t_f = 10,\ E = 1,\ c = 17.5$

$r = 0.71,\ k = 80.5,\ U_{max} = 20$

```
from gekko import GEKKO
import numpy as np
import matplotlib.pyplot as plt
# create GEKKO model
m = GEKKO()
# time points
n=501
m.time = np.linspace(0,10,n)
# constants
E,c,r,k,U_max = 1,17.5,0.71,80.5,20
# fishing rate
u = m.MV(value=1,lb=0,ub=1)
u.STATUS = 1
u.DCOST = 0
x = m.Var(value=70) # fish population
# fish population balance
m.Equation(x.dt() == r*x*(1-x/k)-u*U_max)
J = m.Var(value=0) # objective (profit)
Jf = m.FV() # final objective
Jf.STATUS = 1
m.Connection(Jf,J,pos2='end')
m.Equation(J.dt() == (E-c/x)*u*U_max)
m.Obj(-Jf) # maximize profit
m.options.IMODE = 6  # optimal control
m.options.NODES = 3  # collocation nodes
m.options.SOLVER = 3 # solver (IPOPT)
m.solve() # Solve
print('Optimal Profit: ' + str(Jf.value[0]))
plt.figure(1) # plot results
plt.subplot(2,1,1)
plt.plot(m.time,J.value,'r--',label='profit')
plt.plot(m.time,x.value,'b-',label='fish')
plt.legend()
plt.subplot(2,1,2)
plt.plot(m.time,u.value,'k.-',label='rate')
plt.xlabel('Time (yr)')
plt.legend()
plt.show()
```

Listing A4. Jennings Problem: Minimize Final Time.

Original Form

$\min_{u(t)} t_f$

subject to

$\frac{dx_1}{dt} = u$

$\frac{dx_2}{dt} = \cos(x_1(t))$

$\frac{dx_3}{dt} = \sin(x_1(t))$

$x(0) = [\pi/2, 4, 0]$

$x_2(t_f) = 0$

$x_3(t_f) = 0$

$-2 \leq u(t) \leq 2$

Equivalent Form for GEKKO

$\min_{u(t), t_f} t_f$

subject to

$\frac{dx_1}{dt} = t_f\, u$

$\frac{dx_2}{dt} = t_f \cos(x_1(t))$

$\frac{dx_3}{dt} = t_f \sin(x_1(t))$

$x(0) = [\pi/2, 4, 0]$

$x_2(t_f) = 0$

$x_3(t_f) = 0$

$-2 \leq u(t) \leq 2$

```
import numpy as np
from gekko import GEKKO
import matplotlib.pyplot as plt
m = GEKKO() # initialize GEKKO
nt = 501
m.time = np.linspace(0,1,nt)
# Variables
x1 = m.Var(value=np.pi/2.0)
x2 = m.Var(value=4.0)
x3 = m.Var(value=0.0)
p = np.zeros(nt) # final time = 1
p[-1] = 1.0
final = m.Param(value=p)
# optimize final time
tf = m.FV(value=1.0,lb=0.1,ub=100.0)
tf.STATUS = 1
# control changes every time period
u = m.MV(value=0,lb=-2,ub=2)
u.STATUS = 1
m.Equation(x1.dt()==u*tf)
m.Equation(x2.dt()==m.cos(x1)*tf)
m.Equation(x3.dt()==m.sin(x1)*tf)
m.Equation(x2*final<=0)
m.Equation(x3*final<=0)
m.Obj(tf)
m.options.IMODE = 6
m.solve()
print('Final Time: ' + str(tf.value[0]))
tm = np.linspace(0,tf.value[0],nt)
plt.figure(1)
plt.plot(tm,x1.value,'k-',label=r'$x_1$')
plt.plot(tm,x2.value,'b-',label=r'$x_2$')
plt.plot(tm,x3.value,'g--',label=r'$x_3$')
plt.plot(tm,u.value,'r--',label=r'$u$')
plt.legend(loc='best')
plt.xlabel('Time')
plt.show()
```

References

1. Nyström, R.H.; Franke, R.; Harjunkoski, I.; Kroll, A. Production campaign planning including grade transition sequencing and dynamic optimization. *Comput. Chem. Eng.* **2005**, *29*, 2163–2179. [CrossRef]
2. Touretzky, C.R.; Baldea, M. Integrating scheduling and control for economic MPC of buildings with energy storage. *J. Process Control* **2014**, *24*, 1292–1300. [CrossRef]
3. Powell, K.M.; Cole, W.J.; Ekarika, U.F.; Edgar, T.F. Dynamic optimization of a campus cooling system with thermal storage. In Proceedings of the 2013 European Control Conference (ECC), Zurich, Switzerland, 17–19 July 2013; pp. 4077–4082.
4. Pontes, K.V.; Wolf, I.J.; Embiruçu, M.; Marquardt, W. Dynamic Real-Time Optimization of Industrial Polymerization Processes with Fast Dynamics. *Ind. Eng. Chem. Res.* **2015**, *54*, 11881–11893. [CrossRef]
5. Zhuge, J.; Ierapetritou, M.G. Integration of Scheduling and Control with Closed Loop Implementation. *Ind. Eng. Chem. Res.* **2012**, *51*, 8550–8565. [CrossRef]
6. Beal, L.D.; Park, J.; Petersen, D.; Warnick, S.; Hedengren, J.D. Combined model predictive control and scheduling with dominant time constant compensation. *Comput. Chem. Eng.* **2017**, *104*, 271–282. [CrossRef]
7. Huang, R.; Zavala, V.M.; Biegler, L.T. Advanced step nonlinear model predictive control for air separation units. *J. Process Control* **2009**, *19*, 678–685. [CrossRef]
8. Zavala, V.M.; Biegler, L.T. Optimization-based strategies for the operation of low-density polyethylene tubular reactors: Moving horizon estimation. *Comput. Chem. Eng.* **2009**, *33*, 379–390. [CrossRef]
9. Rall, L.B. *Automatic Differentiation: Techniques and Applications; Lecture Notes in Computer Science*; Springer: Berlin, Germany, 1981; Volume 120.
10. Cervantes, A.; Biegler, L.T. Optimization Strategies for Dynamic Systems. In *Encyclopedia of Optimization*; Floudas, C., Pardalos, P., Eds.; Kluwer Academic Publishers: Plymouth, MA, USA, 1999.

11. Bock, H.; Plitt, K. A Multiple Shooting Algorithm for Direct Solution of Optimal Control Problems*. In Proceedings of the 9th IFAC World Congress: A Bridge Between Control Science and Technology, Budapest, Hungary, 2–6 July 1984; Volume 17, pp. 1603–1608.
12. Lorenz, T. Biegler. *Nonlinear Programming: Concepts, Algorithms, and Applications to Chemical Processes*; Siam: Philadelphia, PA, USA, 2010.
13. Ross, I.M.; Karpenko, M. A review of pseudospectral optimal control: From theory to flight. *Ann. Rev. Control* **2012**, *36*, 182–197. [CrossRef]
14. Qin, S.J.; Badgwell, T.A. A survey of industrial model predictive control technology. *Control Eng. Pract.* **2003**, *11*, 733–764. [CrossRef]
15. Findeisen, R.; Allgöwer, F.; Biegler, L. *Assessment and Future Directions of Nonlinear Model Predictive Control*; Springer: Berlin, Germany, 2007; Volume 358, p. 642.
16. Ellis, M.; Durand, H.; Christofides, P.D. A tutorial review of economic model predictive control methods. *J. Proc. Control* **2014**, *24*, 1156–1178. [CrossRef]
17. Ji, L.; Rawlings, J.B. Application of MHE to large-scale nonlinear processes with delayed lab measurements. *Comput. Chem. Eng.* **2015**, *80*, 63–72. [CrossRef]
18. Würth, L.; Rawlings, J.B.; Marquardt, W. Economic dynamic real-time optimization and nonlinear model predictive control on infinite horizons. *Symp. Adv. Control* **2009**, *42*, 219–224. [CrossRef]
19. Hart, W.E.; Laird, C.; Watson, J.P.; Woodruff, D.L. *Pyomo–Optimization Modeling in Python*; Springer International Publishing: Cham, Switerland, 2012; Volume 67.
20. Dunning, I.; Huchette, J.; Lubin, M. JuMP: A modeling language for mathematical optimization. *SIAM Rev.* **2017**, *59*, 295–320. [CrossRef]
21. Andersson, J.; Åkesson, J.; Diehl, M. CasADi: A symbolic package for automatic differentiation and optimal control. In *Recent Advances in Algorithmic Differentiation*; Springer: Berlin, Germany, 2012; pp. 297–307.
22. Bisschop, J.; Meeraus, A. On the development of a general algebraic modeling system in a strategic planning environment. In *Applications*; Springer: Berlin, Germany, 1982; pp. 1–29.
23. Fourer, R.; Gay, D.; Kernighan, B. *AMPL*; A Modeling Language for Mathematical Programming; Boyd & Fraser Pub. Co.: Danvers, MA, USA, 1993.
24. Barton, P.I.; Pantelides, C. gPROMS-A combined discrete/continuous modelling environment for chemical processing systems. *Simul. Ser.* **1993**, *25*, 25–25.
25. Åkesson, J.; Årzén, K.E.; Gäfvert, M.; Bergdahl, T.; Tummescheit, H. Modeling and optimization with Optimica and JModelica. org—Languages and tools for solving large-scale dynamic optimization problems. *Comput. Chem. Eng.* **2010**, *34*, 1737–1749. [CrossRef]
26. Houska, B.; Ferreau, H.J.; Diehl, M. ACADO toolkit—An open-source framework for automatic control and dynamic optimization. *Opt. Control Appl. Meth.* **2011**, *32*, 298–312. [CrossRef]
27. Ross, I.M. *User's Manual for DIDO: A MATLAB Application Package for Solving Optimal Control Problems*; Tomlab Optimization: Vasteras, Sweden, 2004; p. 65.
28. Patterson, M.A.; Rao, A.V. GPOPS-II: A MATLAB software for solving multiple-phase optimal control problems using hp-adaptive Gaussian quadrature collocation methods and sparse nonlinear programming. *ACM Trans. Math. Softw.* **2014**, *41*. [CrossRef]
29. Rutquist, P.E.; Edvall, M.M. *Propt-Matlab Optimal Control Software*; Tomlab Optimization Inc.: Pullman, WA, USA, 2010; p. 260.
30. Becerra, V.M. Solving complex optimal control problems at no cost with PSOPT. In Proceedings of the 2010 IEEE International Symposium on Computer-Aided Control System Design (CACSD), Yokohama, Japan, 8–10 September 2010; pp. 1391–1396.
31. Bisschop, J. *AIMMS—Optimization Modeling*; Paragon Decision Technology: Kirkland, WA, USA, 2006.
32. Grant, M.; Boyd, S. Graph implementations for nonsmooth convex programs. In *Recent Advances in Learning and Control*; Blondel, V.; Boyd, S.; Kimura, H., Eds.; Lecture Notes in Control and Information Sciences, Springer: Berlin, Germany, 2008; pp. 95–110.
33. Andersen, M.; Dahl, J.; Liu, Z.; Vandenberghe, L. Interior-point methods for large-scale cone programming. *Optim. Mach. Learn.* **2011**, *5583*, 55–83.
34. Löfberg, J. YALMIP: A Toolbox for Modeling and Optimization in MATLAB. In Proceedings of the CACSD Conference, Taipei, Taiwan, 2–4 September 2004.

35. Mitchell, S.; Consulting, S.M.; Dunning, I. *PuLP: A Linear Programming Toolkit for Python*; The University of Auckland: Auckland, New Zealand, 2011.
36. Biegler, L.T. An overview of simultaneous strategies for dynamic optimization. *Chem. Eng. Proc. Proc. Intensif.* **2007**, *46*, 1043–1053. [CrossRef]
37. Hedengren, J.D.; Shishavan, R.A.; Powell, K.M.; Edgar, T.F. Nonlinear modeling, estimation and predictive control in APMonitor. *Comput. Chem. Eng.* **2014**, *70*, 133–148. [CrossRef]
38. De Souza, G.; Odloak, D.; Zanin, A.C. Real time optimization (RTO) with model predictive control (MPC). *Comput. Chem. Eng.* **2010**, *34*, 1999–2006. [CrossRef]
39. Safdarnejad, S.M.; Hedengren, J.D.; Lewis, N.R.; Haseltine, E.L. Initialization strategies for optimization of dynamic systems. *Comput. Chem. Eng.* **2015**, *78*, 39–50. [CrossRef]
40. Waechter, A.; Biegler, L.T. On the Implementation of a Primal-Dual Interior Point Filter Line Search Algorithm for Large-Scale Nonlinear Programming. *Math. Program.* **2006**, *106*, 25–57. [CrossRef]
41. Hedengren, J.; Mojica, J.; Cole, W.; Edgar, T. APOPT: MINLP Solver for Differential and Algebraic Systems with Benchmark Testing. In Proceedings of the INFORMS National Meeting, Phoenix, AZ, USA, 14–17 October 2012.
42. Gill, P.E.; Murray, W.; Saunders, M.A. SNOPT: An SQP algorithm for large-scale constrained optimization. *SIAM Rev.* **2005**, *47*, 99–131. [CrossRef]
43. Hedengren, J.D.; Eaton, A.N. Overview of Estimation Methods for Industrial Dynamic Systems. *Optim. Eng.* **2017**, *18*, 155–178. [CrossRef]
44. Lewis, N.R.; Hedengren, J.D.; Haseltine, E.L. Hybrid Dynamic Optimization Methods for Systems Biology with Efficient Sensitivities. *Processes* **2015**, *3*, 701–729. [CrossRef]
45. Powell, K.M.; Eaton, A.N.; Hedengren, J.D.; Edgar, T.F. A Continuous Formulation for Logical Decisions in Differential Algebraic Systems using Mathematical Programs with Complementarity Constraints. *Processes* **2016**, *4*, 7. [CrossRef]
46. Beal, L.D.; Petersen, D.; Grimsman, D.; Warnick, S.; Hedengren, J.D. Integrated scheduling and control in discrete-time with dynamic parameters and constraints. *Comput. Chem. Eng.* **2018**, *115*, 361–376. [CrossRef]
47. Eaton, A.N.; Beal, L.D.; Thorpe, S.D.; Hubbell, C.B.; Hedengren, J.D.; Nybø, R.; Aghito, M. Real time model identification using multi-fidelity models in managed pressure drilling. *Comput. Chem. Eng.* **2017**, *97*, 76–84. [CrossRef]
48. Park, J.; Webber, T.; Shishavan, R.A.; Hedengren, J.D.; others. Improved Bottomhole Pressure Control with Wired Drillpipe and Physics-Based Models. In Proceedings of the SPE/IADC Drilling Conference and Exhibition, Society of Petroleum Engineers, The Hague, The Netherlands, 14–16 March 2017.
49. Mojica, J.L.; Petersen, D.; Hansen, B.; Powell, K.M.; Hedengren, J.D. Optimal combined long-term facility design and short-term operational strategy for CHP capacity investments. *Energy* **2017**, *118*, 97–115. [CrossRef]
50. Safdarnejad, S.M.; Hedengren, J.D.; Baxter, L.L. Dynamic optimization of a hybrid system of energy-storing cryogenic carbon capture and a baseline power generation unit. *Appl. Energy* **2016**, *172*, 66–79. [CrossRef]
51. Safdarnejad, S.M.; Gallacher, J.R.; Hedengren, J.D. Dynamic parameter estimation and optimization for batch distillation. *Comput. Chem. Eng.* **2016**, *86*, 18–32. [CrossRef]
52. Safdarnejad, S.M.; Hedengren, J.D.; Baxter, L.L. Plant-level dynamic optimization of Cryogenic Carbon Capture with conventional and renewable power sources. *Appl. Energy* **2015**, *149*, 354–366. [CrossRef]
53. DeFigueiredo, B.; Zimmerman, T.; Russell, B.; Howell, L.L. Regional Stiffness Reduction Using Lamina Emergent Torsional Joints for Flexible Printed Circuit Board Design. *J. Electron. Packag.* **2018**. [CrossRef]
54. Valderrama, F.; Ruiz, F. An optimal control approach to steam distillation of essential oils from aromatic plants. *Comput. Chem. Eng.* **2018**, *117*, 25–31. [CrossRef]

processes

Article

Modelling Condensation and Simulation for Wheat Germ Drying in Fluidized Bed Dryer

Der-Sheng Chan [1,2], Jun-Sheng Chan [3] and Meng-I Kuo [1,3,*,†]

1 PhD Program in Nutrition and Food Science, Fu Jen Catholic University, New Taipei City 24205, Taiwan; dschan@ms58.hinet.net
2 Department of Information Technology, Lee-Ming Institute of Technology, New Taipei City 24346, Taiwan
3 Department of Food Science, Fu Jen Catholic University, New Taipei City 24205, Taiwan; stozerchan@gmail.com
* Correspondence: 062998@mail.fju.edu.tw; Tel.: +886-2-2905-2019; Fax: +886-2-2209-3271
† Current Address: 510 Zhong-Zheng Road, Xinzhuang District, New Taipei City 24205, Taiwan.

Received: 10 April 2018; Accepted: 5 June 2018; Published: 9 June 2018

Abstract: A low-temperature drying with fluidized bed dryer (FBD) for wheat germ (WG) stabilization could prevent the loss of nutrients during processing. However, both evaporation and condensation behaviors occurred in sequence during FBD drying of WG. The objective of this study was to develop a theoretical thin-layer model coupling with the macro-heat transfer model and the bubble model for simulating both the dehydration and condensation behaviors of WG during low-temperature drying in the FBD. The experimental data were also collected for the model modification. Changes in the moisture content of WG, the air temperature of FBD chamber, and the temperature of WG during drying with different heating approaches were significantly different. The thermal input of WG drying with short heating time approach was one-third of that of WG drying with a traditional heating approach. The mathematical model developed in this study could predict the changes of the moisture content of WG and provide a good understanding of the condensation phenomena of WG during FBD drying.

Keywords: wheat germ; fluidized bed drying; mathematical model; moisture content; condensation; simulation

1. Introduction

Wheat germ (WG) is one of the wheat milling by-products. It is rich in many nutrients and bioactive compounds, thus, it is widely applied in bread, snack foods, and breakfast cereals [1]. Since WG contains phenolic compounds, it is also used to produce valuable antioxidant supplements [2]. However, fresh WG could easily undergo oxidative rancidity due to a high moisture content (MC), unsaturated fats, and endogenous enzymes, resulting in a limited shelf-life. Controlling the MC within a proper range or inactivating the enzymes is the key to preventing WG from deterioration, and consequently extending its shelf-life.

Several studies have been conducted to develop the suitable methods for WG stabilization, including short wave infrared radiation [3], gamma irradiation [4], roasting [5–7], steaming [8,9], microwave treatment [9–12], toasting [13,14], temperature controlled water-bath [11,12], sourdough fermentation [14], and fluidization method [9,15–18]. However, the processing temperature of these methods was high. High temperature processing not only imparted the color, flavor, and nutrients of WG but also affected the energy efficiency. In our research group, a low-temperature (<100 °C) drying process with a fluidized bed dryer (FBD) was developed for WG stabilization in order to prevent the loss of nutrients during processing [19]. FBD could introduce a high interface area of hot air on material and consequently reduce the drying temperature and time. However, low-temperature WG drying with FBD is complex in evaporation (dehydration at heating operation) and condensation (adsorption at cooling operation) behaviors, leading to an increase in the amount of trial and error in practice. Therefore, developing a mathematical model to simulate the evaporation and condensation of WG during the heating and cooling process is important for optimizing the operation of FBD.

Heat transfer between particles and air is complicated in the FBD [20]. The Davidson's bubble model has been successfully applied to simulate the operational performance of FBD for chemical industry [20,21]. It provides helpful information between the macro fluid (air bubbles) and the micro fluid (suspended particles). The mass transfer equations for the thin-layer drying approach were used to calculate the MC of grain [22–30]. Thin-layer drying models for drying behaviors of biological products can be classified into three different types, namely, theoretical, semi-theoretical, and empirical models [26–28]. The theoretical type is generally derived from the Fick's second law of diffusion with Arrhenius-type temperature dependent diffusivity or simultaneous heat and mass transfer equations [22–25,29]. The semi-theoretical type concerns the approximated theoretical equations, which is valid for the temperature, relative humidity, air velocity, and MC [29]. They can be categorized into the Lewis model, Page's model, two-term model, Henderson and Pabis model, Thomson model, and the Wang and Singh model. An empirical type provides a direct relationship between average MC and time, but it neglects the physical correlation between the drying processes and their parameters [26]. Gili et al. [31] successfully developed an analytical series solution for the microscopic diffusion equations and applied a theoretical thin-layer model coupling with a macro-heat transfer model to predict the MC of WG during drying in the FBD. They also observed that the moisture loss occurred in the falling drying rate period for WG drying in the FBD. However, the condensation at the surface of WG was not investigated in these studies.

The objective of this study was to develop a theoretical thin-layer model coupled with the macro-heat transfer model and the bubble model for simulating both the dehydration and condensation of WG during low-temperature drying in the FBD. The experimental data of inlet air temperature and MC of WG during drying were also collected for the model verification in this study, in order to improve the accuracy of the prediction.

2. Materials and Methods

2.1. Materials

The raw WG was obtained from the local supplier in Taiwan and stored in a walk-in freezer at −20 °C immediately upon arrival. WG was tempered in an environment-controlled storeroom at 25 °C one day before the drying experiment. The particle size distribution of raw WG was measured by using the sieving process with the sieve trays of 50, 40, 30, 20 and 16 mesh sizes.

2.2. Fluidized Bed Dryer

A pilot-scale vertical FBD (Shia Machinery Industrial Co., Ltd., Taichung, Taiwan) with a height of 228 cm and a diameter of 45 cm was used for the drying of WG. The schematic diagram of the FBD is shown in Figure 1A. The FBD is made up of air compressor, heater, sample bin, drying chamber, filter bags, outlet air motor and inbuilt program logical controller. The dimensions of FBD are shown in Figure 1B. The surrounding air (about 25 °C) was first introduced into the heater by air compressor and passed through the bottom of FBD to the sample bin and drying chamber, and then flowed up to the filter bags. Finally, the air was exhausted to atmosphere. There are a total of seven filter bags positioned at the top part of the drying chamber to prevent the samples from escaping the chamber. The inlet air flow velocity was measured with a gas flow transmitter (LABO-FG, GHM Messtechnik GmbH, Erolzheim, Germany).

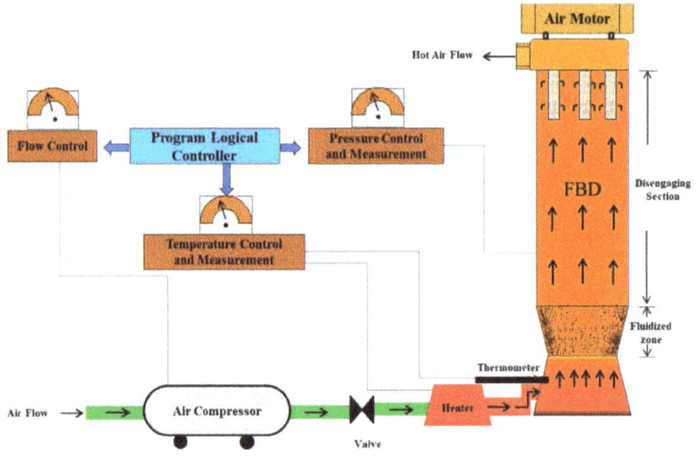

(A)

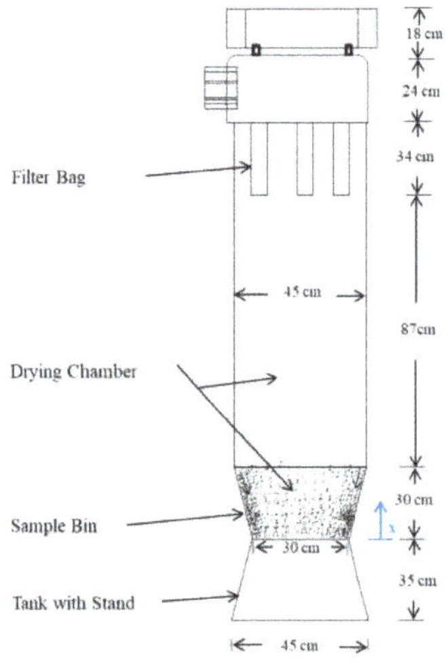

(B)

Figure 1. Schematic diagram (**A**) and dimensions (**B**) of the fluidized bed dryer used in this study.

Once the velocity of air reached 1.2 m/s, the WG particles were loosened from the bottom of the sample bin and started to swirl and expand with air in the FBD. Practically, the 2 kg WG sample was fluidized at a height of 30 cm from the bottom of the sample bin with 1.2 m/s air velocity.

2.3. Drying Experiment

Both short heating time approach (SHTA) and traditional heating approach (THA) were conducted to dry the WG using FBD at 80 °C. The entire drying process was divided into four stages, preheating, sample loading, heating, and cooling (Table 1). The temperature of FDB chamber was controlled at 80 °C and preheated for 10 min to ensure an equilibrium condition. The 2 kg WG sample was spread as a thin layer (5 cm) on the bottom of the sample bin. After loading, the WG sample was heated at 80 °C for two different times, 4 min (SHTA) and 25 min (THA). In order to obtain a final product temperature of around 40 °C, the cooling time of SHTA and THA were 10 min and 4 min, respectively.

Table 1. Time (min) allocation of each stage for wheat germ drying at 80 °C.

Stages	Short Heating Time Approach (SHTA)	Traditional Heating Approach (THA)
Preheating	10	10
Sample Loading	1	1
Heating	4	25
Cooling	10	4

2.4. Analytical Methods

The changes of inlet air temperature, air humidity, and MC of WG during drying process were measured. The air humidity was measured by the thermal hygrometer (Testo 635, Testo Inc., Lenzkirch, Germany). The temperature of inlet air was measured by the K-type thermometer (Tecpel 318, Tecpel Co., Ltd., Taipei, Taiwan). The MC of WG was measured according to the American Association of Cereal Chemists (AACC) Method 44-19 [32]. 2 g ± 1 mg of the sample was dried at 135 °C for 2 h. Moreover, the protein, fat, and ash contents of WG were measured according to the AACC Methods 46-12, 30-25, and 08-03 [32], respectively.

3. Mathematical Modelling

3.1. Governing Equations and Assumptions

Due to the existence of temperature gradient on WG surface, the heat transfer takes place by conduction from the surface to the inside of WG and leads to moisture evaporation. The water vapor diffuses in an opposite direction from the inside to the surface of WG. The water vapor diffusion occurs in the micro pores of WG particles. Accordingly, the bubble model was developed in this study to link the information between the microscopic and the macroscopic heat and mass transfer for WG drying process. This model divided the fine particles distributed within the FBD chamber into three phases, including bubble phase, cloud phase, and emulsion phase. The bubble phase was the gas phase; the cloud phase was the gas phase with thin WG particles; the emulsion phase was the gas phase with dense WG particles [20]. However, WG drying in the FBD is a complex phenomenon because the heat transfer and mass transfer simultaneously occur. Thus, the microscopic assumptions (Terms (1)–(3)) and the macroscopic assumptions (Terms (4)–(10)) are made in this study as follows:

(1) All particles of WG are uniform in size and shape [33].
(2) All particles of WG have the same MC.
(3) Volume shrinkage of WG is negligible during drying [22,23].
(4) Horizontal variation of air temperature and humidity in the FBD are negligible [22,23].
(5) Cloud phase and emulsion phase are within one region [20].
(6) Plug flow model is applied for air and mixed flow model is applied for emulsion phase.
(7) The emulsion phase stays at minimum fluidizing conditions [21].
(8) The size of bubble phase is uniform within the FBD [21].
(9) The difference in temperature between emulsion phase and bubble phase is negligible.
(10) Evaporation mechanism terminates during cooling.

Based on the above assumptions, the heat and mass transfer equations, including initial and boundary conditions are developed to estimate the heat and mass transfers in the FBD during WG drying. The equations for simulation the changes of MC and heat in WG, emulsion phase, and bubble phase during drying are developed in this study as well. In the microscopic point view, the r is the radial coordinate from the center to the surface of WG particle. The x axis is oriented from the base to the top of the FBD (Figure 1B). In macroscopic point of view, the balance of mass and temperature depend on the position x and the drying time t.

3.1.1. Microscopic Energy Balance of Wheat Germ Particles

The energy balance equation for the WG particles is given by:

$$\rho_g C_{P,g} \frac{\partial T_g}{\partial t} = \frac{1}{r^2} \frac{\partial}{\partial r} (r^2 k_g \frac{\partial T_g}{\partial r}) \tag{1}$$

Initial condition:

$$T_g = T_{gi} \text{ for } 0 \leq r \leq r_g \tag{2}$$

Boundary conditions:

$$\frac{\partial T_g}{\partial r} = 0 \text{ for } r = 0 \tag{3}$$

$$k_g \frac{\partial T_g}{\partial r} = h(T_e - T_g) + \lambda D_e \frac{\partial C_{mg}}{\partial r} \text{ for } r = r_g \tag{4}$$

where ρ_g is the density of WG; $C_{p,g}$ is the specific heat of WG; k_g is the thermal conductivity of WG; T_{gi} is the initial WG temperature; r_g is the radius of WG; h is the convective heat transfer coefficient; T_e is the emulsion phase temperature in °C.

3.1.2. Microscopic Mass Balance of Wheat Germ Particles

The mass balance equation for the WG particles is given in Equation (5).

$$\frac{\partial C_{mg}}{\partial t} = \frac{1}{r^2} \frac{\partial}{\partial r} (r^2 D_e \frac{\partial C_{mg}}{\partial r}) \tag{5}$$

Initial condition:

$$C_{mg} = \frac{W_L X_{wi}}{M_w V_T} \text{ for } 0 \leq r \leq r_g \tag{6}$$

Boundary conditions:

$$\frac{\partial C_{mg}}{\partial r} = 0 \text{ for } r = 0 \tag{7}$$

$$D_e \frac{\partial C_{mg}}{\partial r} = K_{de}(C_{mg} - C_e) f_{de} + K_{con}(C_{me} - C_{sat}) f_{con} \text{ for } r = r_g \tag{8}$$

where D_e is the effective diffusivity of water in WG; W_L is the loading of WG; X_{wi} is the initial MC of WG; M_w is the water molecular weight; V_T is the volume of fluidization; K_{de} is the bubble-emulsion mass transfer coefficient for dehydration; C_{mg} is the concentration of moisture in WG; C_e is the equilibrium MC of WG; f_{de} is the step numerical parameter for dehydration; K_{con} is the bubble-emulsion mass transfer coefficient for condensation; C_{me} is the concentration of moisture in emulsion phase; C_{sat} is the saturation concentration; f_{con} is the step numerical parameter for condensation. The equilibrium MC of WG; C_e is modified according to Giner and Calvelo [34] as

$$C_e(RH, T_e) = \frac{1}{100} \left[\frac{-\ln(1 - RH)}{1.2310^{-5}(T_e + 65.8)} \right]^n \tag{9}$$

where RH is the relative humidity in the FBD; n is the constant coefficient.

3.1.3. Macroscopic Energy Balance of Emulsion Phase

The energy balance equation of emulsion phase in the FBD during drying is given in Equation (10).

$$\rho_{em} C_{P,e} (\frac{\partial T_e}{\partial t} + u \frac{\partial T_e}{\partial x}) = \frac{\partial}{\partial x} (k_{em} \frac{\partial T_e}{\partial x}) - A_g(1 - \varphi_b)(k_g \frac{\partial T_g}{\partial r}) \tag{10}$$

Initial condition:

$$T_e = T_i \text{ for } 0 \leq x \leq H_f \tag{11}$$

Boundary conditions:

$$T_e = T_{in} \text{ for } x = 0 \tag{12}$$

$$\frac{\partial T_e}{\partial x} = 0 \text{ for } x = H_f \tag{13}$$

where ρ_{em} is the density of emulsion phase; $C_{P,e}$ is the specific heat of emulsion phase; k_{em} is the thermal conductivity of emulsion phase; A_g is the surface area per unit volume of WG; φ_b is the porosity of fluidized bed; T_i is the initial temperature of emulsion phase; H_f is the height of fluidization; T_{in} is the inlet temperature of emulsion phase.

3.1.4. Macroscopic Mass Balance of Emulsion Phase

The mass balance equation of emulsion phase in the FBD during frying is given by:

$$\frac{\partial C_{me}}{\partial t} = \frac{\partial}{\partial x}(D_v \frac{\partial C_{me}}{\partial x}) - A_g(1 - \varphi_b)(D_e \frac{\partial C_{mg}}{\partial r}) - R_{dl} \tag{14}$$

Initial condition:

$$C_{me} = \frac{RH_i P_{sat}(T_i)}{R_T T_i} \text{ for } 0 \leq x \leq H_f \tag{15}$$

Boundary conditions:

$$C_{me} = \frac{RH_{in} P_{sat}(T_{in})}{R_T T_{in}} \text{ for } x = 0 \tag{16}$$

$$\frac{\partial C_{me}}{\partial x} = 0 \text{ for } x = H_f \tag{17}$$

where D_v is the diffusivity of water vapor; R_{dl} is the rate of mass transfer between bubble and emulsion; RH_i is the initial relative humidity of air in the fluidized bed; $P_{sat}(T_i)$ is the saturated vapor pressure of water and is the function of initial temperature; R_T is the gas constant; T_i is the initial air temperature in the fluidized bed; RH_{in} is the inlet relative humidity of air; $P_{sat}(T_{in})$ is the saturated vapor pressure of water and the function of inlet temperature. $P_{sat}(T)$ is calculated according to Naghavi et al. [25] as

$$P_{sat}(T) = 0.1 \exp(27.0214 - \frac{6887}{T} - 5.31 \ln(\frac{T}{273.15})) \tag{18}$$

The cloud phase and the emulsion phase are considered as perfectly mixed [21]. The rate of mass transfer is modified as

$$R_{dl} = K_{be}\,(C_{me} - C_{mb}) \tag{19}$$

where K_{be} is the mass transfer coefficient between bubble and emulsion; C_{mb} is the concentration of moisture in bubble phase.

3.1.5. Macroscopic Mass Balance of Bubble Phase

The mass balance equation of bubble phase in the FBD during frying is given by:

$$\frac{\partial C_{mb}}{\partial t} + u\frac{\partial C_{mb}}{\partial x} = \frac{\partial}{\partial x}(D_v \frac{\partial C_{mb}}{\partial x}) + R_{dl} \tag{20}$$

Initial condition:

$$C_{mb} = \frac{RH_i P_{sat}(T_i)}{R_T T_i} \text{ for } 0 \leq x \leq H_f \tag{21}$$

Boundary conditions:

$$C_{mb} = \frac{RH_{in}P_{sat}(T_{in})}{R_T T_{in}} \text{ for } x = 0 \tag{22}$$

$$\frac{\partial C_{mb}}{\partial x} = 0 \text{ for } x = H_f \tag{23}$$

3.2. Material Properties of Wheat Germ

3.2.1. Density, Heat Capacity, and Thermal Conductivity Properties of Wheat Germ

The chemical compositions of WG is complex, including moisture, protein, fat, carbohydrate, and ash. In the present study, the effective properties of WG, including density, heat capacity, thermal conductivity, and diffusivity, were estimated using a mass fraction averaged mixing rule [35] based on the compositions and the local temperature (Table 2). The temperature is expressed in °C.

Table 2. Material properties of wheat germ [35].

Description	Equation
Density $\rho_g = \Sigma\rho_j\,\omega_j$	$\rho_w = 9.9718 \times 10^2 + 3.1439 \times 10^{-3}\,T - 3.7574 \times 10^{-3}\,T^2$ $\rho_p = 1.3299 \times 10^3 - 0.51814\,T$ $\rho_{fat} = 9.2559 \times 10^2 - 0.41757\,T$ $\rho_{corb} = 1.5991 \times 10^3 - 0.31046\,T$ $\rho_{ash} = 2.4238 \times 10^3 - 0.28063\,T$
Heat Capacity $C_{P,g} = \Sigma C_{P,j}\,\omega_j$	$C_{P,w} = 4.1762 + 9.0862 \times 10^{-3}\,T - 5.4731 \times 10^{-6}\,T^2$ $C_{P,p} = 2.0082 + 1.2089 \times 10^{-3}\,T - 1.3129 \times 10^{-6}\,T^2$ $C_{P,fat} = 1.9842 + 1.4733 \times 10^{-3}\,T - 4.8008 \times 10^{-6}\,T^2$ $C_{P,carb} = 1.5488 + 1.9625 \times 10^{-3}\,T - 5.9399 \times 10^{-6}\,T^2$ $C_{P,ash} = 1.0926 + 1.8896 \times 10^{-3}\,T - 3.6817 \times 10^{-6}\,T^2$
Thermal conductivity $k_g = \Sigma k_j\,\omega_j$	$k_w = 5.7109 \times 10^{-1} + 1.7625 \times 10^{-3}\,T - 6.7036 \times 10^{-6}\,T^2$ $k_p = 1.788 \times 10^{-1} + 1.1958 \times 10^{-3}\,T - 6.7036 \times 10^{-6}\,T^2$ $k_{fat} = 1.8071 \times 10^{-1} - 2.7604 \times 10^{-3}\,T - 1.7749 \times 10^{-7}\,T^2$ $k_{carb} = 2.014\ 10^{-1} + 1.3874 \times 10^{-3}\,T - 4.3312 \times 10^{-6}\,T^2$ $k_{ash} = 3.296 \times 10^{-1} - 1.401 \times 10^{-3}\,T - 2.9069 \times 10^{-6}\,T^2$

3.2.2. Diffusivity of Moisture in Wheat Germ

The diffusivity of moisture depends on temperature [34]

$$D_e = D_o \exp(-\frac{E_a}{R_g T_g}) \tag{24}$$

where E_a is the activation energy of water diffusion in the WG.

3.3. Thermal Input for Drying Process

When WG was dried in the FBD, the energy must be transferred from the hot air to the surface of WG. The air temperature highly depends on the heater power. The surface temperature of WG is governed by the difference between air temperature and surface temperature of WG and the physicochemical properties of WG. The time–temperature curve of bread baking [36] was used to assess the thermal input (*TI*) in this study. For example, the combination of temperature and time was subjected during drying as

$$TI = \int_0^t T_g dt \tag{25}$$

3.4. Numerical Step Function for Condensation

A numerical approach for the step function was employed to describe the condensation of WG, i.e., the temperature of emulsion phase is smaller than the WG temperature [37]

$$f_{con} = \begin{cases} 0 & \text{if } T_g < T_e \\ 1 & \text{if } T_g \geq T_e \end{cases} \tag{26}$$

Based on the assumption (10), a step function, f_{de}, for evaporation is obtained

$$f_{de} = 1 - f_{con}$$

4. Results and Discussion

4.1. Physical Dimensions and Chemical Composition of Wheat Germ

The particle size distribution of WG is shown in Figure 2. There were about 67% of WG particles within the size range of 643–973 µm. The physical dimensions of WG with the average of 20 particles were 985 µm and 689 µm in long axis and short axis, respectively. The equivalent diameter (D_{eg}) could be calculated based on the results of Figure 2 using a weight fraction averaged mixing rule. Thus, the D_{eg} of WG particles was 810 µm. The diameter of WG particles observed by Gili et al. [30] was in the range between 766 µm and 1223 µm, and their result was similar to our study. According to Srinivasakannan and Balasubramanian [33], the ratio of surface area to volume (A_g) of WG could be calculated by $6/D_{eg}$. Therefore, the A_g of WG particles was 7407 1/m.

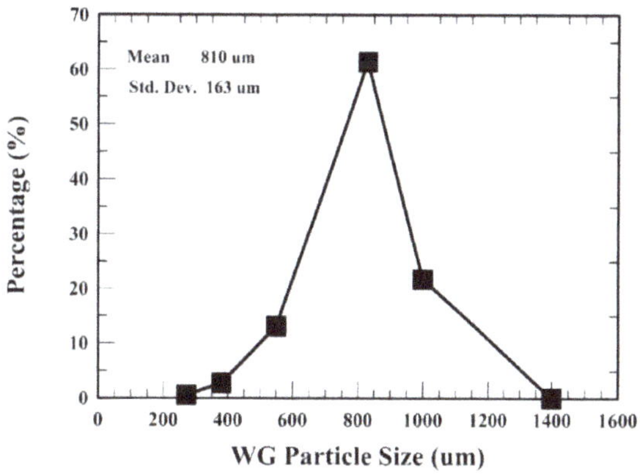

Figure 2. Particle size distribution of wheat germ used in this study.

The chemical composition of WG is shown in Table 3. The WG comprised 13.2% moisture, 9.0% fat, 5.9% ash, 24.1% protein, and 48.2% carbohydrate. The above results are similar to other studies [1,4,38].

Table 3. Chemical composition of raw wheat germ (%, wet basis).

Moisture	Protein	Fat	Carbohydrate *	Ash
13.15 ± 0.28	24.11 ± 0.23	9.02 ± 0.13	48.2	5.91 ± 0.34

Values are mean ± standard deviations of n = 3. * Carbohydrate = 100-Moisture-Protein-Fat-Ash.

4.2. Simulation of the Inlet Air Temperature of FBD

In order to consider the effect of heater on the inlet air temperature of FBD, a function, T_{heater}, is defined. The equations used to simulate the T_{heater} are

$$T_{heater} = S_{heating} \times f_{step} + S_{cooling} \times (1 - f_{step}) \tag{27}$$

$$S_{heating} = T_{start} + (T_s - T_{start}) \times (1.0 - \exp(-2.0t/20)) \tag{28}$$

$$S_{cooling} = 310.15 + 1.01 \times (T_s - 310.15) \times \exp(-0.6 \times (t - t_h)/240) \tag{29}$$

where $S_{heating}$ is the temperature at the heating stage; $S_{cooling}$ is the temperature at the cooling stage; f_{step} is the step function; T_{start} is the measured temperature (67 °C) at the sample loading stage; t_h is the heating time; T_s is the set temperature (80 °C). The step function is expressed as

$$f_{step} = \begin{cases} 1 & \text{if } t < t_h \\ 0 & \text{if } t \geq t_h \end{cases} \tag{30}$$

The experimental data and the predicted values of the inlet air temperature (the heater temperature) during drying at 80 °C in the FBD are shown in Figure 3. The predicted values calculated by the Equations (27)–(30) were consistent well with the experimental data. Therefore, T_{heater} could be applied to employ a boundary condition of heat equations for simulation.

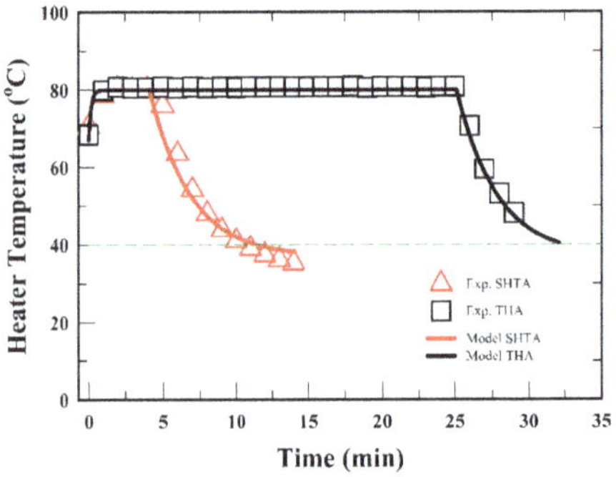

Figure 3. Profiles of the inlet air temperature (the heater temperature) during drying at 80 °C in the FBD. Experimental data, SHTA (Δ) and THA (□); Predicted values, SHTA (**red line**) and THA (**black line**).

4.3. Macroscopic Behavior

4.3.1. Response of the Moisture Content of Wheat Germ

It is important to simulate the changes of the MC of WG during drying with SHTA and THA in the FBD in order to control the quality of the final product. The experimental data and the predicted values of the MC of WG during drying at 80 °C in the FBD are illustrated in Figure 4. The experimental results revealed that the MC of WG drastically decreased to a minimal value of 6.0% at the heating stage with SHTA. Furthermore, the MC of WG slightly increased at the cooling stage. The MC of WG decreased sharply at the early stage of heating with THA and reduced gradually until 25 min of heating. According to Gili et al. [30], the MC of WG should be controlled within 5–8% to avoid spoilage. The SHTA applied in this study required longer cooling time to achieve this goal, but not THA.

The nonlinear partial differential equations presented in the mathematical model at Section 3 were solved by using COMSOL Multiphysics 5.1 with a finite element method. The model parameters listed in Table 4 are given for simulation. The local sensitivity analysis approach was used to verify the model parameter. The criterion of the model verification was calculated by using the minimum of the absolute average deviation (AAD) [31]. The equation is defined as

$$AAD = Min\left(\frac{1}{n}\sum_{j=1}^{n}\left|MC_{\exp,j} - MC_{pre,j}\right|\right) \quad (31)$$

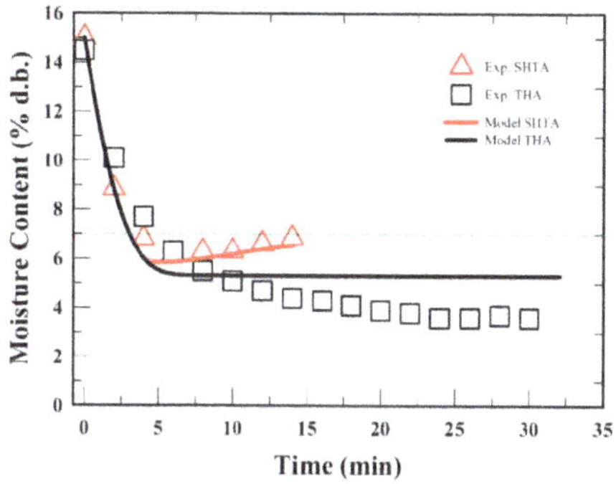

Figure 4. Profiles of the moisture content of wheat germ during drying at 80 °C in the FBD. Experimental data, SHTA (Δ) and THA (□); Predicted values, SHTA (**red line**) and THA (**black line**).

Table 4. Input parameter values.

Measured		Estimated		Set		Theoretical	
Parameter	Value	Parameter	Value	Parameter	Value	Parameter	Value
H_f	0.30 m	A_g	7407 1/m	D_o	2.3×10^{-4} m²/s	M_w	0.018 kg/mol
RH_i	0.15 (-)	r_g	4.05×10^{-3} m	D_v	1.0×10^{-5} m²/s	R_T	0.08205 atm/mol·K
RH_{in}	0.75 (-)	n	0.36	E_a	29.4×10^3 J/mol	R_g	8.314 J/mol·K
T_{gi}	298.15 K			h	32.5 W/m² K	λ	41.4 kJ/mol
T_{start}	340.15 K			K_{con}	4.5×10^{-3} m/s		
W_L	2.0 kg			K_{de}	1.2×10^{-2} m/s		
u	1.2 m/s			K_{be}	1.2×10^{-3} m/s		
V_T	0.214 m³						
X_{wi}	15% d.b.						
φ_b	0.92 (-)						

The mass transfer coefficient for condensation (K_{con}) was the important parameter to predict the MC of WG during drying in the FBD. Therefore, the local impact of K_{con} parameter was evaluated. Table 5 demonstrates the changes in the experimental MC, predicted MC of WG with different values of K_{con} parameter, and the corresponding AAD values during drying. The values of AAD ranging from 0.23 to 0.74 were observed. The minimum value of AAD occurred when the value of K_{con} parameter was 4.5×10^{-3} m/s. Accordingly, this value was selected to predict the MC of WG in this study.

The predicted values of the MC of WG during drying in the FBD were shown in Figure 4. The predicted values were consistent well with the experimental data of SHTA, indicating that the model is able to characterize the physical phenomenon of WG at the heating and cooling stages and predict the changes of the MC of WG during drying very well. Moreover, the predicted values of THA drying were slightly higher than the experimental data. In this study, the equilibrium MC of WG (Equation (9)) was used to fit different operating temperature, but it was not proper to simulate the MC of WG for longer drying time.

Table 5. The experimental moisture content (MC), predicted MC, K_{con} parameter, and absolute average deviation (AAD).

		K_{con} (m/s)				
		0	1.0×10^{-3}	4.5×10^{-3}	9.0×10^{-3}	1.8×10^{-2}
Time (min)	**Experimental MC**	**Predicted MC**				
0	15	15	15	15	15	15
2	8.84	9.56	9.56	9.56	9.56	9.56
4	6.81	6.43	6.43	6.43	6.43	6.43
6	6.15	5.89	5.90	5.9	5.91	5.93
8	6.29	5.89	5.94	6.08	6.25	6.53
10	6.31	5.89	6.01	6.4	6.82	7.46
12	6.61	5.89	6.09	6.71	7.32	8.14
14	6.85	5.89	6.17	6.96	7.68	8.53
AAD		0.48	0.40	0.23	0.43	0.74

4.3.2. Responses of the Air Temperature in the FBD Chamber

The experimental data (at 1/2 height of FBD, about 70 cm above the bottom of bin) and the simulated average values of air temperature in the FBD chamber during drying at 80 °C with SHTA and THA are shown in Figure 5. During the loading of WG, the air temperature in the FBD chamber rapidly dropped from 80 °C, and then it increased gradually because the heat continuously transferred from the heater to the air. After 4-min heating with SHTA, the air temperature reached maximum of 70 °C. During cooling, the air temperatures quickly dropped to a lower level. The air temperature in the FBD chamber with THA drying reached maximum of 72 °C after 25 min. Comparing with Figures 3 and 5, the air temperature in the FBD chamber strongly depends on the inlet air temperature (heater temperature). The simulated average air temperatures were slightly higher than the experimental data at the heating stage and were congruous with the experimental data at the cooling stage (Figure 5). The FBD height used for simulation was 30 cm which was closer to the heater, thus, the simulated air temperature was higher at the heating stage.

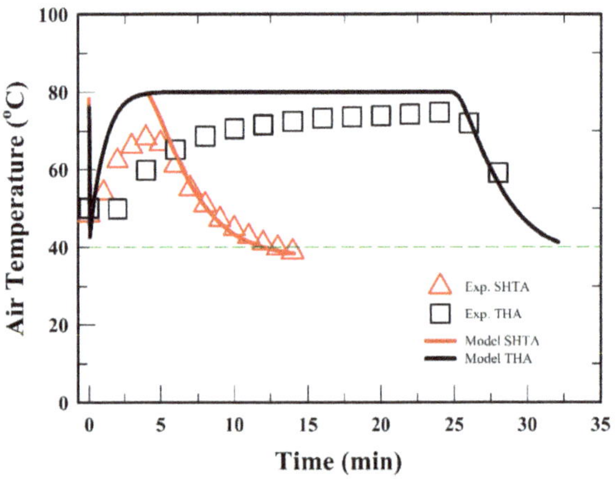

Figure 5. Profiles of air temperature during drying at 80 °C in the FBD chamber. Experimental data, SHTA (Δ) and THA (□); predicted values. SHTA (**red line**) and THA (**black line**).

4.3.3. Responses of the Simulated Thermal Input and the Absolute Humidity in the FBD Chamber

The simulated thermal input during drying at 80 °C in the FBD with SHTA and THA is shown in Figure 6. The thermal input increased with drying time. It was about 0.33 × 10^5 K × min at 8-min drying. At the end of drying process, the thermal input were 0.48 × 10^5 and 1.41 × 10^5 K × min with SHTA and THA, respectively. The thermal input of SHTA is one-third of THA. The above results indicated that the energy applied depended on the heating time, and SHTA could save more energy than THA.

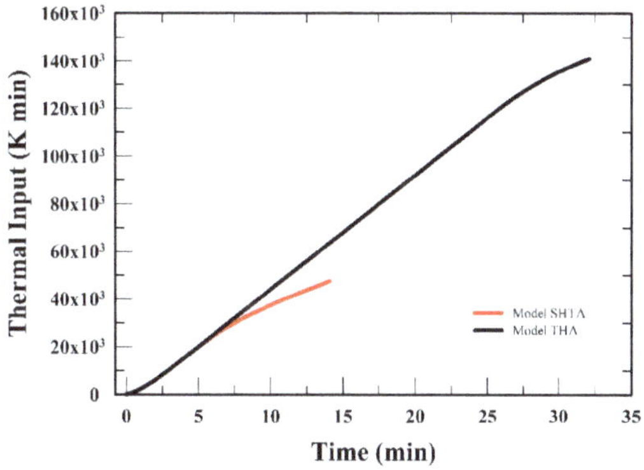

Figure 6. Simulated thermal input of heating germ during drying at 80 °C in the FBD. SHTA (**red line**) and THA (**black line**).

The water vapor in the WG particles would diffuse from the micro pores inside the germ to the air phase of the FBD chamber during drying. The simulated absolute humidity in the FBD chamber during WG drying at 80 °C with SHTA and THA are shown in Figure 7. The absolute humidity increased sharply and then decreased gradually during drying. The maximum absolute humidity was 0.032 and 0.033 kg/kg with SHTA and THA drying, respectively. The time for the absolute humidity in the FBD chamber to reach maximum was about 4.1 min and 4.8 min with SHTA and THA drying, respectively.

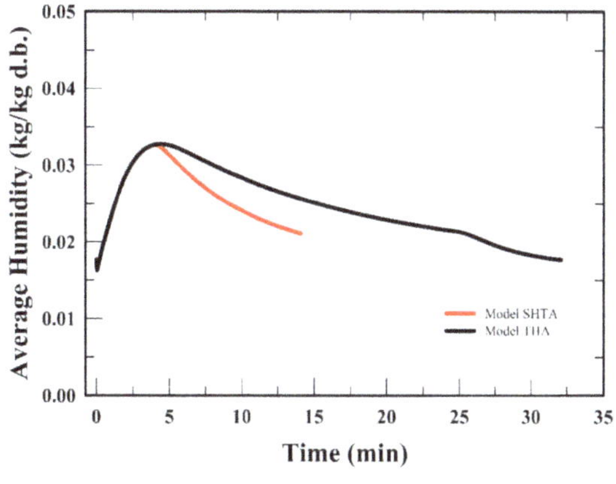

Figure 7. Simulated profiles of absolute humidity of air during drying at 80 °C in the FBD. SHTA (**red line**) and THA (**black line**).

4.4. Microscopic Behavior

4.4.1. Responses of the Simulated Wheat Germ Temperature

Simulation of the local temperature of WG during frying in the FBD with different approaches is important to control the bulk quality. The microscopic behavior of WG during drying could be evaluated by the mathematical model. The simulated WG temperature during drying at 80 °C in the FBD with SHTA and THA is shown in Figure 8. The maximum temperature of WG was 79.0 °C and 80.0 °C with SHTA and THA drying, respectively. The time required to reach the maximum temperature was 4.1 min and 5.8 min with SHTA and THA drying, respectively. The WG temperature slowly dropped at the cooling stage. According to Giner and Calvelo [34], the grain might damage when the heating temperature was above 65 °C. During drying in the FBD with SHTA, the temperature of WG was over 65 °C for 4.4 min. However, the WG temperature was over 65 °C for 25.8 min (1.6–27.4 min) during drying with THA. The above results indicated that the low-temperature drying with SHTA was better for WG stabilization. The product temperature of 40 °C was employed in the industrial drying. Comparing the results between SHTA and THA in Figure 8, the time needed to cool the dried WG to 40 °C was the same.

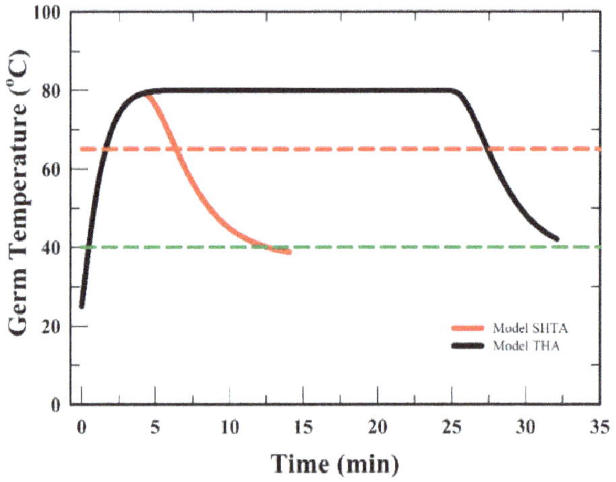

Figure 8. Simulated profiles of wheat germ temperature during drying at 80 °C in the FBD. SHTA (**red line**) and THA (**black line**).

4.4.2. Responses of the Simulated Moisture Distribution in Wheat Germ

The simulated profiles of MC distribution of WG during drying at 80 °C in the FBD are shown in Figure 9. The microscopic changes of MC distribution of WG particles during drying were barely presented in the literature. The MC of WG at different location decreased with drying time, and the rate of dehydration was higher in the germ surface than the center. At 4 min of the heating stage, the MC of WG surface was 5.6%, but the center was 6.9%. At the cooling stage, the MC of WG surface increased due to vapor absorbing, resulting in the condensation of water vapor on the WG surface. Meanwhile, the water vapor still diffused from the center to the surface of WG particle. At the end of cooling stage, the MC of WG surface was higher than that of WG center (Figure 9A). However, the condensation phenomena of WG dried with THA was not observed (Figure 9B).

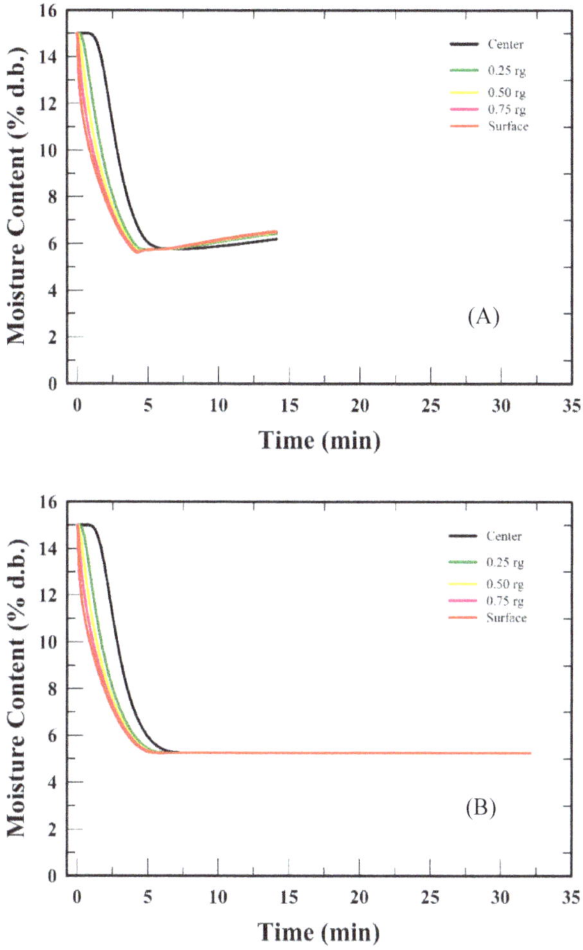

Figure 9. Simulated profiles of MC distribution of wheat germ during drying at 80 °C in the FBD. (**A**) SHTA; (**B**) THA.

According to the results in Figure 9, we postulated that a dynamic equilibrium of water vapor might exist between the WG surface and the emulsion phase (Figure 10). The condensation phenomena of WG dried with SHTA might be more obvious than that of WG dried with THA.

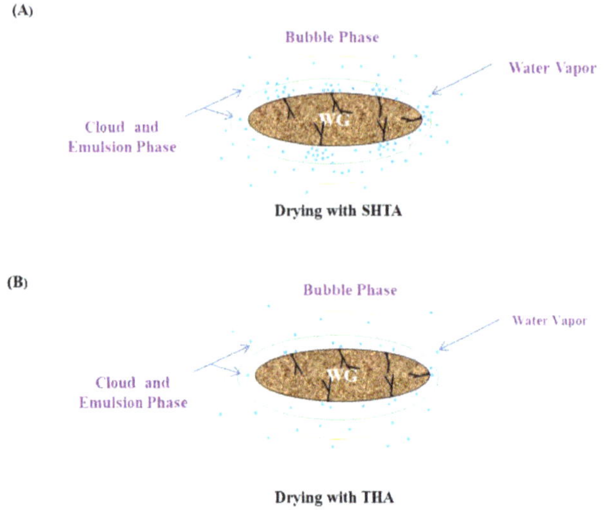

Figure 10. Schematic view of water vapor on the WG surface, emulsion phase, and bubble phase. (**A**) SHTA; (**B**) THA.

5. Conclusions

A mathematical model coupling with the macro-heat transfer model and the bubble model was developed in this study to simulate the dehydration and condensation phenomena during low-temperature wheat germ (WG) drying in the fluidized bed dryer (FBD). Changes in the moisture content (MC) of WG and the air temperature and humidity of FBD chamber during drying with short heating time approach (SHTA) and traditional heating approach (THA) were significantly different. The vapor condensation of WG occurred at the cooling stage of drying with SHTA. SHTA proposed in this study could control both temperature and MC of WG product to 40 °C and 7%, respectively, at the end of drying process. In contrast, the temperature of WG product dried with THA could be controlled at 40 °C while the MC of WG product was 3.5%. The thermal input of the drying process with SHTA was one-third of that of the drying process with THA. Therefore, the low-temperature drying in the FBD with SHTA is the best strategy for the WG stabilization.

Author Contributions: D.-S.C. developed the process model and computational simulations. J.S.C. performed the experiments. M.-I.K. was responsible for the conception and research ideas.

Funding: This research was funded by Texture Maker Enterprise Co., Ltd., Taiwan Contract NO. 600312.

Conflicts of Interest: The authors declare no conflict of interest.

Nomenclature

Symbol	Meaning (units)
A_g	surface area per unit volume of wheat germ (1/m)
C_{mb}	concentration of moisture in bubble phase (mol/m^3)
C_{me}	concentration of moisture in emulsion phase (mol/m^3)
C_{mg}	concentration of moisture in wheat germ (mol/m^3)
C_{sat}	saturation concentration (mol/m^3)
C_e	equilibrium moisture content (mol/m^3)
$C_{p,e}$	effective heat capacity (J/kg K)
$C_{p,j}$	heat capacity of the jth compound (J/kg K)

D_e	effective diffusivity of water in wheat germ (m^2/s)
D_{eg}	equivalent diameter of wheat germ (m)
D_o	pre-exponential factor (m^2/s)
E_a	activation energy of water diffusion (J/mol)
f_{con}	step numerical parameter for condensation
f_{de}	step numerical parameter for dehydration
H_f	height of fluidization (m)
h	convective heat transfer coefficient (W/(m K))
K_{be}	mass transfer coefficient between bubble and emulsion (1/s)
K_{con}	mass transfer coefficient for condensation (m/s)
K_{de}	mass transfer coefficient for dehydration (m/s)
k_g	thermal conductivity of wheat germ (W/m K)
M_w	molecular weight of water (kg/mol)
r_g	radius of wheat germ (m)
R_g	gas constant (J/mol K)
RH	relative humidity of air
T	temperature (K, unless it is specified in °C)
t	time (s)
r	r coordination (m)
x	x coordination (m)
WL	loading of wheat germ (kg)
X_{wi}	initial moisture content of wheat germ (%, d.b.)
Greek Symbols	
κ	thermal conductivity (W/m K)
φ	porosity of bed
ω	mass fraction
ρ	density (kg/m^3)
λ	latent heat of vaporization (J/mol)
Subscript	
ash	ash
b	bed
carb	carbohydrate
con	condensation
e	effective or emulsion
fat	fat
g	germ
h	heating
i	initial
in	inlet
j	jth compound
m	moisture
mb	moisture in bubble phase
me	moisture in emulsion phase
p	protein
sat	saturation
s	set
w	water

References

1. Mahmoud, A.A.; Mohdaly, A.A.A.; Elneairy, N.A.A. Wheat germ: An overview on nutritional value, antioxidant potential and antibacterial characteristics. *Food Nutr. Sci.* **2015**, *6*, 265–277. [CrossRef]

2. Zhu, K.X.; Lian, C.X.; Guo, X.N.; Peng, W.; Zhou, H.M. Antioxidant activities and total phenolic content of various extracts from defatted wheat germ. *Food Chem.* **2011**, *126*, 1122–1126. [CrossRef]
3. Li, B.; Zhao, L.; Chen, H.; Sun, D.; Deng, B.; Li, J.; Liu, Y.; Wang, F. Inactivation of lipase and lipoxygenase of wheat germ with temperature-controlled short wave infrared radiation and its effect on storage stability and quality of wheat germ oil. *PLoS ONE* **2016**, *9*, e0167330. [CrossRef]
4. Jha, P.K.; Kudachikar, V.B.; Kumar, S. Lipase inactivation in wheat germ by gamma irradiation. *Radiat. Phys. Chem.* **2013**, *86*, 136–139. [CrossRef]
5. Zou, Y.P.; Yang, M.; Zhang, G.; He, H.; Yang, T.K. Antioxidant activities and phenolic compositions of wheat germ as affected by the roasting process. *J. Am. Oil Chem. Soc.* **2015**, *92*, 1303–1312. [CrossRef]
6. Murthy, K.V.; Ravi, R.; Bhat, K.K.; Raghavarao, K.S.M.S. Studies on roasting of wheat using fluidized bed roaster. *J. Food Eng.* **2008**, *89*, 336–342. [CrossRef]
7. Jurkovic, N.; Colic, I. Effect of thermal processing on the nutritive value of wheat germ protein. *Die Nahrung.* **1993**, *37*, 538–543. [CrossRef] [PubMed]
8. Ferrara, P.J.; Ridge, R.D.; Benson, J.T. Method of producing shelf stable wheat germ. US Patent 5063079, 5 November 1991.
9. Srivastava, A.K.; Sudha, M.L.; Baskaran, V.; Leelavathi, K. Studies on heat stabilized wheat germ and its influence on rheological characteristics of dough. *Eur. Food Res. Technol.* **2007**, *224*, 365–372. [CrossRef]
10. Xu, B.; Zhou, S.L.; Miao, W.J.; Gao, C.; Cai, M.J.; Dong, Y. Study on the stabilization effect of continuous microwave on wheat germ. *J. Food Eng.* **2013**, *117*, 1–7. [CrossRef]
11. Xu, B.; Wang, L.K.; Miao, W.J.; Wu, Q.F.; Liu, Y.X.; Sun, Y.L.; Gao, C. Thermal versus microwave inactivation kinetics of lipase and lipoxygenase from wheat germ. *J. Food Process Eng.* **2015**, *39*, 247–255. [CrossRef]
12. Kermasha, S.; Bisakowski, B.; Ramaswamy, H.; Van De Voort, B. Comparison of microwave, conventional and combination heat treatments on wheat germ lipase activity. *Int. J. Food Sci. Technol.* **1993**, *28*, 617–623. [CrossRef]
13. Ali, S.; Usman, S.; Nasreen, Z.; Zahra, N.; Nazir, S.; Yasmeen, A.; Yaseen, T. Nutritional evaluation and stabilization studies of wheat germ. *Pak. J. Food Sci.* **2013**, *23*, 148–152.
14. Marti, A.; Torri, L.; Casiraghi, M.C.; Franzetti, L.; Limbo, S.; Morandin, F.; Quaglia, L.; Pagani, M.A. Wheat germ stabilization by heat-treatment or sourdough fermentation: Effects on dough rheology and bread properties. *LWT-Food Sci. Technol.* **2014**, *59*, 1100–1106. [CrossRef]
15. Fernando, W.J.N.; Hewavitharana, L.G. Effect of fluidized bed drying on stabilization of rice bran. *Drying Technol.* **1993**, *11*, 1115–1125. [CrossRef]
16. Shingare, S.P.; Thorat, B.N. Fluidized bed drying of sprouted wheat (*Triticum aestivum*). *Int. J. Food Eng.* **2014**, *10*, 29–37. [CrossRef]
17. Martinez, M.L.; Marin, M.A. Optimization of soybean heat-treating using a fluidized bed dryer. *J. Food Sci. Technol.* **2013**, *50*, 1144–1150. [CrossRef] [PubMed]
18. Yondem-Makascıoglu, F.; Gurun, B.; Dik, T.; Kıncal, N.S. Use of a spouted bed to improve the storage stability of wheat germ followed in paper and polyethylene packages. *J. Sci. Food Agric.* **2005**, *85*, 1329–1336. [CrossRef]
19. Hung, J.M. Investigation on the Effect of Fluidized-Bed Drying Processing on the Storage Stability of Wheat Germ. Master's Thesis, Fu-Jen Catholic University, Taiwan, 2017.
20. Yang, W.C. *Handbook of Fluidization and Fluid–Particle System*; Marcel Dekker: New York, NY, USA, 2003.
21. Levenspiel, O. *Chemical Reaction Engineering*, 3rd ed.; John Wiley & Sons, Inc.: New York, NY, USA, 1999.
22. Hemis, M.; Singh, C.B.; Jaya, D.S.; Bettahar, A. Simulation of coupled heat and mass transfer in granular porous media: Application to the drying of wheat. *Dry Technol.* **2011**, *29*, 1267–1272. [CrossRef]
23. Srivastava, V.K.; John, J. Deep bed grain drying model. *Energy Convers. Manag.* **2002**, *43*, 1689–1708. [CrossRef]
24. Giner, S.A.; De Michelis, A. Evaluation of the thermal efficiency of wheat drying in fluidized beds: Influence of air temperature and heat recovery. *J. Agric. Eng. Res.* **1988**, *41*, 11–23. [CrossRef]
25. Naghavi, Z.; Moheb, A.; Ziaei-Rad, S. Numerical simulation of rough rice drying in a deep-bed dryer using non-equilibrium model. *Energy Convers. Manag.* **2010**, *51*, 258–264. [CrossRef]
26. Vijayaraj, B.; Saravanan, R.; Renganarayanan, S. Studies on thin layer drying of bagasse. *Int. J. Energy Res.* **2007**, *31*, 422–437. [CrossRef]

27. Zare, D.; Minaei, S.; Mohamad Zadeh, M.; Khoshtaghaza, M.H. Computer simulation of rough rice drying in a batch dryer. *Energy Convers. Manag.* **2006**, *47*, 3241–3254. [CrossRef]
28. Proietti, N.; Adiletta, G.; Russo, P.; Buonocore, R.; Mannina, L.; Crescitelli, A.; Capitani, D. Evolution of physicochemical properties of pear during drying by conventional techniques, portable-NMR, and modeling. *J. Food Eng.* **2018**, *230*, 82–98. [CrossRef]
29. Mohapatra, D.; Rao, P.S. A thin layer drying model of parboiled wheat. *J. Food Eng.* **2005**, *66*, 513–518. [CrossRef]
30. Gili, R.D.; Martín Torrez Irigoyen, R.; Cecilia Penci, M.; Giner, S.A.; Ribotta, P.D. Physical characterization and fluidization design parameters of wheat Germ. *J. Food Eng.* **2017**, *212*, 29–37. [CrossRef]
31. Gili, R.D.; Martín Torrez Irigoyen, R.; Cecilia Penci, M.; Giner, S.A.; Ribotta, P.D. Wheat germ thermal treatment in fluidised bed. Experimental study and mathematical modelling of the heat and mass transfer. *J. Food Eng.* **2018**, *221*, 11–19. [CrossRef]
32. American Association of Cereal Chemists. *Approved Methods of Analysis*, 10th ed.; AACC International PRESS: St. Paul, MN, USA, 2000.
33. Srinivasakannan, C.; Balasubramanian, N. An Analysis on Modeling of Fluidized Bed Drying of Granular Material. *Adv. Power Technol.* **2008**, *19*, 73–82. [CrossRef]
34. Giner, S.A.; Calvelo, A. Modeling of wheat drying in fluidized bed. *J. Food Sci.* **1987**, *52*, 1358–1363. [CrossRef]
35. Romeo, T.T. *Fundamentals of Food Process Engineering*, 3rd ed.; Springer Asia Limited: Hong Kong, China, 2007.
36. Purlis, E. Baking process design based on modelling and simulation: Towards optimization of bread baking. *Food Control.* **2012**, *27*, 45–52. [CrossRef]
37. Christie, J.G. *Transport Processes and Separation Process Principles (Includes Unit Operations)*, 4th ed.; Goodwin, B., Ed.; Prentice Hall: Upper Saddle River, NJ, USA, 2003.
38. Brandolini, A.; Hidalgo, A. Wheat germ: Not only a by-product. *Int. J. Food Sci. Nutr.* **2012**, *63*, 71–74. [CrossRef] [PubMed]

processes

MDPI

Article

Model Development and Validation of Fluid Bed Wet Granulation with Dry Binder Addition Using a Population Balance Model Methodology

Shashank Venkat Muddu [1], Ashutosh Tamrakar [1], Preetanshu Pandey [2,†] and Rohit Ramachandran [1,*]

1 98 Brett Rd., Department of Chemical and Biochemical Engineering, Rutgers-The State University of New Jersey, Piscataway, NJ 08854, USA; shashank1992.venkat@gmail.com (S.V.M.); tamrakar.ashutosh@gmail.com (A.T.)
2 Drug Product Science and Technology, Bristol-Myers Squibb, New Brunswick, NJ 08854, USA; preetanshu@gmail.com
* Correspondence: rohit.r@rutgers.edu; Tel.: +1-848-445-6278; Fax: +1-732-445-2581
† Currently at Kura Oncology Inc., Drug Product Development, 3033 Science Park Rd., San Diego, CA 92121, USA.

Received: 15 June 2018; Accepted: 21 August 2018; Published: 1 September 2018

Abstract: An experimental study in industry was previously carried out on a batch fluid bed granulation system by varying the inlet fluidizing air temperature, binder liquid spray atomization pressure, the binder liquid spray rate and the disintegrant composition in the formulation. A population balance model framework integrated with heat transfer and moisture balance due to liquid addition and evaporation was developed to simulate the fluid bed granulation system. The model predictions were compared with the industry data, namely, the particle size distributions (PSDs) and geometric mean diameters (GMDs) at various time-points in the granulation process. The model also predicted the trends for binder particle dissolution in the wetting liquid and the temperatures of the bed particles in the fluid bed granulator. Lastly, various process parameters were varied and extended beyond the region studied in the aforementioned experimental study to identify optimal regimes for granulation.

Keywords: fluid bed granulation; heat and mass balance; population balance model; binder dissolution; kernel development

1. Introduction and Objectives

Fluid bed granulation (FBG) is a process widely used in the pharmaceutical industry to manufacture solid dosage products and its importance to the industry has resulted in it being extensively studied over the last few decades [1–3]. During an FBG unit operation, size enlargement of circulating primary powder particles takes place within the fluid bed chamber due to the wetting of particles by the sprayed binder solution and the resulting collisions between wetted particles. The particle wetting rates in an FBG are highly dependent on the particle flow pattern which in turn is affected by the flow field of the fluidizing gas [4]. Moreover, the flow behavior of the powder within the granulator also depends on the geometry of the vessel. Due to such inherent complexity, there has been a lack of science-based mathematical models that can be applied to better design and operate FBG processes at scale-up [5]. Several attempts are being made to understand the granular flow pattern, but many aspects of it (i.e., a more detailed study of the vessel geometry, fluid flow field, collision frequencies etc.) using sophisticated modeling tools and techniques are yet to be developed [6]. As the pharmaceutical industries are transitioning towards implementation of US Food and Drug

Administration (FDA) recommended Quality by Design (QbD) guidelines, there is a growing effort related to the development of practical and predictive models of particulate processes over the last decade to incorporate discrete particle simulations in the study of particle and particle–fluid flow.

Recently, fundamental modeling approaches such as discrete element modeling (DEM) and computational fluid dynamics (CFD) have become exceedingly popular in capturing particle level physics [4,6–15]. One such coupled DEM-CFD model for simulating a fluid bed granulation has been elucidated in detail in Tamrakar et al. [16]. However, employing such complex multi-phase simulations requires a steep computational cost. In DEM simulation which models the solid particle phase, for example, the computational algorithm needs to solve a set of momentum equations for each and every particle being handled in the system for very small time intervals (around 10^{-6} s scale) while also tracking their interactions and spatial movements. Similarly, the discretization of the fluid flow field region in CFD simulations into large number of cells to solve Navier Stokes energy and mass conservation equations requires substantial computational power. The high computational cost incurred for implementing a computationally complex model such as DEM-CFD renders it grossly inefficient while developing models for real scale geometries and for simulating actual time length of unit operations. Another significant challenge in the implementation of the mechanistic DEM-CFD model arises from its inability and inefficiency to simulate particle size and property changes resulting from the sub-processes in wet granulation. Since these mechanistic models do not directly relate the captured particle-scale properties back to meso-scopic/macro-scopic quality attributes traditionally utilized by the pharmaceutical standards (such as changes in particle size distribution, average liquid content, etc.), there is a disparity between the needs of the industry and the aims of academic projects. Industrial practitioners generally prefer relations that merely incorporate correction factors and predict properties like bed temperature and bed moisture without simulating the specific system. The same has been described in detail by Gupta in the chapter 'Fluid Bed Granulation and Drying' [17].

In order to deal with these limitations, an integrated modular meso/macro-scale model development is proposed that will contain the mechanics for binder dissolution, energy balances of the FBG, moisture balance of the granulator and the particulate formation mechanics. This model would be computationally inexpensive too as it does not take any data from DEM/CFD coupled models. This model can be tailor made to adapt to different unit operations and scenarios. The most general rate-based modeling technique used for simulating granulation processes are population balance models (PBMs) that can accommodate various granulation subprocesses (e.g., aggregation, consolidation and breakage) that result in changes to the particle property including size, porosity, liquid content, etc. [10,18–24]. PBM relies heavily on the different aggregation, breakage, consolidation, etc. rate expressions—also called rate kernels—which describe the sub-mechanisms involved in the size enlargement phenomena during granulation. By summing the effect of different sub-mechanisms for individual size class of particles, PBM is able to offer a net change information on the size of particles within the system in response to different operating conditions. PBMs are thus used to gain insight into the underlying granule growth/attrition rate processes. Dosta et al. [25], for instance, used a one-dimensional PBM to mathematically model FBG where simple fluid bed dynamics were coupled with heat and mass transfer. Traditionally, however, stochastic PBM methods model processes without particle transport, implying that the system is perfectly mixed. Tan et al. [10], on the other hand, developed a kinetic energy (EKE) kernel for PBM to describe the evolution of granule size distributions in fluidized bed melt granulation from kinetic theory of granular flow (KTGF). For heterogeneous granulation behaviors observed in fluidized beds, where all powder particles are not wetted to the same degree, the assumption of uniformity breaks down. In such circumstances, compartmental PBMs have been proposed to account for the heterogeneous nature of liquid addition and powder mixing [26,27]. Another important thing to note for PBM development is that the mechanistic information, used in calculation of aggregation and breakage kernels (i.e., effect of particle properties, spatial and size effect, degree of particle wetting etc.) in different size classes, are usually derived from empirical relations or particle-level models. PBM models, thus, cannot independently capture the detailed granulation behavior without empirical parameters inherent in its kernels. As a result, integrating PBM with binder dissolution mechanics and heat balances using first principles drying mechanics in a commercial programming interface would calculate the parameters in every step of the process and enable one to tune the empirical parameters to available experimental data that can be then used to validate and predict data at other operating conditions.

Objectives

The first goal of this work is to validate the available industrial data on mass fraction based particle size distributions (PSDs) and the derived geometric mean diameters (GMDs). Along with the mass balance equations, the energy balance relations have been developed in this work in order to get estimates on the temperature of the granulation bed over time. The changes in viscosity due to addition of liquid with pre-dissolved binder to the granular bed consisting of dry binder particles have been modeled by employing the binder dissolution model and viscosity prediction model that has been integrated into the aggregation mechanism of the PBM. Upon estimation of the empirical parameters in the model, the simulation design space has been extended beyond the region studied by the industrial experimenters. Conditions have been explored that lead to desired granulation with higher median diameter values (D_{50}) and undesired granulation regimes with a higher amount of fines despite having the optimum D_{50}.

2. Materials and Experimental Methods

The materials and experimental methods were performed by Pandey et al. [28] at Drug Product Science and Technology, Bristol-Myers Squibb, New Brunswick, NJ, USA and have been summarized below.

2.1. Materials

The experimental formulation consisted of active pharmaceutical ingredient (API) Lamivuidine, micro-crystalline cellulose (MCC-Avicel PH 102) as excipient, sodium starch glycolate (SSG) as the glidant, hydroxypropyl cellulose (HPC) as the binder and, lastly, magnesium stearate (MgSt) as the extra-granular lubricant. Of these components, Lamivuidine, MCC, SSG and HPC were present in the granulation blend and therefore have been considered for the modeling simulation study in this work. Although it was a multi-component granulation, the multiple components have been modeled as a singular granular blend solid in the subsequent section. In the granulation process, half of the HPC was pre-blended with the other excipients before granulation in the fluid bed; the other half was added as part of the sprayed binder solution.

2.2. Batch Design

Two formulations were evaluated using FBG: one containing 5% *w*/*w* SSG, and the other containing 10% *w*/*w* SSG. In order to account for the different SSG content in different designs of experiments' (DOE) settings, the content of the MCC excipient was adjusted accordingly (Table 1).

Table 1. Formulation compositions for fluid-bed wet granulation batches.

Formulation Ingredient	5% SSG % *w/w*	7.5% SSG % *w/w*	10% SSG % *w/w*
	Intra-granular		
Lamivudine (API)	70.00	70.00	70.00
MCC (Excipient)	20.25	17.75	15.25
SSG (super-disintegrant)	2.50	3.75	5.00
HPC (binder-dry)	2.00	2.00	2.00
HPC (binder-soln)	2.00	2.00	2.00
	Extra-granular		
SSG (super-disintegrant)	2.50	3.75	5.00
Mg Stearate (glidant)	0.75	0.75	0.75
Total	100	100	100

The mass percentages of the components in the dry granular pre-blend and after the incorporation of the dissolved wet binder onto the granules were accordingly computed as shown (Table 2).

Table 2. Intra-granular blend components' compositions.

Formulation	5% SSG		7.5% SSG		10% SSG	
Ingredient	% *w*/*w*		% *w*/*w*		% *w*/*w*	
	Dry blend	All HPC bound	Dry blend	All HPC bound	Dry blend	All HPC bound
Lamivudine	73.88	72.35	74.87	73.30	75.88	74.27
MCC	21.37	20.93	18.98	18.59	16.53	16.18
SSG	2.64	2.58	4.01	3.93	5.42	5.31
HPC	2.11	4.13	2.14	4.19	2.17	4.24
Total	100	100	100	100	100	100

A $2^{(4-1)}$ fractional factorial design of experiemnts (DOE) with two center points (10 experiments) was used to study the effect of one formulation factor (level of super-disintegrant) and three process factors (inlet air temperature, spray rate and atomization air pressure) on the moisture profile and granule growth rates of the FBG batches. The study design is listed in Table 3.

Table 3. Various design of experiemnts (DOE) settings for fluid bed granulation (FBG) batches.

Batch #	Atomization Air Pressure (Bar)	Disintegrant Amount % *w/w*	Inlet Air Temp. (°C)	Binder Spray Rate (g/min)
1	1	10	30	16
2	1	5	30	8
3	2.5	7.5	40	12
4	4	5	50	8
5	4	10	50	16
6	4	5	30	16
7	1	5	50	16
8	2.5	7.5	40	12
9	4	10	30	8
10	1	10	50	8

The system was modeled only for nozzle atomization pressure of 1 bar as PBMs cannot incorporate the effect of changing nozzle atomization pressures. Coupled CFD studies would be required for the same that incorporate the effect of liquid atomization pressures on the liquid flow patterns. Therefore, the PSD data of experimental runs 1, 2, 7 and 10 were used for calibrating the empirical parameters and validating the experimental data (75% training/calibration & 25% validation) in this study. The parameters obtained were then used for simulating the PSD growth behavior cases for which experimental data were not available at settings beyond the ones covered in the experimental DOE as shown in (Table 4). The study was expanded to include seven L/S ratios, three disintegrant amounts, three inlet air temperatures and three liquid spray rates. Therefore, the total number of runs in the extended simulations was 7 × 3 × 3 × 3 which is a total of 189 simulation cases.

Table 4. The DOE settings for other modeled FBG cases at 1 bar nozzle atomization pressure.

Varied Parameter	Parameter Values
Ratio of total liquid added to initial solid bed mass	0.1334, 0.2001, 0.2668, 0.3335, 0.4002, 0.4669 and 0.5336
Total SSG amt in formulation	5%, 7.5% and 10%
Inlet air temperature during liquid addition	30 °C, 40 °C and 50 °C
Liquid spray rate	8 g/min, 12 g/min and 16 g/min

2.3. Fluid Bed Granulation

The experimentation performed by the research group Pandey et al. [28] on the fluidized bed has been described in this section. Fluid-bed wet granulation was performed in a GPGC-2 (Glatt Air Techniques Inc., Ramsey, NJ, USA) with an integrated peristaltic pump. Inlet air temperature and velocity were controlled during processing. A total of 667 g of binder solution was delivered to the blend through a 1 mm spray nozzle positioned in the top spray configuration. Intra-granular components were pre-blended in a 10 L bin blender for 125 revolutions (5 min at 25 rpm). An initial 2 kg mass of the pre-mixed blend was charged into the fluid bed, and pre-warmed using inlet air at the target temperature for 5 min at an initial air flow velocity of 40 m^3/h. The addition of binder solution spray was started after this pre-warm time. During the fluid bed granulation process, airflow velocity was increased by 10 m^3/h for every 100 g of binder solution added. When the desired amount of binder solution had been sprayed, inlet air temperature was increased to 60 °C to facilitate drying of the granulation. Drying continued until the granule moisture content was below 1.5%.

2.4. Particle Size Distribution

The particle size distributions of the initial and final granular material blends were measured by sieve analysis. The screen sizes used for analysis were as follows: 885 μm, 505 μm, 335 μm, 213 μm, 141 μm, 79 μm and 26.5 μm. The results were reported as mass percentages of material retained on each of the meshes.

3. Mathematical Model Development

3.1. Particle Grid Configuration

As the granular bed contains both solid formulation particles and liquid particles in varying proportions, the particles have been defined on the basis of their solid and liquid contents which are the two key coordinates to describing all of a particle's dimensions and physical properties such as density, wetted area, etc. The size of the particles have been defined and tracked by the available solid volume content s and liquid volume content l in each different class or 'bin' of particles. The solid content and liquid content have been varied independently in a geometric progression. The ratio of the progression has been kept roughly the cubed value of the ratio of sieve size between two consecutive meshes. As eight individual volume amounts of solid and liquid each have been chosen that defined a particle's contents, there are distinctly 16 different classes of particles in the model and each class is defined by the respective solid and liquid volume content. Therefore, the grid of particles is a two-dimensional (2D) grid and the resulting population balance model developed too is a 2D PBM.

3.2. Particle Wetness Classification

The particles have been modeled as solid spheres consisting of the granular blend composition materials with liquid adhered onto the solid surface. Depending on the volumetric amount of liquid to solid ratio in each particle/bin class, the liquid on the solid has been been modeled either as a concentric layer or as a small circular patch on the spherical solid particle. The wetting behavior of the liquid drop on the solid sphere is determined by the contact angle value. An expression for the radius of the wet patch in case of liquid addition to powder blend consisting of pre-mixed dry binder particles previously used in [29] has been adapted in this work. The expression for the base radius of the wet patch is given as:

$$R_{wet} = \left(\frac{3\Pi(\theta)l}{\pi}\right)^{\frac{1}{3}}, \tag{1}$$

where $\Pi(\theta)$ is a function of the contact angle θ of the liquid on the solid particle, and l is the liquid volume of the particle defined by the grid location (s, l). The expression for the contact angle function is given as:

$$\Pi(\theta) = \left(\frac{\sin^3\theta}{2 - 3\cos\theta + \cos^3\theta}\right). \tag{2}$$

The area of the wet patch is therefore given as:

$$Area_{wet} = \pi \times R_{wet}^2. \tag{3}$$

The fraction of the particle surface covered by liquid has been mathematically as follows in terms of the total external particle surface area and the area of the corresponding wet liquid patch on the particle. It is given as follows:

$$\alpha_{wet} = \frac{Area_{wet}}{Area_{particle}}, \tag{4}$$

where $Area_{particle}$ is the external surface area of the particle. It must be noted that the wet-patch model calculating the radius of the wet patch on the solid is applicable only to particles in those grids where corresponding R_{wet} is less than the radius of the solid sphere of the particles. In the cases where the condition doesn't hold true, the particles have been modeled assuming that the liquid covers the solid particle entirely as a concentric layer. In these cases, it can be seen that the fractional wet area of the particle becomes equal to one ($\alpha_{wet} = 1$).

3.3. PBM Configuration

The material balance in the fluid bed granulator (FBG) has been modeled in terms of population balance models (PBMs) which track the change in particle size over time for different classes. PBMs are the established framework for particulate systems with distinct and evolving particle size distributions (PSDs). Moreover, PBMs are nearly indispensable where the rate processes depend on the particle sizes and compositions. Therefore, a semi-mechanistic PBM model has been developed in which the equations were developed from the first principles, but some experimental data and/or material properties were used as input parameters to the rate equations. The granulation process took into account the rate processes of the following phenomena: aggregation of the smaller particles into larger granules, liquid addition due to binder spraying and liquid evaporation due to hot air drying. A basic 2D PBM equation depending on the particle properties (solid and liquid contents) has been formulated as follows:

$$\frac{\partial N(s,l,t)}{\partial t} + \frac{\partial}{\partial l}\left[N(s,l,t)\frac{dl}{dt}\right] = \Re_{agg}(s,l,t), \tag{5}$$

where $N(s, l, t)$ is the number of particles in bin class (s, l) at time point t, $\frac{dl}{dt}$ accounts for change in liquid content of the particle due to liquid addition and evaporation and $\Re_{agg}$ is the aggregation rate of particles of class (s, l, t) at time point t. The net aggregation for a general particle of bin class (s, l) at time t is defined as follows:

$$\Re_{agg}(s,l,t) = \Re_{agg}^{form}(s,l,t) - \Re_{agg}^{dep}(s,l,t), \tag{6}$$

where $\Re_{agg}(s,l,t)$ is the net aggregation rate of any particular particle class, $\Re_{agg}^{form}(s,l,t)$ is the rate of formation of a particle due to aggregation during a binary collision of particles, and $\Re_{agg}^{dep}(s,l,t)$ is the rate of depletion of particles due to collision with other particles. The rate of formation of a particle of class (s,l) due to aggregation of two smaller particles at a given time instant t is given similar treatment as a kinetic reaction and thus has been defined as follows:

$$\Re_{agg}^{form}(s,l,t) = \frac{1}{2}\sum_{s'=s_{min}}^{s}\sum_{l'=l_{min}}^{l}\beta(s',s-s',l',l-l',t)N(s',l',t)N(s-s',l-l',t), \tag{7}$$

where s_{min} and l_{min} are the respective minimum solid and liquid volumes for a particle, and $\beta(s',s-s',l',l-l',t)$ indicates the specific aggregation rate of any two particles whose net respective bin volumes equate to (s,l). The above equation takes into account all the possible ordered pair combinations such that the net volume of the colliding particles is equal to the volume of the particle use formation rate is being tracked. The product of the aggregation kernel and the number of particles of each colliding ordered pair is multiplied by half as each possible combination is counted twice while performing the ordered pair multiplication. The rate of depletion of a particle due to aggregation of the particle with another at a given time instant t is given as follows:

$$\Re_{agg}^{dep}(s,l,t) = N(s,l,t)\sum_{s'=s_{min}}^{s}\sum_{l'=l_{min}}^{l}\beta(s',s,l',l,t)N(s',l',t). \tag{8}$$

It can be seen from Equation (6) that the aggregation of particles is overall a second order process, and is directly proportional to the number/quantity of each colliding particle bin/ size class.

3.4. Aggregation Kernel Development

The aggregation rate kernel is akin to a kinetic reaction rate constant. However, in the PBM model developed in our study, the kernel rather depends on the properties of the colliding particles/granules. The kernel that was developed in [29] has been adapted in this work and has been described as follows:

$$\beta(s',s-s',l',l-l',t) = \beta_0 \times \psi_{physical}(s',s-l',l',l-l',t) \times \psi_{geometric}(s',s-s',l',l-l'), \tag{9}$$

where β_0 is an empirical rate constant, $\psi_{physical}(s',s-s',l',l-l',t)$ is a physical success parameter and $\psi_{geometric}(s',s-s',l',l-l')$ is a geometric success parameter for aggregation to occur. The physical and geometric success parameters have been developed along the similar lines as in [7]. The physical success factor for aggregation depends on the Stokes number and critical Stokes number of the system. The Stokes number for any ordered pair of particles was defined by [30] as follows:

$$St_{ij} = \frac{u_o\rho}{18\mu}\left(\frac{dia_i dia_j((dia_i+dia_j)^2)}{dia_i^3+dia_j^3}\right), \tag{10}$$

where St_{ij} is the Stokes' Number for an ordered pair of colliding particles i and j, respectively, u_o is the magnitude of the velocity of the colliding particles, μ is the viscosity of the liquid binder film wetting/covering the surface of the particles, ρ is the density of the solid component of the particles, and dia_i and dia_j are the diameters of the solid spherical parts of the colliding particles i and j, respectively. The magnitude of the velocity of the particles u_o is given by the expression from [7] as:

$$u_o = \frac{3}{2}\sqrt{\pi \times T_{bed}}, \tag{11}$$

where T_{bed} is the average temperature of the bed of particles in the fluid bed granulator. The critical Stokes' Number has been defined as follows:

$$St^* = \left(1 + \frac{1}{e_{rest}}\right) ln \left(\frac{h_{o,i} + h_{o,j}}{h_{a,red}}\right), \tag{12}$$

where e_{rest} is the coefficient of restitution of the colliding particles, $h_{o,i}$ and $h_{o,j}$ are the depth of the liquid on the solid particle surfaces for the ordered pair of particles i and j, respectively, and, lastly, $h_{a,red}$ is the harmonic mean of the asperity values of the particles i and j. The value of the physical success factor $\psi_{physical}$ is binary and is determined by the comparison of the Stokes' number St_{ij} of the ordered pair of particles with the critical Stokes's number St^* for the same ordered set. It has been defined as follows:

$$\psi_{physical} = \begin{cases} 1, & St \leq St^*, \\ 0, & otherwise. \end{cases} \tag{13}$$

The geometric success factor for aggregation upon collision of two particles i and j depends upon the fractional wetted surface area α_{wet} in accordance with [7] as follows:

$$\psi_{geometric} = 1 - (1 - \alpha_i) \times (1 - \alpha_j). \tag{14}$$

The above expression in Equation (14) takes into account the possibility of collisions in all the combinations of directions: wet surface on wet surface and wet surface on dry surface of particles.

3.5. Liquid Balance of the Particle Bed

The liquid balance for the particles due to liquid addition and aggregation has been developed as follows:

$$-\frac{\partial}{\partial l}\left[N(s,l,t)\frac{dl}{dt}\right] = N_{l,spray}(s,l,t) + N_{l,evap}(s,l,t). \tag{15}$$

Here, $N_{l,spray}$ is the rate of particles being formed due to spray liquid addition into the bin (s,l) and $N_{l,evap}$ is the rate of particles being depleted due to liquid evaporation from the the bin (s,l) at time t. The net rate of particles being formed due to liquid addition into the bin (s,l) can be described as follows:

$$N_{l,spray}(s,l,t) = N_{l,spray,form}(s,l,t) - N_{l,spray,dep}(s,l,t), \tag{16}$$

where $N_{l,spray,form}$ is the rate of particles formed in bin (s,l) and $N_{l,spray,dep}$ is the rate of particles lost from bin (s,l) at time t due to the liquid sprayed. The rate of particles being formed in bin (s,l,t) due to liquid addition is given as follows:

$$N_{l,spray,form}(s,l,t) = \frac{\dot{v}_{spray}}{(v_l(l) - v_l(l-1))} \times \frac{N(s,l-1,t-1)v_s(s)}{\sum v_s(s)N(s,:,t-1)}, \tag{17}$$

where the particles are being formed in bin (s,l) due to liquid addition in bin $(s,l-1)$. Here, it is assumed that the spray liquid is being distributed according to the volume of the solid content in the bin $(s,l-1)$. It is to be noted that the number of particles $N(s,l-1,t-1)$ is used, as the calculations for the time point t are done using the data available at time point $(t-1)$. This is in accordance with the 1st order forward difference principles being used in the discretization of all the partial differential equations in the modeling of the system. The discretization has been described further in Section 3.8. The rate of particles being depleted in bin (s,l,t) due to liquid addition is given in a similar vein as follows:

$$N_{l,spray,dep}(s,l,t) = \frac{\dot{v}_{spray}}{(v_l(l+1) - v_l(l))} \times \frac{N(s,l,t-1)v_s(s)}{\sum v_s N(t-1)}, \tag{18}$$

where particles are moving from bin (s, l) to bin $(s, l+1)$ due to liquid addition. The net rate of particles being formed due to liquid evaporation from the bin (s, l) can be described as follows:

$$N_{l,evap}(s,l,t) = N_{l,evap,form}(s,l,t) - N_{l,evap,dep}(s,l,t), \tag{19}$$

where $N_{l,evap,form}$ is the rate of particles formed in bin (s, l) and $N_{l,evap,dep}$ is the rate of particles lost from bin (s, l) at time t due to the liquid being evaporated from convective heat transfer from the surrounding air in the granulator. The rate of particles being formed in bin (s, l, t) due to liquid evaporation is given as follows:

$$N_{l,evap,form}(s,l,t) = \frac{\dot{m}_{evap}(s,l+1)}{\rho(s,l+1)} \times \frac{N(s,l+1,t-1)}{(v_l(l+1) - v_l(l))}, \tag{20}$$

where the particles are being formed in bin (s, l) due to liquid evaporation from bin $(s, l+1)$. Here, $\dot{m}_{evap}(s, l+1)$ is the evaporation rate of liquid from one particle of bin $(s, l+1)$. The rate of particles being formed in bin (s, l, t) due to liquid evaporation is given as follows:

$$N_{l,evap,dep}(s,l,t) = \frac{\dot{m}_{evap}(s,l)}{\rho(s,l)} \times \frac{N(s,l,t-1)}{(v_l(l) - v_l(l-1))}, \tag{21}$$

where the particles are being depleted from bin (s, l) into the bin $(s, l-1)$ due to liquid evaporation from bin (s, l) The bed evaporation rate $(\dot{m}_{evap})$ is calculated as [31]:

$$\dot{m}_{evap} = A_{wet} k \rho_{air}(x_{sat} - x_{out}), \tag{22}$$

where ρ_{air} is the density of air, x_{sat} and x_{out} are the saturation moisture content and moisture content of the air leaving the compartment, respectively. It is to be noted that, technically, the evaporation rate would be proportional to the difference between the saturated moisture content and the average moisture content in the air inside the fluid bed granulator. However, in this work, the outlet air moisture content has been assumed to be the same as the average moisture content inside the granulator. This is due to the lumped vessel approach used for modeling where the spatial variation of the granule and air properties have not been considered.

The saturated moisture content of the vapor in the air is calculated from the saturated pressure equation as follows:

$$x_{sat} = \frac{0.621}{1013325 - P_{sat}}, \tag{23}$$

where it has been assumed that the fluid bed granulator is operated at 1 atm pressure and P_{sat} is the saturated pressure of water vapor in Pascals at temperature T_{air}. The saturated pressure of water vapor P_{sat} is given as a relation of the air temperature T_{air} by Antoine's relation as follows:

$$P_{sat} = exp\left(21.8261 - \frac{3130.7984}{T_{air} - 70.6573}\right). \tag{24}$$

The Sherwood Number Sh of water in a fluidised bed is related to the Schmidt number Sc and Reynolds number by the following relation given in [31]:

$$Sh = 2 + 0.6(Sc)^{1/3}(Re)^{1/2}, \tag{25}$$

where Sc is the Schmidt number and Re is the Reynolds number. Furthermore, at low fluidization velocity, the Sherwood number can be assumed to be 2 in value. The Sherwood number is physically defined as the ratio of convective mass transfer rate to the diffusive mass transfer rate. It is given in the following expression:

$$Sh = \frac{dia_p k}{D_{water}}, \tag{26}$$

where dia_p is the characteristic length for diffusion i.e., the diameter of a particle in the fluidized bed, k is the drying coefficient of water and D_{water} is the diffusion of coefficient of water. Therefore, the coefficient of mass transfer of the moisture from the particle to the surrounding air due to diffusion is given from the above expressions as:

$$k(s,l) = \frac{2D_{water}}{dia_p(s,l)}. \tag{27}$$

D_{water} is the diffusivity of water and dia_p the particle diameter of the corresponding bin (s,l).

3.6. Integration of Heat Balance and Liquid Evaporation Due to the Heating of Particle Bed

The heat and liquid evaporation model coupled with population balance model helps in tracking the variation in bed temperature and outlet air temperature in the fluid bed granulation. This model is developed by considering the energy balance of bed along with the energy balance of the system air inside the granulator. In order to estimate drying or evaporation rates under different temperature and relative humidty conditions, the concept of "drying coefficient" is applied [32]. This model predicts the bed temperature, air outlet temperature and air outlet moisture content as a function of time in the fluid bed granulator. The equations were developed by using the film model as used in [14,25,33]. The rate of heat transfer from the bed to the spray liquid due to the cooling effect of the binder Q_{spray} is given as:

$$Q_{spray} = \dot{m}_{spray} C_{p,liquid}(T_{spray} - T_{bed}), \tag{28}$$

where $C_{p,liquid}$ is the heat capacity of the binder spray liquid and $(T_{spray} - T_{bed})$ is the difference in temperature between the binder spray liquid and the particles, respectively. The temperature of the inlet spray liquid T_{spray} is a given input parameter constant at 25 °C (298 K). The heat lost due to cooling by the binder spray liquid is included subsequently in the modeling of particle bed temperature in Equation (36). The expression for the heat transferred from the surrounding air inside the granulator to the bed of particles is given as follows:

$$Q_t = hA_{bed}(T_{air} - T_{bed}), \tag{29}$$

where A_{bed} is the total external surface area of all the particles in the bed, T_{air} is the temperature of the air inside the granulator and h is the convective heat transfer coefficient. The convective heat transfer is calculated from the diffusive bulk heat transfer coefficient in a manner similar to the mass transfer coefficient.The expression for the same is given as follows:

$$h = \frac{2k_t \sum\sum (N(s,l,t). \times dia_p(s,l)^2)}{\sum\sum (N(s,l,t). \times dia_p(s,l)^3)}, \tag{30}$$

where $2k_t$ is the heat transfer conductivity of moisture in air. The conductivity is described as a function of the air moisture content x_{air}:

$$k_t = 0.0243\sqrt{\frac{x_{air}}{273.2}}. \tag{31}$$

The total mass balance equation of the dry part of the air in the fluid bed granulator can be expressed as:

$$\begin{aligned}\frac{dM_{air,dry}}{dt} &= \dot{m}_{air,dry}[(1-x_{in})-(1-x_{air})] \\ &\quad - \sum_{all\ s}\sum_{all\ l}\left(\frac{dN}{dt}-N_{l,evap}\right)\rho_{air}(1-x_{air}),\end{aligned} \tag{32}$$

where $\dot{m}_{air,dry}$ is the mass flow rate of the air flowing through the granulator, and ρ_{air} is the density of the air at the temperature T_{air}. x_{in} is the moisture content of the inlet air and x_{air} is the moisture content of the air inside the granulator and the outlet air. Here, the moisture content of the air inside the system is equivalent to the moisture content of the air flowing out owing to the lumped system approach applied in the energy balance equations. The mass balance expression for the total water vapor present in the air inside the granulator is given similarly as:

$$\begin{aligned}\frac{dM_{air,wet}}{dt} &= \dot{m}_{air,wet}[x_{in}-x_{air}] \\ &\quad - \sum_{all\ s}\sum_{all\ l}\left(\frac{dN}{dt}-N_{l,evap}\right)\rho_{air}x_{Air}+M_{evap}.\end{aligned} \tag{33}$$

Here, the term M_{evap} is the total evaporation rate of the moisture from the bed that is obtained by summing the product of evaporation rate m_{evap} and the number of particles in the corresponding bins N over all the bins. The expression for the density of air ρ_{air} is given from the ideal gas approximation as:

$$\rho_{air} = \frac{(101.325\times 28.96)}{8.314\times T_{air}}. \tag{34}$$

The total energy balance equation of the air in the fluid bed granulator can be expressed as:

$$\begin{aligned}\frac{dT_{air}}{dt} &= \frac{\dot{m}_{air}(C_{p,air}((1-x_{in})T_{in}-(1-x_{air})T_{air}))-Q_t}{(M_{air,dry}C_{p,air}+M_{air,wet}C_{p,water\ vap})} \\ &\quad -\left(\frac{dM_{air,dry}}{dt}C_{p,air}+\frac{dM_{air,wet}}{dt}C_{p,water\ vap}\right) \\ &\quad \times\frac{T_{air}}{M_{air,dry}C_{p,air}+M_{air,wet}C_{p,water\ vap}},\end{aligned} \tag{35}$$

where $C_{p,air}$ is the heat capacity of the solid components in the particle bed and $C_{p,water\ vap}$ is the heat capacity of water vapor in the system. The total energy balance equation for the particle bed of the fluidized granulation system is as follows:

$$\begin{aligned}\frac{dT_{bed}}{dt} &= \frac{Q_t+Q_{spray}-M_{evap}\lambda_{vap}}{\sum\left(m_{particle,dry}NC_{p,solid}\right)+\sum\left(m_{particle,wet}NC_{p,water}\right)} \\ &\quad -\left(\frac{dN}{dt}m_{particle,dry}C_{p,solid}+\frac{dN}{dt}m_{particle,wet}C_{p,water}\right) \\ &\quad \times\frac{T_{bed}}{\sum\left(m_{particle,dry}NC_{p,solid}\right)+\sum\left(m_{particle,wet}NC_{p,water}\right)},\end{aligned} \tag{36}$$

where λ_{vap} is the latent heat of vaporization of water, $C_{p,solid}$ is the heat capacity of the solid components in the particle bed and $C_{p,liquid}$ is the heat capacity of the liquid water.

3.7. Binder Dissolution Module

The dissolution model for the dry binder particles in liquid solvent was developed according to the work published in [29,34]. The expression for the change in radius of the binder due to dissolution is given as follows:

$$\frac{dRad_{binder}}{dt} = -\frac{D_{HPC}}{\rho_{HPC}}(C_{sat,HPC} - C_{bulk,HPC})\left(\frac{1}{y_{binder}} + \frac{1}{Rad_{binder}}\right), \quad (37)$$

where Rad_{binder} is the radius of the binder, D_{HPC} is the diffusion coefficient of the binder component HPC in water, ρ_{HPC} is the density of the HPC binder, $C_{sat,HPC}$ is the maximum possible concentration of the HPC binder in the liquid solvent under saturation conditions, $C_{bulk,HPC}$ is the current concentration of the HPC binder in the liquid solvent and y_{binder} is the diffusion layer thickness for the HPC binder.

As the system experimented upon and modeled contains half of the binder dissolved in the liquid solvent prior to granulation, the expression for the bulk concentration of the binder is therefore formulated as:

$$C_{bulk,HPC} = \frac{0.5 \times m_{HPC,initial} + \rho_{HPC} N_{HPC} \frac{4}{3}\pi(Rad^3_{binder,0} - Rad^3_{binder})}{Liquid_{total}}, \quad (38)$$

where $m_{HPC,initial}$ is the initial mass of HPC binder in the granular bed, N_{HPC} is the estimated number of binder particles, $Rad_{binder,0}$ is the initial radius of the binder particles and $Liquid_{total}$ is the total amount of liquid present in the system at a particular instance.

3.8. Numerical Techniques

While the aggregation and liquid phenomena happen during liquid addition and wet granulation, the growth rate of the mass of particles is linear as the liquid was sprayed at a constant linear rate during the experimental conditions. However, as mentioned in Section 3.1, the grid of particle sizes used for developing the PBM is exponential in nature. Therefore, the number of particles was allocated in the appropriate grid locations, and the cell average technique as developed in [35] was employed where lever rule techniques are used to distribute particles in nearest grid locations by linear interpolations.

Since the various equations formulated in the model for emulating the system consist of differential equations, an appropriate discretization technique must be utilized in order to solve them over time given the initial conditions. Therefore, the solution comes down to an initial-value problem of a few coupled differential equations. The differential equations were discretized using Euler's first order finite forward difference method where the value of the unknown property at a subsequent time-step is given by the value at the present time-step, the rate expression governing the property and the time-step chosen.

The issue arising during development of the numerical solution of the coupled differential equations over the process time is the value of the time-step chosen for the simulations. An adaptive time-step method was chosen where the time-step or interval of discretization is computed according to the CFL (Courant-Fredrichs-Lewis) condition. The time-step chosen was such that the interval was small enough to come well within the bounds of the CFL conditions for all the differential equations involved in the computation.

4. Results and Discussion

4.1. Training and Validation of the Model to Experimental Data

The experimental runs simulated are No. 1, 2 and 10 for training the model to the data and consequently arrive at the optimal parameters for the aggregation kernel constant β_o, the coefficient of restitution for binary collision of particles e_{rest} and the contact angle θ. The values of the various constants and parameters used in the integrated model have been given below in Table 5.

Table 5. The values of various constants and parameters.

Symbol	Meaning	Parameter Value
θ	Contact Angle	42.7 °
β_o	Aggregation kernel constant	6.87 Number$^{-1}\cdot$s^{-1}
e_{rest}	Coefficient of restitution	0.326
$C_{p,air}$	Heat Capacity of air	1006 Jkg$^{-1}\cdot$K^{-1}
$C_{p,liquid}$	Heat Capacity of liquid water	4187 Jkg$^{-1}\cdot$K^{-1}
$C_{p,watervap}$	Heat Capacity of water vapor	1996 Jkg$^{-1}\cdot$K^{-1}
$C_{p,solid}$	Heat Capacity of solids	1000 Jkg$^{-1}\cdot$K^{-1}
λ_{evap}	Latent heat of evaporation for water	2.27×10^{6} J$\cdot$kg^{-1}
D_{HPC}	Diffusivity of HPC	8.3×10^{-9} m$^{2}\cdot$s^{-1}
ρ_{HPC}	Density of HPC	500 kg$\cdot$m^{-3}
$C_{sat,HPC}$	Saturated concentration of HPC	330 kg$\cdot$m^{-3}
$Rad_{binder,o}$	Initial radius of HPC binder particles	49.375 μm
x_{in}	Moisture content of inlet air	1.5% w.r.t. dry air

The model PSD and GMD results have been compared against their experimental counterparts for the aforementioned cases.

The prediction of the PSDs against the corresponding experimental values have been shown in Figure 1. During experimentation, the granules were sampled at 200 g (red curves), 400 g (pink curves) and 667 g (blue curves) of liquid added. The granules were sieved and the mass fraction based PSDs were obtained accordingly.

It is seen from the figures that the mass fraction of material in larger bins were overpredicted in the model when compared to the experimental data. However, the PSD trends were closer to the experimentally reported values in the validation Set 7 (see Figure 2) for the same tunable model parameters. This leads us to infer that the overprediction in the training experimental data (Set 1, 2 and 10) may plausibly be due to under-measurement of the experimental PSD data given by the authors in [28]. However, it must be noted that all four of these sets were run only once without any duplicates/triplicates by the authors responsible for the generation of experimental data in Pandey et al. [28].

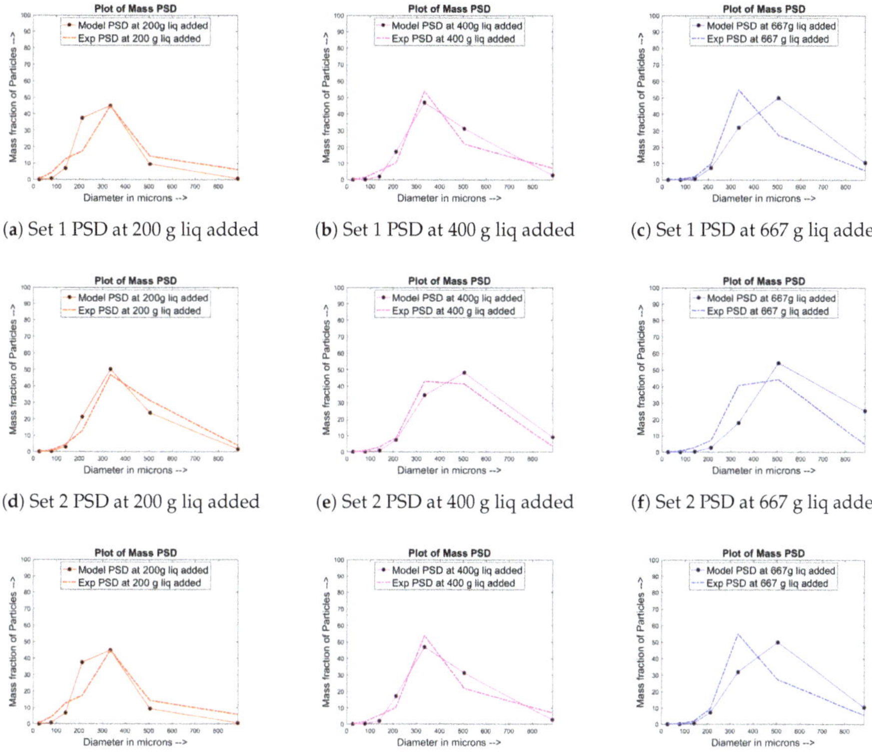

(**a**) Set 1 PSD at 200 g liq added (**b**) Set 1 PSD at 400 g liq added (**c**) Set 1 PSD at 667 g liq added

(**d**) Set 2 PSD at 200 g liq added (**e**) Set 2 PSD at 400 g liq added (**f**) Set 2 PSD at 667 g liq added

(**g**) Set 10 PSD at 200 g liq added (**h**) Set 10 PSD at 400 g liq added (**i**) Set 10 PSD at 667 g liq added

Figure 1. Experimental vs. model particle size distributions (PSDs) for Sets 1, 2 and 10 that were used for training the model.

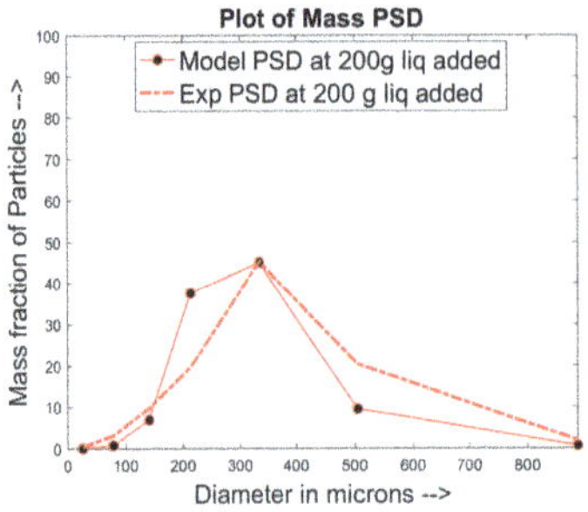

(**a**) Mass fraction based PSD at 200 g liq added

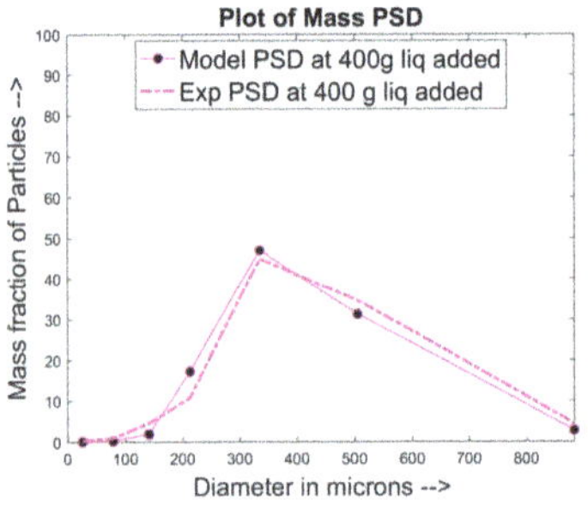

(**b**) Mass fraction based PSD at 400 g liq added

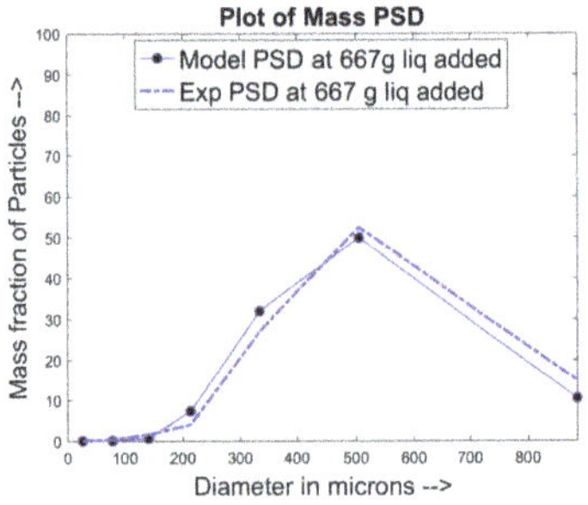

(**c**) Mass fraction based PSD at 667 g liq added

Figure 2. Experimental vs. model PSDs for validation Set 7.

The previously shown distributions were one-dimensional distributions where the evolution of particle size trends were shown with respect to their sizes at various points of liquid addition. Another interesting approach to look at these results would be to classify them both with particle size and liquid content. The charts in Figure 3 showcase the same for the readers' interest.

From Figure 3a,b, it can be that less particles are formed containing higher liquid mass content. This may probably have been due to the higher spray rate of the granulation liquid as the particles received more liquid coming onto them each time.

On the other hand, the particle mass frequency in the lower spray rate conditions were higher in the lower liquid content regions. Therefore, we observe from these figures that one must ideally aim for lower liquid spray rates to have lower liquid content at the end of granulation.

The simulated GMDs were also compared against their model counterparts and have been shown in the Figures 4 and 5.

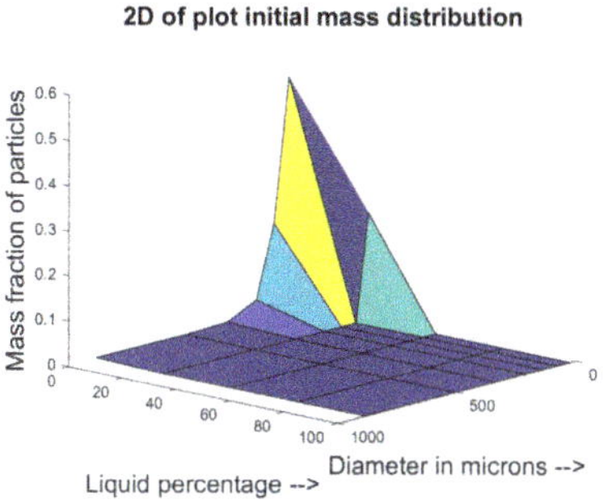

(**a**) Normalised mass frequency distribution of blend before granulation

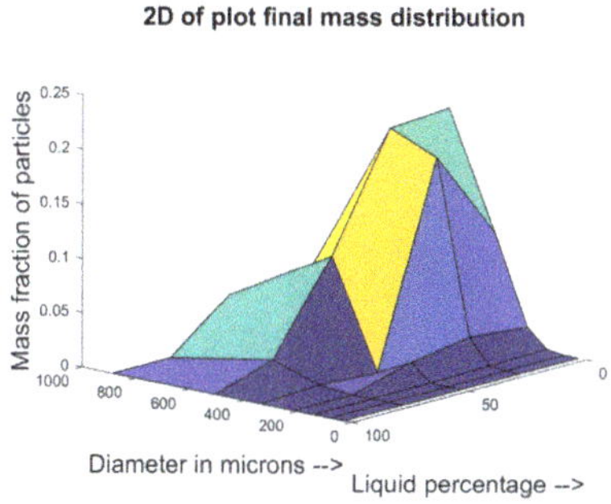

(**b**) Normalised mass frequency of blend after granulation for Set 1

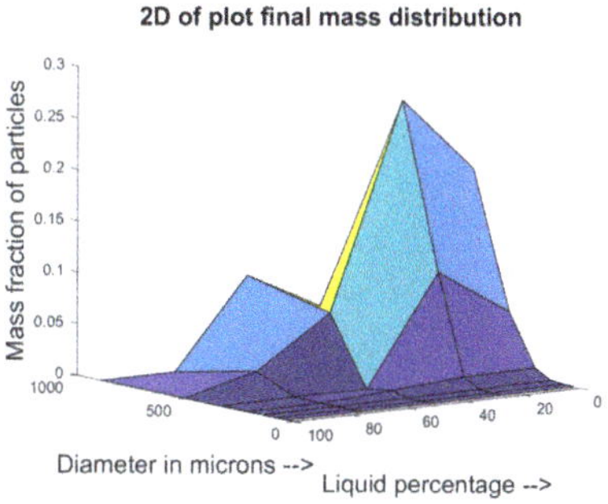

(**c**) Normalised mass frequency of blend after granulation for Set 2

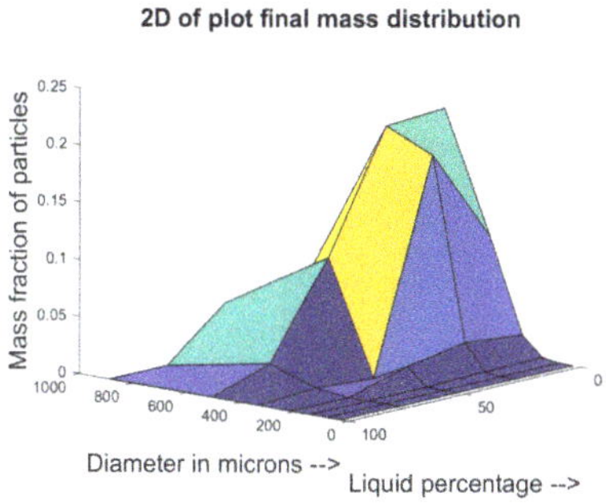

(**d**) Normalised mass frequency of blend after granulation for Set 7

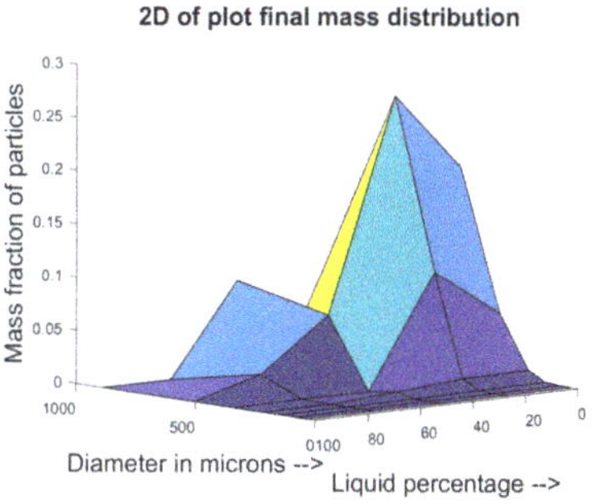

(**e**) Normalised mass frequency of blend after granulation for Set 10

Figure 3. 2-dimesional mass frequency of particles with respect to both diameter and moisture content.

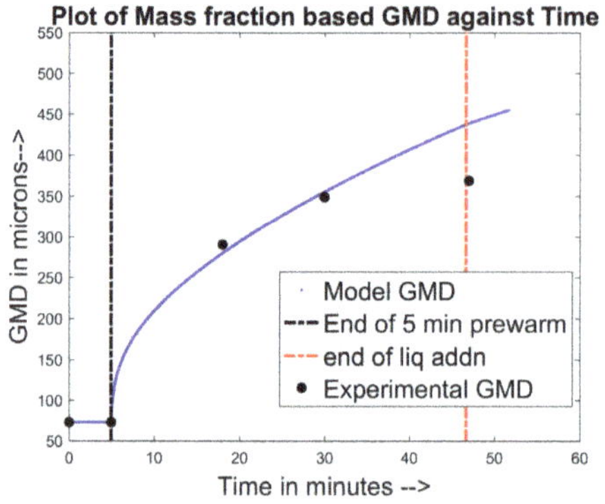

(a) Mass fraction based GMD for Set1

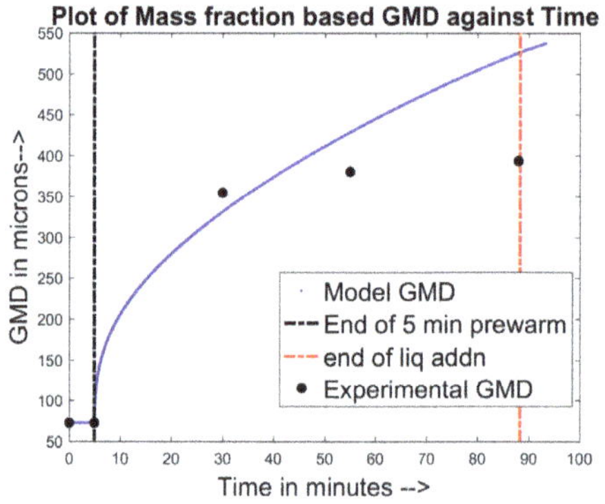

(b) Mass fraction based GMD for Set2

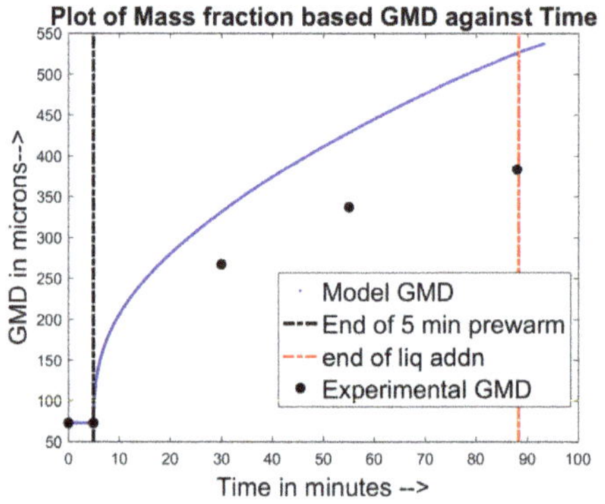

(c) Mass fraction based GMD for Set10

Figure 4. Experimental vs. model PSDs for Sets 1, 2 and 10 that were used for training the model.

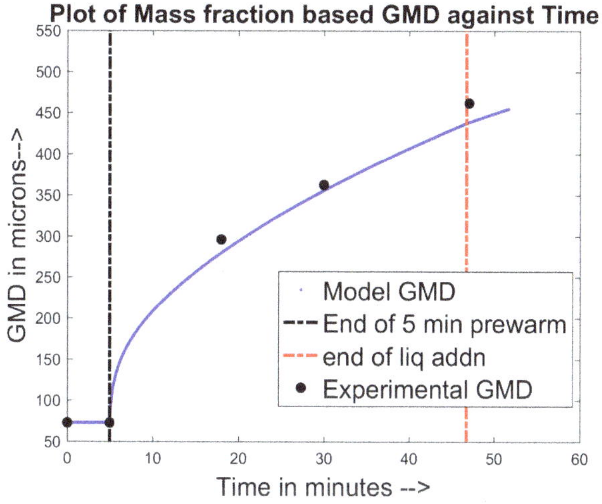

Figure 5. Experimental vs. model geometric mean diameters (GMDs) for validation Set 7.

From Figure 4a,b, one can see that the model predicted the growth trend fairly well until the point where 400 g of liquid had been added. However, beyond this point, the model results were more than the data obtained experimentally. It should be noted that only experimental points at 667 g liquid added were available to compare the difference. It might be plausible that they were erroneous points/outlier points in their respective data sets. However, the model overpredicted the experimental GMD at even the point when just 200 g liquid was added. From Figure 5, it was seen that the model predicted the trends for Set 7 fairly well. The explanation for the same is beyond the scope of this work.

Similar results were obtained when the training and validation sets were mixed and matched. Therefore, the estimated values for the empirical aggregation rate constant, coefficient of restitution and contact angle were accepted to be true for the purpose of this work and were used for further simulation studies. This was done in order to predict the system behavior at points in the expanded DOE. The estimation was carried out by minimizing the sum of the relative errors between the experimental and model predicted PSD values at all the sieve sizes for 200 g liquid added, 400 g liquid added and 667 g liquid added time points. This was carried out in MATLAB (R2016b, Mathworks, Inc., Natick, MA, USA) using the inbuilt *fminsearch* function.

The overall minimisng objective function can be written as follows in Equation (39):

$$SSRE = \sum_{200\ g\ liqadded}^{667\ g\ liq\ added} \left(\sum_{all\ sieves} \left(\frac{Mass\ frac_{exp} - Mass\ frac_{model}}{Mass\ frac_{exp}} \right)^2 \right). \tag{39}$$

The veracity of the model results has been tested by computing the ratio of rate of formation of particles due to aggregation to the rate of depletion of particles due to aggregation. Theoretically, the ratio should always be '0.5' because the aggregation of particles was modelled to be due to binary collisions between combination pairs. It is described as follows in Equation (40):

$$Ratio = \sum \left(\frac{\Re_{agg}^{form}(s,l,t)}{\Re_{agg}^{dep}(s,l,t)} \right). \tag{40}$$

From Figure 6a, it can be seen that the range of values on the y-axis is very small. This is indicative of the fact that the developed model was correct for the working precision of the MATLAB software and that the aggregation mechanism is computed properly throughout the iterative steps despite being closely integrated with the heat and mass balance of the moisture content and the binder dissolution mechanics. Along with the ratio, the total amount of liquid added over time to the system is an indication that the rate processes are being calculated correctly over every time step (see Figure 6b).

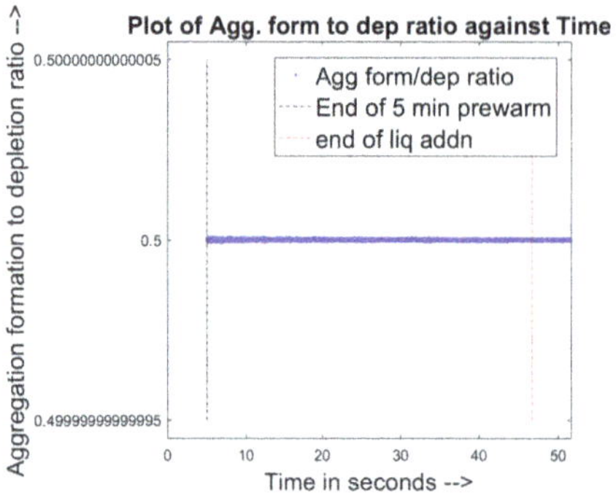

(**a**) Ratio of aggregation formation to depletion

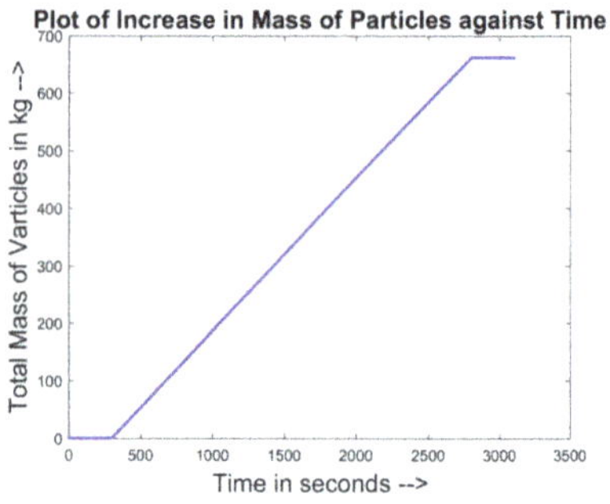

(**b**) Total mass of particles (initial solid + liquid added) over process time

Figure 6. Plots showcasing the veracity of the results obtained from the integrated population balance, energy balance, binder dissolution and viscosity prediction models.

4.2. Prediction of Other Data for the Experimental Sets

The simulated bed temperatures for the four experimental sets have been recorded and shown as follows in Figure 7.

It can be seen from the temperature plots in Figure 7 that the inlet temperature of the air (shown in the charts as cyan colour and labelled as DOE temp.) and the binder liquid addition rate are most significant in determining the temperature profile of the particle bed over time. It can be seen that, during pre-warm stage, the temperature of the bed (solid pink curve) rises until it reaches closer to the inlet air temperature. Then, during liquid addition/granulation stage, the temperature initially falls approaching the temperature of the binder liquid spray. The temperature rises whenever the airflow rate is increased due to increased heat transferred from the air to the bed of particles. It can be seen from Figure 7a,b that the rate of temperature decrease is the same when the temperature of the inlet air is 30 °C. Moreover, the temperature of the bed doesn't rise after reaching the spray temperature of 25 °C in these experiments. This is because the heat provided by the inlet air is not sufficient to raise the temperature of the bed after negating the cooling effects of the binder spray liquid. It can be seen that, once the addition of binder spray liquid has stopped, the temperature of the bed increases until it reaches the inlet temperature of air during drying i.e., 60 °C. From the the four charts in Figure 7, it can be seen that the predicted bed temperature for Set 10 (i.e., 8 g/min and 50 °C) was the highest. This may possibly be due to the combination of higher inlet air temperature during granulation and the lower binder addition rate that would lead to the bed particles losing lesser heat as the amount of liquid coming in an instant of time is too small to cause the bed temperature to fall/not rise.

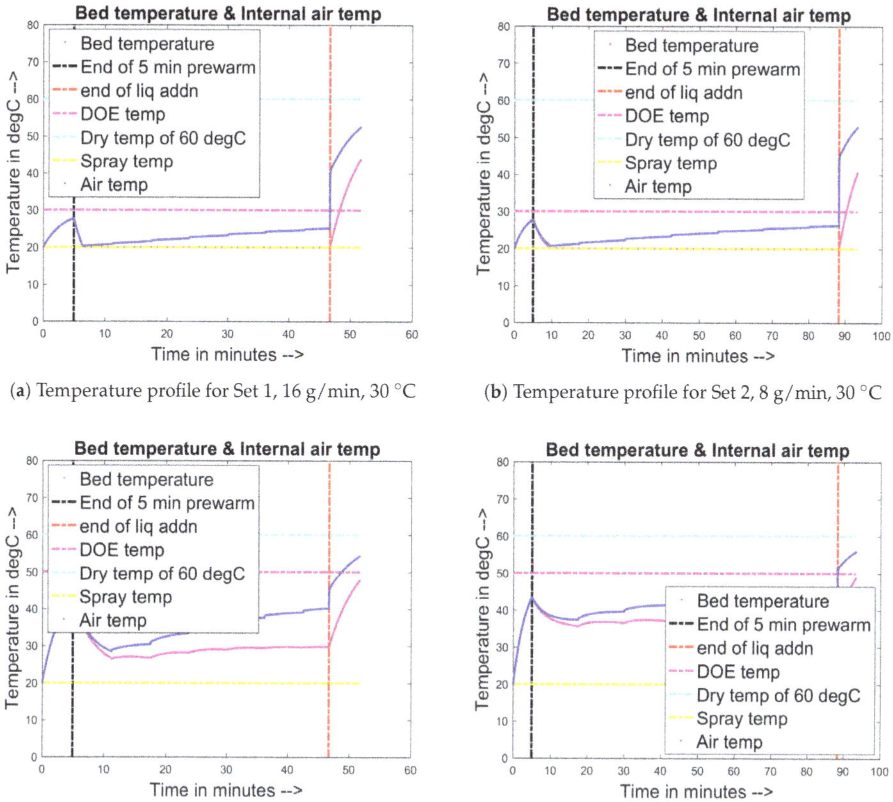

(**a**) Temperature profile for Set 1, 16 g/min, 30 °C

(**b**) Temperature profile for Set 2, 8 g/min, 30 °C

(**c**) Temperature profile for Set 7, 16 g/min, 50 °C

(**d**) Temperature profile for Set 10, 8 g/min, 50 °C

Figure 7. Plots showcasing the temperature profiles of the particle bed for the various experimental cases.

The various charts in Figure 8 show the results obtained from the integrated binder dissolution models in the overall system PBM. It can be seen from Figure 8a that, during pre-warm stage of the granulation, the binder concentration remained constant. Upon addition of binder initially at 5 min, the binder concentration increased and reached its maximum value. This may possibly due to the initial rapid dissolution of the binder particles in the minimal amount of liquid that has been sprayed until this time. The binder concentration remained at the highest, saturated concentration for another 5 min during which all of the binder particles dissolved completely into the spraying liquid until they reached the theoretical minimum radius. The binder concentration started decreasing after the binder dissolution was over, during which only dilution of the liquid occurs due to the addition of further liquid into the system.

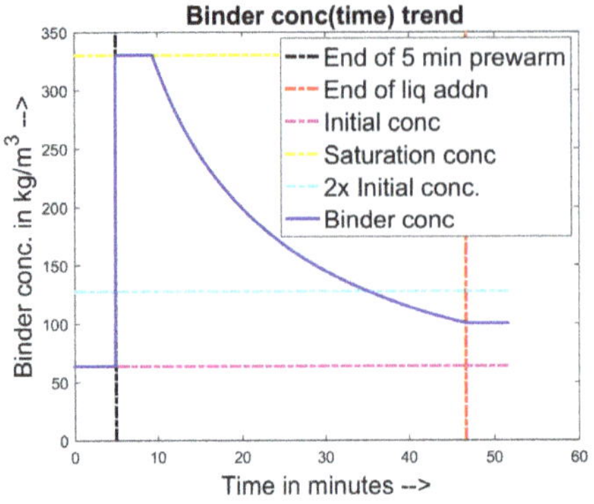

(a) Binder concentration for Set7

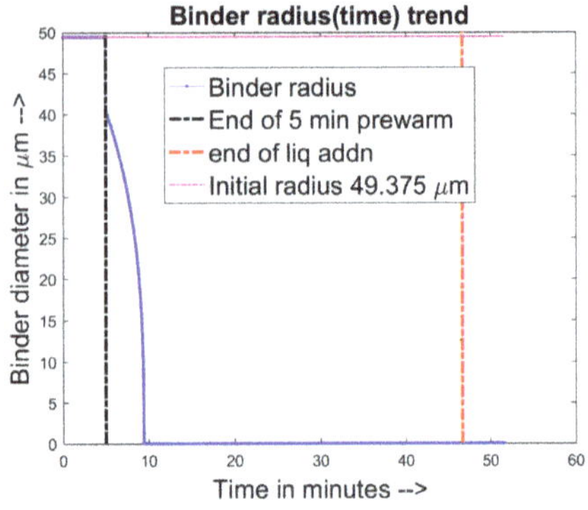

(b) Binder radius for Set7

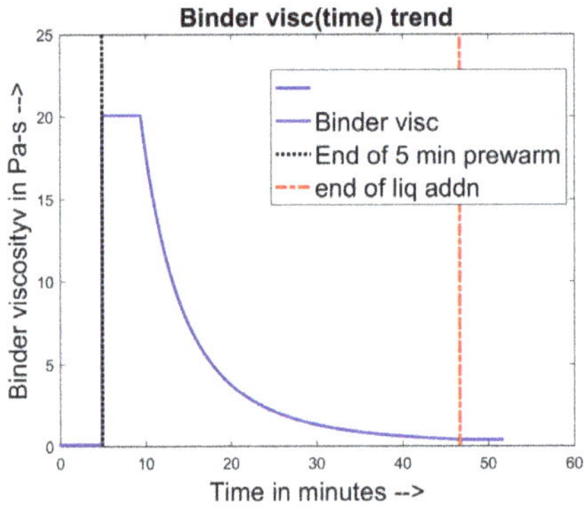

(c) Binder liquid viscosity for Set7

Figure 8. Simulated data for the binder liquid obtained by integrating the binder dissolution and viscosity prediction models into the main mass and energy balance model of the system.

4.3. Prediction of Regimes That Lead to Good Granulation

The identification of optimal regimes for granulation were identified by running all of the one-hundred eighty-nine (189) cases in the extended DOE as described previously in Table 4. The resulting 189 PBM trials were carried out using the *parfor* feature of MATLAB in a 16 core PC, where any 16 trials were run simultaneously, and the resulting PSDs were compared in terms of the span and the D50 of the particles with respect to the initial dry powder charge PSD. The expression for the span of a bed of particles is given as follows in Equation (41):

$$Span = \frac{D_{90} - D_{10}}{D_{50}}, \tag{41}$$

where D_{90}, D_{10} and D_{50} are the diameter sizes below which 90%, 10% and 50% of the particles in the granulator bed exist at any point of time. Upon examining the results of the extended DOE model, it was found that the higher amount of total liquid added, the lower amount of SSG in the blend and the lower inlet temperature yielded the greatest increase in D_{50} while keeping the fines to a minimum extent as possible. The total amount of liquid added during granulation seems to have had a significant influence on the PSDs. This may be explained by the fact that the addition of liquid would lead to a shift of a greater number of particles to the bins of higher sizes. This outweighs the dampening effect on aggregation by the lowering of viscosity of the system, thereby leading to higher Stokes' criterion numbers for the particles. The effect of lower inlet air temperature on the greater growth of particles is more intuitive to grasp as the lowering of temperature would cause lower moisture evaporation rates from the bed, thereby leading to formation of bigger particles.

On the other hand, conditions have been observed for cases where the D_{50} remains in the same ballpark as in the optimal case, but the span is larger thereby leading to a wider distribution and greater amount of fines. The identification of the DOE setting has been achieved by implementing an *fminsearch* algorithm. The initial guess for the DOE setting was given as the one that gave optimal final PSD of the granule, and the objective function was defined as the sum of relative error in the D_{50} with respect to the D_{50} obtained at the optimum settings and the inverse of the span. The expression for the objective function for the identification of granulation regimes leading to wider PSD is given as follows in Equation (42):

$$F_{obj} = abs\left(\frac{D_{50} - D_{50,optimal}}{D_{50,optimal}}\right) + \frac{1}{Span}. \tag{42}$$

It was observed that a greater amount of fines was obtained in the simulations of the systems having greater SSG amount and higher inlet air temperature. The cumulative PSDs for initial powder bed, optimal run setting and the run yielding a greater amount of fines have been shown below in Figure 9.

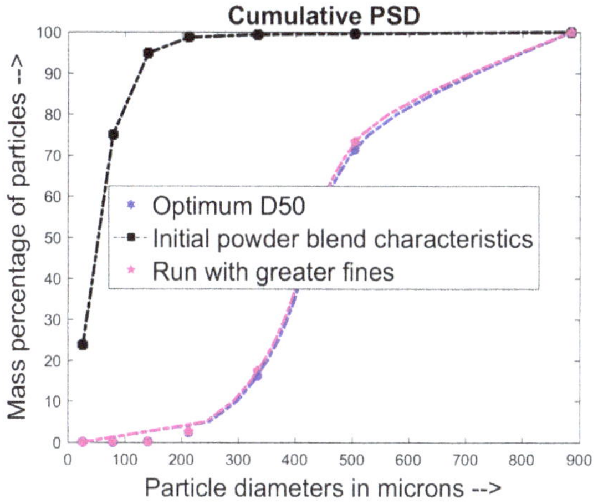

Figure 9. Mass PSDs for inital powder blend, final granular blends at optimum D50 run settings and at setting generating more fines.

5. Conclusions

An attempt has been made to validate the model results with the available experimental results limited in both quantity and quality. This goal has been achieved by developing a semi-mechanistic population balance model with aggregation and liquid addition mechanisms that tracked the growth rate of smaller dry particles into larger wet particles. Furthermore, the mass balance equations have been integrated with the energy balance equations of the system with the goal of profiling the temperature of the bed particles over time.The second objective has been achieved by modeling the bed and the air as two individual lumped systems with no spatial variations of temperature for any of them. The experimental PSD and GMD industrial data have been compared. Furthermore, regimes were found that led to poor granulation, indicated by wider spans and lower D10 values.

Thus, the overall model could also simulate the binder dissolution trend and change in kinematic viscosity over time due to the interplay of different mechanisms taking place in the system. With more detailed experimentation, data such as temperature of particle beds at various locations of the bed along the height and the flow patterns at different time points would enable one to discretize the model even along the height and observe the effects of preferential wetting and preferential drying, if any is present.

Author Contributions: S.V.M. and R.R. contributed theory and modeling methods. P.P. provided the experimental data. S.V.M. modeled the system, performed the simulations and wrote the paper. A.T. wrote the introduction section of the paper.

Funding: This work is supported by the National Science Foundation CAREER program through grant no. 1350152.

Acknowledgments: The authors would like to thank Anik Chaturbedi and Indu Mutancheri (affiliated with Dept. of Chem. and Biochem Engg., Rutgers, the State University of New Jersey, USA) for their valuable help in providing their insights of the model formulation. We would also like to thank Christopher Lewins, Steve Pafiakis, Brian Zacour, Dilbir Bindra, Jade Trinh, David Buckley, Shruti Gour, Shasad Sharif and Howard Stamato (from Drug Product Science and Technology, Bristol-Myers Squibb, New Brunswick, NJ, USA) for having generated the experimental data used in this work [28]. Moreover, special thanks is extended to Maksym Dosta at the Dept. of Chem. Engg, TU Hamburg, Germany for having helped with his expertise in the modeling of the energy balance component of the fluidized bed model.

Conflicts of Interest: The authors declare no conflicts of interest.

References

1. Boerefijn, R.; Hounslow, M. Studies of fluid bed granulation in an industrial r&d context. *Chem. Eng. Sci.* **2005**, *60*, 3879–3890.
2. Haapaniemi, H. Development and Validation of Near Infrared Method and Manufacture of Tablet Calibration Set by Fluidized Bed Granulation. Master's Thesis, Aalto University, Espoo, Finland, October 2017. Available online: https://aaltodoc.aalto.fi/handle/123456789/28499 (accessed on 22 August 2018).
3. Abouzaid, A.S.; Salem, M.Y.; Elzanfaly, E.S.; el Gindy, A.E.; Hoag, S.; Ibrahim, A. Screening the fluid bed granulation process variables and moisture content determination of pharmaceutical granules by nir spectroscopy. *Eur. J. Chem.* **2017**, *8*, 265–272. [CrossRef]
4. Fries, L.; Antonyuk, S.; Heinrich, S.; Palzer, S. DEM-CFD modeling of a fluidized bed spray granulator. *Chem. Eng. Sci.* **2011**, *66*, 2340–2355. [CrossRef]
5. Lourenco, V.; Lochmann, D.; Reich, G.; Menezes, J.C.; Herdling, T.; Schewitz, J. A quality by design study applied to an industrial pharmaceutical fluid bed granulation. *Eur. J. Pharm. Biopham.* **2012**, *81*, 438–447. [CrossRef] [PubMed]
6. Sen, M.; Barrasso, D.; Singh, R.; Ramachandran, R. A multi-scale hybrid cfd-dem-pbm description of a fluid–bed granulation process. *Processes* **2014**, *2*, 89–111. [CrossRef]
7. Rajniak, P.; Stepanek, F.; Dhanasekharan, K.; Fan, R.; Mancinelli, C.; Chern, R. A combined experimental and computational study of wet granulation in a wurster fluid bed granulator. *Powder Technol.* **2009**, *189*, 190–201. [CrossRef]
8. Drumm, C.; Attarakih, M.M.; Bart, H.-J. Coupling of CFD with DPBM for an RDC extractor. *Chem. Eng. Sci.* **2009**, *64*, 721–732. [CrossRef]
9. Yan, W.; Luo, Z.-H.; Guo, A.-Y. Coupling of CFD with PBM for a pilot-plant tubular loop polymerization reactor. *Chem. Eng. Sci.* **2011**, *66*, 5148–5163. [CrossRef]
10. Tan, H.; Goldschmidt, M.; Boerefijn, R.; Hounslow, M.; Salman, A.; Kuipers, J. Building population balance model for fluidized bed melt granulation: Lessons from kinetic theory of granular flow. *Powder Technol.* **2004**, *142*, 103–109. [CrossRef]
11. Yuu, S.; Umekage, T.; Johno, Y. Numerical simulation of air and particle motions in bubbling fluidized bed of small particles. *Powder Technol.* **2000**, *110*, 158–168. [CrossRef]
12. Bokkers, G.; Annaland, M.V.; Kuipers, J. Mixing and segregation in a bidisperse gas–solid fluidised bed: A numerical and experimental study. *Powder Technol.* **2004**, *140*, 176–186. [CrossRef]

13. Fries, L.; Antonyuk, S.; Heinrich, S.; Dopfer, D.; Palzer, S. Collision dynamics in fluidised bed granulators: A DEM-CFD study. *Chem. Eng. Sci.* **2013**, *86*, 108–123. [CrossRef]
14. Dosta, M.; Antonyuk, S.; Heinrich, S. Multiscale simulation of the fluidized bed granulation process. *Chem. Eng. Technol.* **2012**, *35*, 1373–1380. [CrossRef]
15. Dosta, M.; Antonyuk, S.; Heinrich, S. Multiscale simulation of agglomerate breakage in fluidized beds. *Ind. Eng. Chem. Res.* **2013**, *52*, 11275–11281. [CrossRef]
16. Tamrakar, A.; Devarampally, D.R.; Ramachandran, R. Advanced multiphase hybrid model development of fluidized bed wet granulation processes. *Comput. Aided Chem. Eng.* **2018**, *41*, 159–187.
17. Pandey, P.; Bharadwaj, R. *Predictive Modeling of Pharmaceutical Unit Operations*; Woodhead Publishing: Sawston, UK, 2016.
18. Barrasso, D.; Walia, S.; Ramachandran, R. Multi-component population balance modeling of continuous granulation processes: A parametric study and comparison with experimental trends. *Powder Technol.* **2013**, *241*, 85–97. [CrossRef]
19. Ramachandran, R.; Immanuel, C.D.; Štěpánek, F.; Litster, J.D.; Doyle, F.J., III. A mechanistic model for breakage in population balances of granulation: Theoretical kernel development and experimental validation. *Chem. Eng. Res. Des.* **2009**, *87*, 598–614. [CrossRef]
20. Heinrich, S.; Peglow, M.; Ihlow, M.; Henneberg, M.; Mörl, L. Analysis of the start-up process in continuous fluidized bed spray granulation by population balance modelling. *Chem. Eng. Sci.* **2002**, *57*, 4369–4390. [CrossRef]
21. Tan, H.; Salman, A.; Hounslow, M. Kinetics of fluidized bed melt granulation–ii: Modelling the net rate of growth. *Chem. Eng. Sci.* **2006**, *61*, 3930–3941. [CrossRef]
22. Vreman, A.; van Lare, C.; Hounslow, M. A basic population balance model for fluid bed spray granulation. *Chem. Eng. Sci.* **2009**, *64*, 4389–4398. [CrossRef]
23. Cryer, S.A. Modeling agglomeration processes in fluid–bed granulation. *AIChE J.* **1999**, *45*, 2069–2078. [CrossRef]
24. Chaudhury, A.; Niziolek, A.; Ramachandran, R. Multi-dimensional mechanistic modeling of fluid bed granulation processes: An integrated approach. *Adv. Powder Technol.* **2013**, *24*, 113–131. [CrossRef]
25. Dosta, M.; Heinrich, S.; Werther, J. Fluidized bed spray granulation: Analysis of the system behaviour by means of dynamic flowsheet simulation. *Powder Technol.* **2010**, *204*, 71–82. [CrossRef]
26. Börner, M.; Peglow, M.; Tsotsas, E. Derivation of parameters for a two compartment population balance model of wurster fluidised bed granulation. *Powder Technol.* **2013**, *238*, 122–131. [CrossRef]
27. Liu, H.; Li, M. Two-compartmental population balance modeling of a pulsed spray fluidized bed granulation based on computational fluid dynamics (CFD) analysis. *Int. J. Pharm.* **2014**, *475*, 256–269. [CrossRef] [PubMed]
28. Pandey, P.; Levins, C.; Pafiakis, S.; Zacour, B.; Bindra, D.; Trinh, J.; Buckley, D.; Gour, S.; Sharif, S.; Stamato, H. Enhancing tablet disintegration characteristics of a highly water-soluble high-drug-loading formulation by granulation process. *Pharm. Dev. Technol.* **2016**, *23*, 1–9. [CrossRef] [PubMed]
29. Chaturbedi, A.; Bandi, C.; Reddy, D.; Pandey, P.; Narang, A.; Bindra, D.; Tao, L.; Zhao, J.; Li, J.; Hussain, M.; et al. Compartment based population balance model development of a high shear wet granulation process via dry and wet binder addition. *Chem. Eng. Res. Des.* **2017**, *123*, 187–200. [CrossRef]
30. Ennis, B.J.; Tardos, G.; Pfeffer, R. A microlevel-based characterization of granulation phenomena. *Powder Technol.* **1991**, *65*, 257–272. [CrossRef]
31. Hu, X.; Cunningham, J.; Winstead, D. Understanding and predicting bed humidity in fluidized bed granulation. *J. Pharm. Sci.* **2008**, *97*, 1564–1577. [CrossRef] [PubMed]
32. Stakic, M.; Stefanovic, P.; Cvetinovic, D.; Skobalj, P. Convective drying of particulate solids–packed vs. fluid bed operation. *Int. J. Heat Mass Transf.* **2013**, *59*, 66–74. [CrossRef]
33. Fries, L.; Dosta, M.; Antonyuk, S.; Heinrich, S.; Palzer, S. Moisture distribution in fluidized beds with liquid injection. *Chem. Eng. Technol.* **2011**, *34*, 1076–1084. [CrossRef]

34. Wang, J.; Flanagan, D.R. General solution for diffusion-controlled dissolution of spherical particles. 1. Theory. *J. Pharm. Sci.* **1999**, *88*, 731–738. [CrossRef] [PubMed]
35. Chaudhury, A.; Kapadia, A.; Prakash, A.V.; Barrasso, D.; Ramachandran, R. An extended cell-average technique for a multi-dimensional population balance of granulation describing aggregation and breakage. *Adv. Powder Technol.* **2013**, *24*, 962–971. [CrossRef]

processes

MDPI

Article

Modeling On-Site Combined Heat and Power Systems Coupled to Main Process Operation

Cristian Pablos [1,2,*], Alejandro Merino [3] and Luis Felipe Acebes [1]

[1] Systems Engineering and Automatic Control Department, University of Valladolid, 47011 Valladolid, Spain; felipe.acebes@eii.uva.es
[2] Institute of Sustainable Processes, University of Valladolid, 47011 Valladolid, Spain
[3] Electromechanical Engineering Department, University of Burgos, 09006 Burgos, Spain; alejandromg@ubu.es
* Correspondence: cristian.pablos@uva.es

Received: 27 February 2019; Accepted: 9 April 2019; Published: 16 April 2019

Abstract: Many production processes work with on-site Combined Heat and Power (CHP) systems to reduce their operational cost and improve their incomes by selling electricity to the external grid. Optimal management of these plants is key in order to take full advantage of the possibilities offered by the different electricity purchase or selling options. Traditionally, this problem is not considered for small cogeneration systems whose electricity generation cannot be decided independently from the main process production rate. In this work, a non-linear gray-box model is proposed in order to deal with this dynamic optimization problem in a simulated sugar factory. The validation shows that with only 52 equations, the whole system behavior is represented correctly and, due to its structure and small size, it can be adapted to any other production process working along a CHP with the same plant configuration.

Keywords: Combined Heat and Power; gray-box model; utility management; CHP legislation; optimization

1. Introduction

In a world facing continuous challenges from an economic and environmental point of view, a lot of industrial processes rely on on-site Combined Heat and Power (CHP) systems to increase their global energy efficiency. The main feature of CHP is the possibility of obtaining heat and electricity simultaneously, from a single power source. This has many advantages, such as an increased reliability, fewer energy generation related costs, a reduction in the amount of emissions and, of course, an improvement in the energy efficiency of the process. More information about CHP systems and their operation can be found in [1].

When it comes to the management of industrial CHP plants, the idea is to exploit them in such a way that the operational costs related to heat and power generation are reduced to a minimum. This is a difficult problem due to the number of things that must be considered at the same time. Among others, the electricity purchase and selling options, the legislation related to cogeneration systems, and the uncertainty inherent to industrial processes, make essential the use of advanced decision-making tools, to help industry managers and operators to manage this kind of systems in the best possible way.

From the literature, it is clear that the traditional way to deal with this problem is to apply a sequential approach, where the main process operation is decided first, then the energy needs are estimated, and finally the CHP plant management is optimized. Thus, the main process and cogeneration are treated independently. This problem is known as the short-term operation planning on CHP systems. Some examples are [2–6]. For these problems, decisions can be taken at two different levels:

1. At the plant level, decisions are related to the quantity of electricity that must be bought from the external grid, or generated on site in order to sell a surplus if possible. Here, different contracts (Base load, Tarif of Use (TOU)), and electricity markets (Day-ahead market, intra-day market) are usually considered, and the price of the electricity takes on a special relevance.
2. At the equipment level, once the quantity of electricity to be generated has been decided, the next step is to decide what equipment is going to be used; this is known as the Unit-Commitment (UC) problem. Then, the load generation of each of the selected pieces of equipment is decided. This is the Economic Dispatch (ED) problem.

In [7], the main research lines related to the short-term operation planning problem are set out. One of the most important branches focuses on the formulation of the model used to later carry out the optimization. This is a critical step in obtaining a good solution with the least computational effort possible. This way, although CHP plants are typically non-linear, they are usually linearized due to the computational advantages involved, being Mixed Integer Linear Programming (MILP) models the approach traditionally selected to solve this problem [5,8,9]. However, works such as [4] and [3] claim that sometimes this kind of model leads the optimization to suboptimal or infeasible solutions, and non-linearities should be considered, turning the problem into Mixed Integer Non-Linear Programing (MINLP).

If the main process and the CHP plant are part of the same company, instead of following the sequential approach, an integrated one could be taken, where the operational planning of the CHP is obtained along with the best process operation strategy. Here, the main idea is to simultaneously adapt the way energy is produced and how production is carried out, according to such external factors as the electricity prices. This approach is taken in [10], leading to a bigger optimization model than if a sequential approach were considered, but obtaining better optimization results.

Although this approach is not so well studied as the sequential one, it is very close to the one proposed for solving the industrial Demand-Side Management problem (iDSM) [11]. There, industrial processes try to adapt the load profile, and therefore the production, according to some financial incentives given by the grid operator, in order to maintain grid stability if emergencies occur. These incentives can be in many forms, one of them being in the electricity prices, encouraging facilities to adapt their production schedule to them. Therefore, in this problem, the decisions taken, as in the integrated approach, affect the production strategy and energy management.

In spite of the fact that, in iDSM problems, only the way electricity is consumed and bought is usually considered, some studies have also explored the possibility of selling electricity to the external grid. This is the case in [12], where the authors developed an MILP model to optimize the production scheduling of an energy-intensive part of the steel production process with energy awareness. Thus, the optimization problem was able to decide whether it was convenient to buy electricity, or generate it in an on-site plant in order to reduce electricity demand at peak time, or even sell it to the external grid. However, the nature of the on-site generation plant was not given, and in this case the load-commitment curve with the grid operator had previously been fixed. In [13], the authors went one step further, developing a stochastic mixed-integer linear programming model capable of obtaining at the same time, the scheduling of a continuous energy-intensive process and its load-commitment curve, considering two different sources of uncertainty.

As can be seen in the literature regarding the kind of models used, in both the sequential and integrated approach, the mixed-integer formulation of the problems come from the possibility of deciding between different equipment, either in the main process or in the CHP plant. On the other hand, the assumptions made concerning the equipment make the problem linear or non-linear, and with respect to dynamics, they are usually neglected. This is due to the fact that changes in the operation are usually caused by changes in the electricity prices, and the latter typically have a much slower dynamic.

Regarding the case studies considered, in the sequential approach bibliography, the CHP plants present different configurations, where energy can be obtained using different equipment and heat can sometimes be stored. However, they do not cover all the possible cases. There are many small or medium sized industries that work with simple CHP units, made up of a small number of boilers and turbines designed to fill the main process energy demands. In these cases, if the sequential approach is used to optimize the CHP management, it is found that neither the equipment nor the plant level decisions can be modified. This is due to the fact that, typically, all the equipment (boilers and turbines) that forms the cogeneration unit is the same, and is being used at the same capacity level, without any spare, so the choice of equipment is not necessary. On the other hand, since these CHP systems are designed to fill the heat demand of the main process they are attached to, and with the heat generated, produce the electricity needed by the main process with a small surplus, buying electricity from the external grid usually makes no sense. Furthermore, the fact that, typically, the excess of power generated is not very big, and that it can vary sharply depending on the main process operation, means that the possibility of selling electricity to the external grid is seldom considered.

Now, if the integrated approach is considered for these cases, two different scenarios may appear, depending on how the main process energy demand can be modified. If it can be adapted without changing the production rate, then a close problem to the one presented by [10] or by the iDSM community could be proposed. Thus, the production would be adapted to generate a bigger electricity surplus when electricity prices are more convenient. However, if this is not the case, the only way to change the energy demanded is by changing the production rate. If this way of working is considered, the fact that it is still an open issue in the literature must be taken into account. Here, the dynamic of the CHP plant is determined by the dynamic of the main process, so it could be said that the generation of electricity is coupled to the production rate. If the main process dynamic is close to that of the electricity prices, the problem can no longer be static, and dynamics will have to be considered explicitly in the optimization model.

This work focuses on this last scenario, where the production rate is adapted in order to operate the whole system (main process and CHP plant) in the best possible way, from an economic point of view. Considering the importance of the dynamics of the main process, and the absence of scheduling decisions, dynamic optimization will be the approach selected to solve this problem. With this technique, a dynamic model of the system is used to find the best control action sequence over time. Typically, depending on how the dynamic model is used during the optimization, two methods can be found to solve these problems: sequential or simultaneous approach. In the sequential approach, the control variables are discretized over time, and an external simulation of the model is used to evaluate the performance of the solution given by the optimizer each iteration. This way of working brings some advantages, like a good accuracy when solving the model. However, there are also some disadvantages, like difficulties when calculating the gradients needed for gradient based optimization solvers, or a high time spent for evaluating each solution in the simulator if the model is too complex. These facts often make the use of this approach infeasible. In the simultaneous method, the whole model is discretized and incorporated to optimization problem as a set of constraints. With this method, the optimization problem increases its size, but if the discretization is done in an appropriate way, it is usually solved faster than the sequential approach. Furthermore, convergence problems in the simulation are avoided as no simulator is being used, and path constraints can be more easily added to the optimization problem [14].

Since it has been explained, models in dynamic optimization are a key part of the problem. Concerning the modeling of energy generation systems, many references can be found. A good review on the modeling of energy generation can be found in [15]. On the other hand, many tools, designed for modeling and simulating different energy simulation systems are available in the market; in [16], an extensive review of the available tools can be found. Some of the developed models found in the literature are built based on model libraries offered by third parties, such is the case for Modelica [17], EcosimPro [18] or Aspenplus [19]. In addition, detailed models for the individual components constituting CHP plants such as boilers [20–22] or steam turbines [23] can be found.

The models presented so far are in general complex models comprised by many equations which represent the behavior of the system with great accuracy. The objective of these models is usually to test different operating policies in simulation, training or control design, but in general are not useful for optimization. This is because its complexity makes it hard to find the solution of the optimization problem. Therefore, simpler models are usually searched for optimization, trying at the same time, to represent the main behavior of the system in the best possible way. In this work, a dynamic gray-box model is proposed for one of the most commonly used CHP configurations in industry, able to be adapted to any main process that uses it. In order to be as accurate as possible, the model will be non-linear where the complexity of the process is too high. However, in order to make it computationally tractable, it will have linear equations based on experimental data where linearity is demonstrated. As a case study, a simulated sugar plant is used as the real plant to obtain experimental data and validate the model proposed. It is based on the experience of the working team in the sugar factory process and previous works in sugar factory simulators [18,24,25], and has been thoroughly validated [26].

The rest of this paper is organized as follows. In the next section, a brief description of the sugar factory problem is given. The third section presents obtaining the gray-box model for the case study considered. The fourth section examines the validity of the model and a discussion about the results. The fifth section explains how to extend the model for other case studies. Lastly, the paper finishes with some conclusions, limitations and future work with the model developed.

2. Case Study

A sugar factory with a CHP configuration like the one shown in Figure 1, is to be operated according to the Day-Ahead electricity market, so as to improve its competitiveness in the sugar market. Currently, due to the complexity of the electricity market considered and the small surplus of power generated in turbines, no electricity is being sold and the plant is operating isolated from the external grid. To benefit from some advantages set by European legislation [27,28], rewarding efficient cogeneration processes, some energy indexes must be considered during CHP operation.

In the factory considered, there are three steam boilers working in parallel to produce the amount of heat needed by the main process. These boilers burn natural gas to obtain heat from boiling water. As a result, superheated high pressure (HP) steam is obtained in this process. In the boilers, there is a control system that allows operators to select the desired steam pressure and temperature. The steam needed by the process must have low pressure (LP) and be saturated, so the expansion of the obtained steam must be carried out. To do so, there are two different possibilities: passing it through three similar steam turbines, or through a bypass. During normal process operation, almost all the steam is passed through the steam turbines to obtain the electricity needed by the main process, which is able to obtain a maximum of 11 MW. However, the process usually needs less electricity than that which can be generated with the steam required, so an electricity surplus can be sold to the external grid. If the possibility of selling electricity is not considered, steam is recirculated through the bypass valve, where its expansion takes place using the valve pressure drop. Once the LP steam leaves the expansion zone, it must be saturated. For that purpose, a saturator, where water absorbs steam energy, is used, and the saturated LP steam is sent to the process.

Regarding the sugar process, it is a continuous process that uses sugar beet to obtain white sugar using the heat and power provided by the CHP system. Before being processed, beet is stored in big piles outside the factory. The storage time is crucial, as the quality of the sugar decreases because of rotting. Inside the factory, sucrose is extracted from the sugar beet, obtaining a juice that has to be purified and then concentrated in an evaporation section. Finally, sucrose is crystallized obtaining sugar crystals. Each of these processes needs heat and power to operate; however, one of the main features of the sugar industry is its energy integration. Evaporation needs LP steam obtained directly from the CHP system to remove water from the sugar solution. This stage is carried out concurrently by different evaporation effects, six in the case study considered. In this process, vapor is obtained as a byproduct, and it is used by the rest of the consumers of the plant to obtain the heat they need. To ensure its availability, the pressure inside the evaporators must be monitored and maintained between bounds.

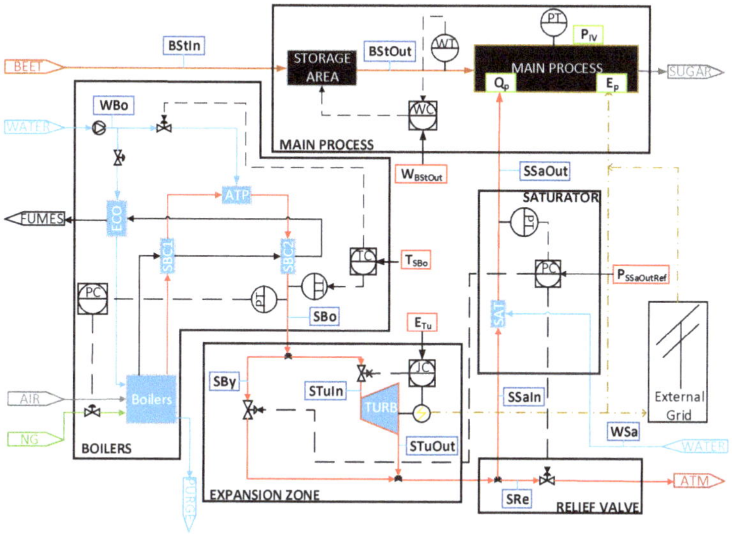

Figure 1. Schematic of the whole system.

The pressure of the steam that leaves the cogeneration is one of the most important variables of the CHP system, and it must be maintained between limits. If the pressure were higher than expected it could caramelize the sugar in the evaporation stage, and if it were lower, evaporation could not take place, compromising the rest of the process. To control the LP steam pressure, a split range controller is used. The manipulated variables are the openings of the bypass valve and the relief valve located between the expansion zone and the saturator. Thus, if the pressure is too low, the bypass valve will open and let live steam mix with the steam expanded in the turbines; and if it is too high, the relief valve will let it escape to the atmosphere. More information about the sugar industry can be found in [29] or [30].

Following the discussion set out in the first section, in this case study, the management optimization of the CHP system cannot be treated independently from the main process operation. Therefore, an integrated approach must be considered. If the modeling literature is reviewed, some approaches have been found that deal with sugar plants working with CHP systems [31–33]. However, the purpose is to study different aspects of process operation, without solving an optimization problem. Furthermore, they are focused on sugar cane factories, where bagasse, the residue of juice extraction from sugar cane, is used as fuel in the boilers, changing considerably the energy problem. In our case, since no process scheduling can be done to modify the energy consumption of the factory, the production rate must be changed to optimize the whole system operation. Due to the inherent complexity to the explained process, the setting time when changes are made in the production rate is around three hours. This contrasts with the quick response of the CHP plant, where changes in the electricity generation, can be done in less than 10 min. Since the Day-Ahead electricity market is considered, and prices there change hourly, in this case the dynamics of the main process must be considered explicitly.

In the next section, the process to obtain the dynamic model, and how it can be extended to other processes with the same CHP configuration, is shown.

3. Obtaining the Model

The first step to obtaining a model, is to make clear its objectives. In our case, a dynamic model is sought to optimize the sugar production process and CHP management according to electricity prices. To do so, an index that evaluates the cost of the energy needed to process a determined amount of beet, will be optimized.

$$Cost(t) = \int_0^{T_N} \frac{\frac{W_G(t)}{\rho_G} \cdot HCV_G \cdot Pr_G - \frac{(E_{Tu}(t) - E_P(t)) \cdot Pr_E}{3600}}{W_{BStOut}(t)} \delta t \qquad (1)$$

In (1), the numerator presents the difference between the cost of natural gas used in boilers and the incomes obtained by selling electricity to the external grid. In the denominator, the beet process rate is found. The function can be interpreted as the integral of the specific energy cost of the process. A full description of the model variables and parameters can be found in Appendix A.

Next, the inputs and outputs of the model can be set. Regarding the inputs, since the aim is to adjust the production rate to electricity generation, both must be considered. On the other hand, manipulating the thermodynamic conditions of the steam generated in the boilers can greatly modify the steam consumption of the whole system and, therefore, its manipulation will be indispensable to achieve the objectives marked by the European legislation mentioned in the previous section. To do so, the temperature of the superheated steam obtained in the boilers and the pressure of the saturated steam delivered to evaporation are the two most influential variables. With that in mind, a list of the model inputs and its working range, is shown below:

1. Beet production rate (W_{BStOut}) [370–430 T/h].
2. Electricity power generation (E_{Tu}) [5000–11,000 kW].
3. Evaporation working pressure (P_{SSaOut}) [2.2–3.0 barA].
4. Superheated steam temperature obtained in boilers (T_{SBo}) [360–420 °C].

Regarding the outputs of the model, the natural gas used in the boilers and the electricity power consumption of the process are essential for computing an operation policy. On the other hand, according to the description of the process, the beet accumulation, the pressure inside the evaporators and the legislation indexes will be needed to assure its feasibility.

With respect to beet accumulation, this can be measured in two different ways. One is measuring the remaining beet in the storage zone, and the other is computing the average residence time the beet spends inside the storage zone. The last is defined as the accumulated beet over the beet production rate, and if this is too high, the beet can rot. On the other hand, it shows how long the process can be operated at a specified production rate without running out of beet, if the beet input drops to zero. Since the residence time gives more information about the beet storage, and it is easier to constrain, this is used in the optimization problem and will therefore be an output of the model.

Concerning the pressure inside the evaporators, as mentioned in the previous section, in the evaporation stage, the pressures are variables that must be monitored continuously during the sugar process operation. Since the crystallization section demands a great part of the steam generated, it is considered the most important steam consumer. Given that the quality of the sugar obtained depends greatly on this last stage, sufficient steam for its operation must be guaranteed. The evaporation effect that feeds the crystallization section with steam may vary for different factories. In the one considered, crystallization is fed by the fourth evaporation effect, so in order to ensure that there will be enough steam to operate that section, the steam pressure of this stage must be maintained above a specified minimum. Having controlled the pressure of the steam that goes into the evaporation and monitored and maintained the pressure of the fourth effect between bounds, it is enough to ensure that the evaporation is running smoothly, so the pressure from other evaporators is not needed.

In relation to the calculation of the European legislation indexes, they have to be measured once per year. However, to introduce them into the optimization problem, they are computed for each instant. Thus, if they are always above the limits, it will be considered that they are respected.

On the other hand, since one of the objectives of any CHP system is to generate the amount of heat energy demanded by the main process, the heat energy consumed by the sugar factory will also be considered as an output of the model.

To sum up, a list of the desired output variables is shown below:

1. Electricity consumption of the sugar factory (E_p).
2. Natural gas mass flow needed to operate the whole process (W_G).
3. Average time beet spends in the storage zone (τ_{St}).
4. Steam pressure inside the fourth effect of the evaporation (P_{IV}).
5. European legislation Indexes.
6. Heat energy consumption of the sugar factory (Q_p).

Once the input and outputs of the model have been specified, the relationships between them are sought. As direct relations are sometimes very difficult or even impossible to find, the whole process is divided into different control volumes in order to make the modeling process easier. The division taken is shown in Figure 1, where the control volumes are emphasized with black lines. They have been selected from knowledge of the process behavior. However, this is not the only possibility, and other control volumes could have been chosen. With the division made, the equations and experimental relationships that represent the behavior of each control volume must be found.

System identification is the technique that has been selected to obtain the experimental relationships [34,35]. This is due to the fact that it is quite a common technique in the process industry and makes the extension to other case studies easier. Thus, since it is explained in Section 5, for other plants with cogeneration systems, if the structure of the CHP is the same, simply identifying the relation between some variables, and using its own parameters, the model proposed can easily be extended. Otherwise, for cases where the structure of the CHP is different and the model developed cannot be used directly, the work described in this paper will serve as a guide for building models for other CHP configurations.

In order to make the optimization model simpler, the possibility of using linear identification is studied first and applied for each case if possible. Among the different types of models available for linear identification, state-space is selected. This is because relationships in this form are written as a system of first-order differential equations depending on time, and this is very convenient for incorporating them later in any simulation or optimization environment. Furthermore, they can deal with Multiple-Input Multiple-Output (MIMO) systems and delays can easily be incorporated, which in our case will be very useful.

3.1. Main Process

The beet sugar plant is understood as a black box system, where beet is transformed into white sugar using a determined amount of heat and electricity power. From this control volume, heat and power energy consumption, pressure inside the fourth evaporation effect, and available beet in the storage zone are sought.

3.2. Heat Consumption

Although the main process heat consumption depends on many variables, it has been assumed that the most influential ones are the specified production rate (W_{BStOut}), and the evaporation working pressure (P_{SSaOut}) due to modifications in the performance of the evaporators [36]. To avoid modeling the whole process, a direct relationship between the heat consumption of the main process and the pressure and production described above, is searched.

To do so, the step response for both inputs, shown in Figure 2, must first be analyzed. A second order dynamic response, with a delay of almost one hour and a settling time of three hours, can be discerned for W_{BStOut}. Regarding P_{SSaOut}, things are more complicated to deduct, due to the step response during the first instants of the transitory period. This behavior is due to the aggressive tuning of the split range pressure controller, which tries to take the pressure to the new set point in the less possible time. This is done in order to send steam to the system as quickly as possible if required. A settling time of one hour can be observed, and there is no delay in the response. If the first part of the transient is ignored, the step response behavior is close to a typical linear first order system, with no delay and a settling time of one hour.

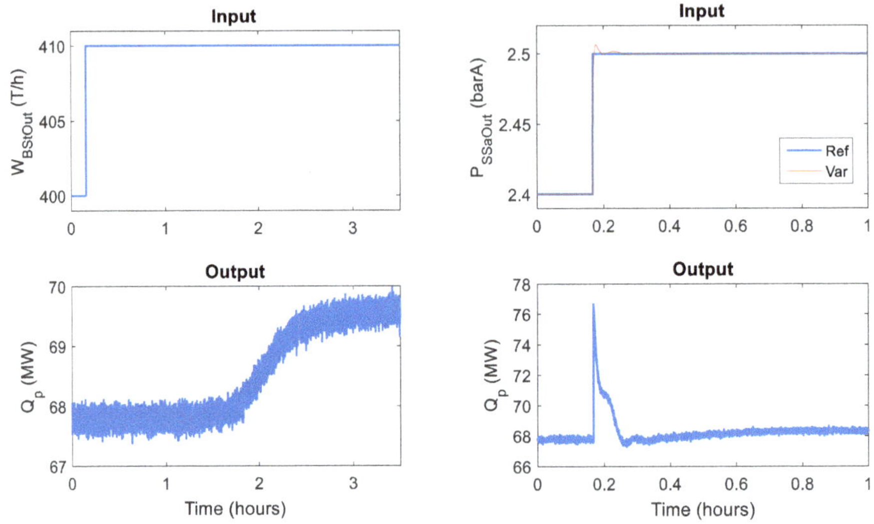

(a) (b)

Figure 2. Dynamic response of Q_p. (**a**) Dynamic response of Q_p for a step in W_{BStOut}; (**b**) Dynamic response of Q_p for a step in P_{SSaOut}.

Once the dynamic response for both inputs has been studied, steady state linearity between inputs and outputs is considered. With this in mind, some experiments are carried out trying different scenarios for both variables throughout its working range, and measuring the heat demand for each case in the stationary state. The experiments have been carried out independently for each input variable, maintaining the other one constant. Therefore, it is assumed that the superposition principle is valid for all the operational range of the variables, and this will be checked in the validation section. This has been done for any identification model where more than one input is involved. The results of these experiments are shown in Figure 3.

As can be seen, for the range of operation considered, the relationship of pressure and production with heat consumption can be assumed to be linear in the steady state. For the optimization problem considered, capturing the steady state is essential, but transients are also important if they expand in time. In the case of the heat energy consumption with respect to the evaporation working pressure, the transient overshoot is damped relatively quickly, so the model proposed for its identification should not be focused on this short response.

Therefore, a second order state-space model is proposed for identification in order to capture the stationary and essential response of the transient. To obtain it, the MATLAB® system identification toolbox [37] is used, and the model obtained is shown in Equation (2). The validity of this model is shown in the next section along with the rest of the model.

$$
\begin{aligned}
&Q_p(t) = Q_{p_{eq}} + \Delta Q_p(t)\\
&\Delta Q_p(t) = 172.400 \cdot Q_{p_{x1}}(t) + 1.744 \cdot Q_{p_{x2}}(t) + 5.629 \times 10^{-5} \cdot Q_{p_{x3}}(t-3312) + 1.744 \cdot Q_{p_{x4}}(t-3312)\\
&\dot{Q}_{p_{x1}}(t) = -1.077 \cdot Q_{p_{x1}}(t) - 8.815 \times 10^{-2} \cdot Q_{p_{x2}}(t) + 256 \cdot (P_{SSaOut}(t) - P_{SSaOut_{eq}})\\
&\dot{Q}_{p_{x2}}(t) = 6.25 \times 10^{-2} \cdot Q_{p_{x1}}(t)\\
&\dot{Q}_{p_{x3}}(t) = -5.521 \times 10^{-4} \cdot Q_{p_{x3}}(t) - 3.252 \times 10^{-4} \cdot Q_{p_{x4}}(t) + 0.250 \cdot (W_{BStOut}(t) - W_{BStOut_{eq}})\\
&\dot{Q}_{p_{x4}}(t) = 4.883 \times 10^{-4} \cdot Q_{p_{x3}}(t)
\end{aligned} \tag{2}
$$

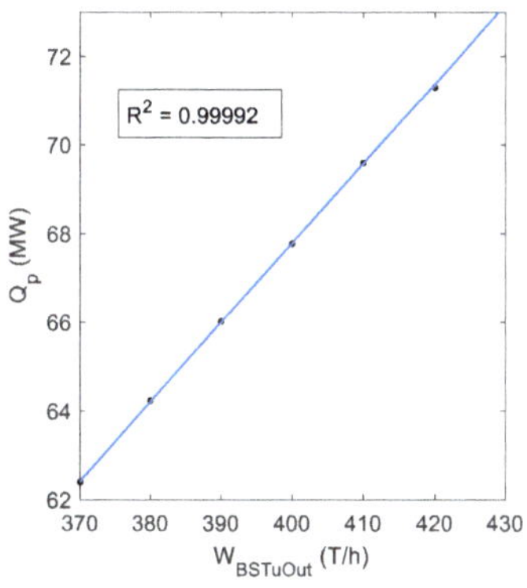

Figure 3. Experiments carried out to show linearity dependence between Q_p and W_{BStOut} (**left**) and P_{SSaOut} (**right**).

3.3. Electricity Power Consumption

According to [36,38], it has been assumed that for the operational range considered, among all the process variables, electricity power consumption depends mainly on the beet processing rate in a linear way. The step response of E_p with respect to W_{BStOut} is shown in Figure 4.

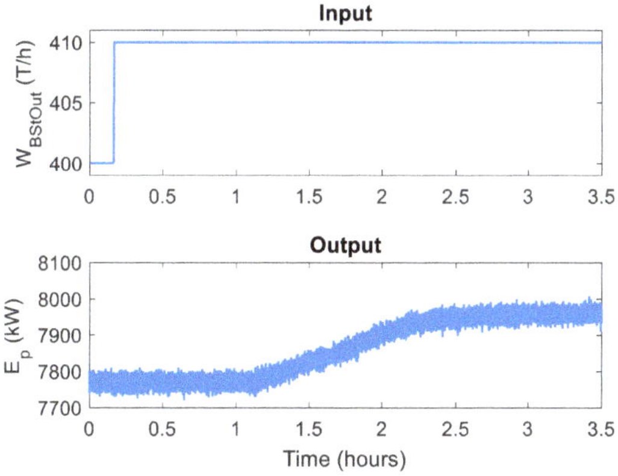

Figure 4. Dynamic response of E_p for a step in W_{BStOut}.

From Figure 4, a second order dynamic response can be inferred. Again, a delay of almost one hour appears between the input and the output, and a settling time of approximately three hours can be discerned. A state-space second order dynamic system is obtained and shown in Equation (3).

$$
\begin{aligned}
&E_p(t) = E_{p_{eq}} + \Delta E_p(t) \\
&\Delta E_p(t) = 2.423 \times 10^{-2} \cdot E_{p_{x1}}(t-3312) + 0.1273 \cdot E_{p_{x2}}(t-3312) \\
&\dot{E}_{p_{x1}}(t) = -1.021 \times 10^{-3} \cdot E_{p_{x1}}(t) - 8.359 \times 10^{-4} \cdot E_{p_{x2}}(t) + 0.125 \cdot (W_{BStOut}(t) - W_{BStOut_{eq}}) \\
&\dot{E}_{p_{x2}}(t) = 4.883 \times 10^{-4} \cdot E_{p_{x1}}(t)
\end{aligned} \quad (3)
$$

3.4. Pressure Inside Fourth Effect

Regarding pressure inside the fourth evaporation effect, since modeling the whole evaporation section would be inefficient for the purpose described, a simple expression is sought based on experimental data that represents how pressure inside this effect varies during the process operation. Firstly, the causes that affect its value are identified. In the factory studied, these are essentially the beet production rate and the evaporation working pressure. If production is increased and the evaporation working pressure is maintained constant, the water removed from the juice will not be enough to cope with the process steam demand, and eventually the pressure of every evaporation effect will drop. Instead, if the working pressure is increased, but the evaporation input juice flow remains constant, more water than necessary will be removed and the pressure of the evaporators will increase. Other variables may affect pressure inside the evaporators, but they are rarely changed during normal process operation and its effect is not very significant, so they will not be considered.

In Figure 5, the step responses of P_{IV} when changes are introduced in W_{BStOut} and P_{SSaOut} are shown. Regarding the beet production rate, when it is increased, the pressure inside the fourth effect rises following a typical linear second order response. The delay and settling time are the same as before with Q_p and E_p. With respect to P_{SSaOut}, a very similar response to the one seen for Q_p is found. Again, if the first part of the transient state is ignored, the response of the system is close to a first order dynamic. There is no delay and the settling time is one hour.

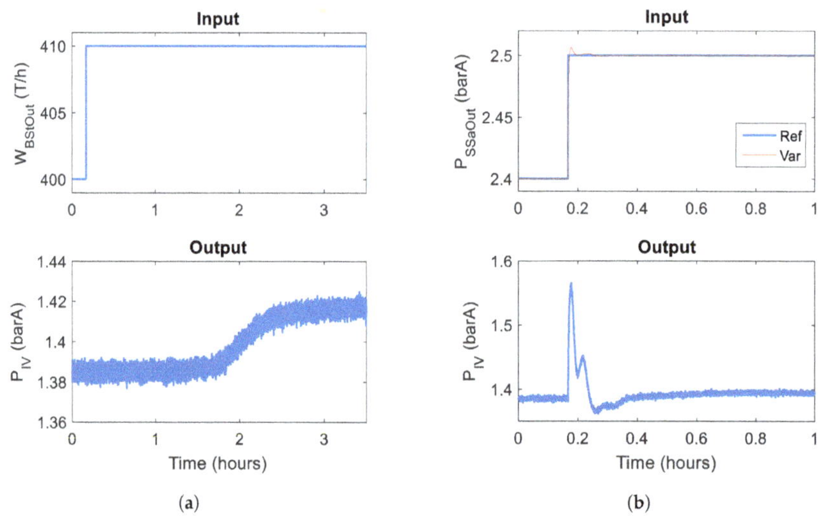

Figure 5. Dynamic response of P_{IV}. (**a**) Dynamic response of P_{IV} for a step in W_{BStOut}; (**b**) Dynamic response of P_{IV} for a step in P_{SSaOut}.

In order to find the linearity of P_{IV} with both inputs, different experiments are carried out with several values of W_{BStOut} and P_{SSaOut}. Again the superposition principle has been assumed as valid. The results of these experiments are shown in Figure 6.

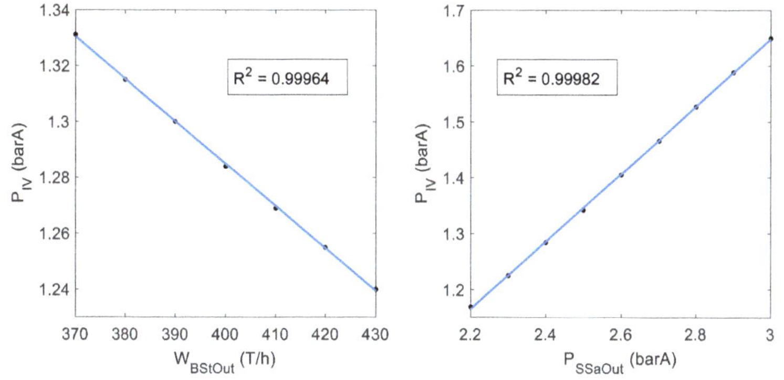

Figure 6. Experiments carried out to show linearity dependence between P_{IV} and W_{BStOut} (**left**) and P_{SSaOut} (**right**).

Figure 6 shows that, for the operational range considered, it can be assumed that both inputs are related to the fourth effect pressure in a linear way. With the same reasoning as that used for the identification of Q_p, a MIMO second order space-estate model is searched using identification techniques. The model obtained can be seen in Equation (4).

$$\begin{aligned}
&P_{IV}(t) = P_{IV_{eq}} + \Delta P_{IV}(t)\\
&\Delta P_{IV}(t) = 9.397 \times 10^{-2} \cdot P_{IV_{x1}}(t) + 0.122 \cdot P_{IV_{x2}}(t) - 1.449 \times 10^{-7} \cdot P_{IV_{x3}}(t-3312) - 5.935 \times 10^{-4} \cdot P_{IV_{x4}}(t-3312)\\
&\dot{P}_{IV_{x1}}(t) = -0.309 \cdot P_{IV_{x1}}(t) - 5.434 \times 10^{-2} \cdot P_{IV_{x2}}(t) + 0.25 \cdot (P_{SSaOut}(t) - P_{SSaOut_{eq}})\\
&\dot{P}_{IV_{x2}}(t) = 6.25 \times 10^{-2} \cdot P_{IV_{x1}}(t)\\
&\dot{P}_{IV_{x3}}(t) = -7.163 \times 10^{-4} \cdot P_{IV_{x3}}(t) - 3.915 \times 10^{-4} \cdot P_{IV_{x4}}(t) + 9.766 \times 10^{-4} \cdot (W_{BStOut}(t) - W_{BStOut_{eq}})\\
&\dot{P}_{IV_{x4}}(t) = 4.883 \times 10^{-4} \cdot P_{IV_{x3}}(t)
\end{aligned} \tag{4}$$

3.5. Storage Zone

The storage area, where beet is accumulated until it can be processed, has also been incorporated into this control volume. The calculation of the remaining beet is necessary to know if a determined control sequence leads to a situation where the beet is completely used up, or on the contrary, the storage zone is overflowing. To model it, a mass balance is considered between its input and output. Since the beet input and output flow are measured in [T/h] and the rest of the model variables are in seconds, a change of units must be done, dividing the difference by 3600.

$$\dot{m}_{St}(t) = \frac{W_{BStIn}(t) - W_{BStOut}(t)}{3600} \tag{5}$$

To calculate the output of this mass balance, the beet input must be known or predicted beforehand. As mentioned before, instead of the accumulated beet in the storage zone, the average residence time beet spends there is the desired output. It is defined as follows:

$$\tau_{St}(t) = \frac{m_{St}(t)}{W_{BStOut}(t)} \tag{6}$$

With the equations described so far, outputs number one, three, four and six from the list enumerated previously in this section are accomplished. The others are related to the CHP plant. Output number two concerns the quantity of natural gas needed during normal process operation that will be used to generate a determined amount of steam demanded by the main process or by the expansion turbines.

To calculate it, an experimental relationship between the natural gas consumed, and the steam used by the main process and turbines is necessary. A priori, it could be thought that both steam mass flows should be close or the same; however, due to the use of the split range controller, this does not always have to be true, and its presence makes the system complex and highly nonlinear. On the other hand, using a first principles model to describe the behavior of the whole CHP plant does not seem wise, since it would yield a large model not suitable for optimization. These problems make use of a different approach advisable, where the natural gas is obtained using a combination of first principle equations and experimental direct relationships, taking advantage of both methods. To do so, the other control volumes depicted in Figure 1 will be used.

3.6. Boilers

Within this control volume, a relationship is sought between the natural gas used in boilers and the superheated steam obtained. In this way, the amount of natural gas needed to generate the steam demanded by both the main process and the turbines could be known. However, it would still be necessary to determine the individual steam consumption of each one. This control volume includes the preheating water system, the boilers themselves, and the overheating of the saturated steam obtained. Modeling the system in detail makes no sense if the aim is to make this model useful for other industries with other types of boiler. Thus, an experimental relation between natural gas and steam must be found.

When some experiments were carried out, we found that, in most of the cases, the relationship between their mass flows is almost linear. However, this is only true if the temperature of the superheated steam remains constant. For example, more natural gas will sometimes be needed to obtain less superheated steam, but with a higher temperature. Therefore, a variable that includes both mass flow and temperature information for the superheated steam is needed. That variable is the energy heat of the current Q_{SBo}.

The dynamic relation between Q_{SBo} and W_G is studied. Since the heat of a current cannot be manipulated directly, a negative step change has been made in W_{BStOut} and a positive one in the power energy generated in the turbine (E_{Tu}) to move it. They are the most influential input variables.

Figure 7 shows the similar behavior presented by both variables. Since a proper step cannot be carried out in this case, apart from the lack of delay between them, little information can be gathered from the dynamic behavior of W_G.

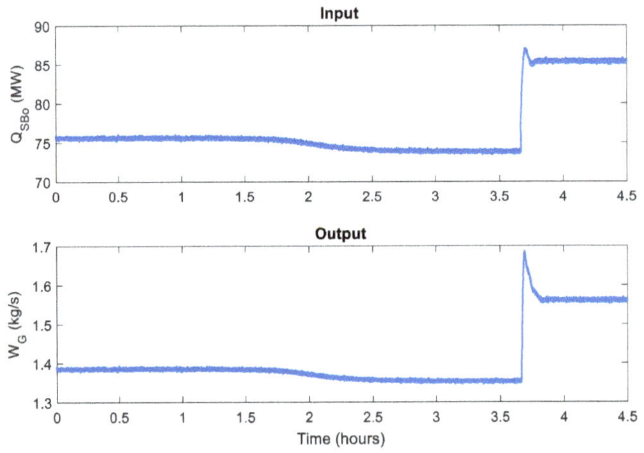

Figure 7. Dynamic response of W_G for changes in Q_{SBo}.

To study the linearity between both variables, several experiments were carried out using the simulator. Different combinations of the input variables, W_{BStOut} and E_{Tu}, were tested to determine if linearity remains for all the operational range considered. The outcome of the experiments is shown in Figure 8.

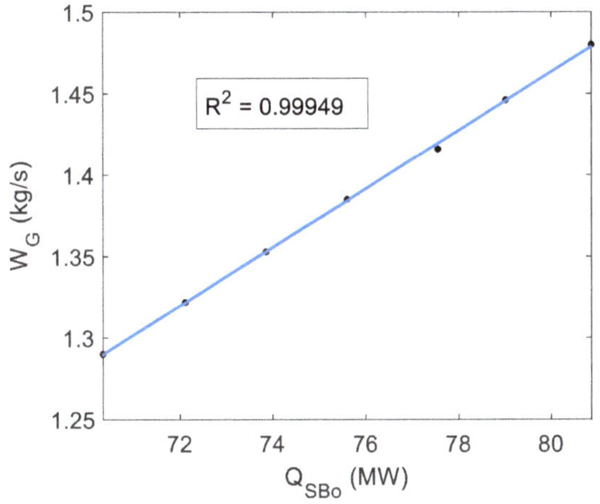

Figure 8. Linearity between Q_{SBo} and W_G.

From these results, it can be inferred that the relation between the superheated steam heat and the natural gas flow is almost linear in a stationary state. Hence, a second order state-space model is identified and the result is shown in Equation (7).

$$\begin{aligned} &W_G(t) = W_{NG_{eq}} + \Delta W_G(t) \\ &\Delta W_G(t) = 1.038 \times 10^{-3} \cdot W_{NG_{x1}}(t) + 2.308 \times 10^{-4} \cdot W_{NG_{x2}}(t) + 2.101 \times 10^{-6} \cdot (Q_{SBo}(t) - Q_{SBo_{eq}}) \\ &\dot{W}_{NG_{x1}}(t) = -7.528 \times 10^{-2} \cdot W_{NG_{x1}}(t) - 2.828 \times 10^{-2} \cdot W_{NG_{x2}}(t) + 1.953 \times 10^{-3} \cdot (Q_{SBo}(t) - Q_{SBo_{eq}}) \\ &\dot{W}_{NG_{x2}}(t) = -1.563 \times 10^{-2} \cdot W_{NG_{x1}}(t) \end{aligned} \quad (7)$$

Since the heat energy of a current is a variable that it is not measured in process industry, it must be calculated using Equation (8).

$$Q_{SBo}(t) = W_{SBo}(t) \cdot H_{SBo}(t) \quad (8)$$

To solve Equation (8), both the mass flow and the specific enthalpy of the superheated steam obtained in the boilers are needed. The enthalpy can be calculated using steam property tables [39], but to incorporate its calculation within the model, a linear function interpolating the data from the table is used instead. The obtained function (Equation (9)) is valid for every pressure, when the temperature is higher than 283 °C.

$$H_{SBo}(t) = 2355 - 1.490 \cdot P_{SBo} + 2.291 \cdot T_{SBo}(t) \quad (9)$$

In Equation (9), both the pressure and temperature of the overheated steam are required. The first can be chosen freely thanks to the pressure control system, but as it has almost no effect on any of the desired outputs, it will remain constant. On the other hand, the temperature will be a decision variable of the optimization problem, and hence it will also be known.

Since the boilers model should be as simple as possible, no more equations are added to this control volume. The steam mass flow generated in the boilers (W_{SBo}) is still needed to obtain the natural gas flow consumed. Therefore, other control volumes must be used.

3.7. Expansion Zone

The steam expansion control volume contains three identical turbines working in parallel and a bypass recirculation for steam. To simplify the problem, only one large steam turbine has been modeled, in such a way that it will consume the same amount of steam to generate the same amount of power as the sum of the other three. The first equation considered in this control volume is a mass balance, where the steam mass flow generated in the boilers must be equal to the steam mass flow that goes through the turbine and the bypass valve.

$$W_{SBo}(t) = W_{SBy}(t) + W_{STuIn}(t) \tag{10}$$

The objective now is to find both mass flows in order to calculate W_{SBo}. The amount of steam used by the turbine can be calculated using the following expression obtained from [40], where K_{Tu} is a parameter that must be adjusted from experimental data.

$$W_{STuIn}(t) = \frac{K_{Tu}}{\sqrt{T_{STuIn}(t)}} \cdot \sqrt{P_{STuIn}(t)^2 - P_{STuOut}(t)^2} \tag{11}$$

It has been assumed that the pressure drop due to the pipes and the saturator is negligible, so the steam pressure at the output of the turbines is assumed to be equal to the evaporation working pressure.

$$P_{STuOut}(t) = P_{SSaOut}(t) \tag{12}$$

To obtain the steam pressure at the input of the turbines, P_{STuIn}, a relation between the enthalpy of this current and its pressure and temperature can be found using the relation shown in Equation (9), but applied to this current.

$$H_{STuIn}(t) = 2355 - 1.49 \cdot P_{STuIn}(t) + 2.291 \cdot T_{STuIn}(t) \tag{13}$$

Valves have been assumed to be completely insulated, so they can be considered as adiabatic. If so, the enthalpy before and after them must remain constant. This property can be used to obtain the steam specific enthalpy at the input of the turbine, which must be equal to that of the overheated steam obtained from the boilers, which was calculated using Equation (9).

$$H_{STuIn}(t) = H_{SBo}(t) \tag{14}$$

Regarding the temperature at the input of the turbine (T_{STuIn}), assuming that the steam expansion through a turbine is isentropic and adiabatic, the following expression is used to obtain the steam temperature at the input of the turbines, where k is the polytrophic expansion factor for steam.

$$\frac{T_{STuOut}(t) + 273.15}{T_{STuIn}(t) + 273.15} = \frac{P_{STuOut}(t)}{P_{STuIn}(t)}^{\left(\frac{k-1}{k}\right)} \tag{15}$$

At this point, since T_{STuIn} is needed to calculate P_{STuIn} and vice versa, a nonlinear algebraic loop appears affecting Equations (13) and (15). To solve Equation (15), the temperature at the output of the turbine is needed (T_{STuOut}). This can be obtained using the relation between the specific enthalpy of this current and its pressure and temperature.

$$H_{STuOut}(t) = 2355 - 1.49 \cdot P_{STuOut}(t) + 2.291 \cdot T_{STuOut}(t) \tag{16}$$

However, as H_{STuOut} is still unknown, more equations must be added to find it. For example, the electricity generated in the turbines, which can be expressed as a relationship between the specific enthalpy of its input and output current.

$$E_{Tu}(t) = \mu_{Tu} \cdot W_{STuIn}(t) \cdot (H_{STuIn}(t) - H_{STuOut}(t)) \tag{17}$$

In Equation (17), μ_{Tu} is the efficiency of the steam turbine, which must be known beforehand, and E_{Tu} is the electricity generated in the turbine. It has been assumed that changes in the power generated by the turbines are very quick compared to the dynamic of the main process, so the equation used to represent the electricity generation is static. For the same reason, the dynamic of the power controller has also been ignored. Thus, changes in the set-point of the power controller are considered to be direct changes to the generated power.

With Equations (11) to (17), the full behavior of the large turbine can be explained, and only the pressure at the output of the saturator (P_{SSaOut}) is needed to obtain the mass flow of steam at the input of the turbines (W_{STuIn}). If these equations are analyzed carefully, it can be seen how a new nonlinear algebraic loop appears affecting all of them.

On the other hand, to solve Equation (10), the steam mass flow passing through the bypass valve must also be calculated. To do so, the following expression is used, where the rated valve coefficient (Kv_{By}) of the bypass valve must be known beforehand.

$$W_{SBy}(t) = Kv_{By} \cdot \frac{Ap_{By}(t)}{100} \cdot \sqrt{P_{SBo}(t)^2 - P_{SSaOut}(t)^2} \tag{18}$$

In Equation (18), everything is known except the opening of the bypass valve (Ap_{By}) and P_{SSaOut}. In Figure 1, it can be seen that this valve is an actuator of the split range controller used to control the evaporation working pressure. In order to obtain the opening of this valve, the output signal of that controller needs to be calculated. In conclusion, to obtain W_{SBo}, there are now two unknown variables, so more equations must be used.

3.8. Saturator

The two variables needed are related to the steam used in the evaporation section, so the saturator control volume is used to find relationships between these two variables and the ones calculated before. This contains the split-range pressure controller and the piece of equipment known as the saturator, where the exhausted steam obtained from the expansion zone is saturated with a water flow before using it in the evaporation section.

In order to calculate the opening of the bypass valve, the entire PI controller has been modeled with the same equations and parameters used in the simulated plant. As the parameters of the controller should always be known, the same approach could be extended to any real factory, using its own equations.

$$\begin{aligned}
Ap_{By}(t) &= max(0, min(100, \frac{v_{max}}{v_{max} - 45} \cdot (v(t) - 45))) \\
Ap_{Re}(t) &= max(0, min(100, \frac{-v_{max}}{45} \cdot (v(t) - 45))) \\
v(t) &= kp \cdot e(t) + v_i(t) \\
\dot{v}_i(t) &= \frac{kp}{Ti} \cdot e(t) \\
e(t) &= P_{SSaOutRef}(t) - P_{SSaOut}(t)
\end{aligned} \tag{19}$$

As can be seen in Equation (19), the opening of the bypass valve depends on the control signal calculated by the split range controller, which in turn also depends on P_{SSaOut}. So if this pressure is obtained, the whole model could be solved.

To do so, the relation between the saturated steam pressure and its specific enthalpy is used. Since it is saturated, the specific enthalpy can be computed using only one thermodynamic property. As mentioned for Equation (9), the specific enthalpy can be computed using thermodynamic tables, but as the operation range is not going to change greatly, an interpolation linear function that works in a pressure interval from 1 to 3 bar, is used instead.

$$H_{SSaOut}(t) = 24.35 \cdot P_{SSaOut}(t) + 2656 \tag{20}$$

To obtain the specific enthalpy of the saturated steam at the output of the saturator (H_{SSaOut}), an energy balance is used.

$$W_{SSaOut}(t) \cdot H_{SSaOut}(t) = W_{SSaIn}(t) \cdot H_{SSaIn}(t) + W_{WSa}(t) \cdot H_{WSa} \tag{21}$$

In Equation (21), H_{WSa} is the specific enthalpy of the water used to saturate the steam, which is assumed to have atmospheric conditions, and can therefore be easily computed. However, the steam mass flow that leaves the saturator (W_{SSaOut}), the one that enters (W_{SSaIn}), its specific enthalpy (H_{SSaIn}) and the mass flow of water needed to saturate the steam (W_{WSa}), are unknown variables which must be calculated. One relation between them can easily be found by using a mass balance in the saturator.

$$W_{WSa}(t) = W_{SSaOut}(t) - W_{SSaIn}(t) \tag{22}$$

W_{WSa} can be computed with Equation (22), while a relationship between W_{SSaOut} and the process heat energy consumed can be obtained with Equation (2), so the following expression can be found.

$$Q_P(t) = W_{SSaOut}(t) \cdot H_{SSaOut}(t) \tag{23}$$

With all the equations expressed above, there are still two unknown variables, W_{SSaIn} and H_{SSaIn}, so the last control volume related to the relief valve, must be used.

3.9. Relief Valve

This control volume includes the bifurcation that leads the expanded steam to the saturator or to the reflief valve, which is used for reducing an excess of pressure in the system, if necessary. The first equation considered to find the unknown variables is a mass balance between the steam mass flow through the bypass, the steam turbine, the relief valve and the saturator.

$$W_{SSaIn}(t) = W_{SBy}(t) + W_{STuIn}(t) - W_{SRe}(t) \tag{24}$$

In this equation, the mass flow that goes through the relief valve appears, and it will be calculated using a similar expression to (18).

$$W_{SRe}(t) = Kv_{Re} \cdot \frac{Ap_{Re}(t)}{100} \cdot \sqrt{P_{SSaOut}(t)^2 - P_{atm}(t)^2} \tag{25}$$

The Kv of the valve and the atmospheric pressure are known parameters, so the opening of the relief valve is the only unknown variable, which can be computed using Equation (19). Therefore, at this point, only the specific enthalpy of the team that goes into the saturator (H_{SSaIn}) is required to solve the entire model.

To obtain this, an energy balance is used with the same streams as in Equation (24). As a disjunction occurs, the specific steam enthalpy at the input of the relief valve is the same as at the input of the saturator. Furthermore, the bypass valve is considered to be isenthalpic, so the specific enthalpy of this current can be assumed to be the same as that of the superheated steam that leaves the boilers.

$$\begin{aligned} H_{SRe}(t) &= H_{SSaIn}(t) \\ H_{SBy}(t) &= H_{SBo}(t) \\ W_{SBy}(t) \cdot H_{SBo}(t) + W_{STuIn}(t) \cdot H_{STuOut}(t) &= W_{SRe}(t) \cdot H_{SSaIn}(t) + W_{SSaIn}(t) \cdot H_{SSaIn}(t) \end{aligned} \tag{26}$$

With Equations (7) to (26), the consumption of natural gas can be computed and, since the behavior of the CHP has been explained, the European indexes can also be computed in the next subsection. In Figure 9, a scheme of the inputs and outputs of the different control volumes used to obtain the model can be found. As it can be observed, a new nonlinear algebraic loop appears for calculating the opening of the relief valve.

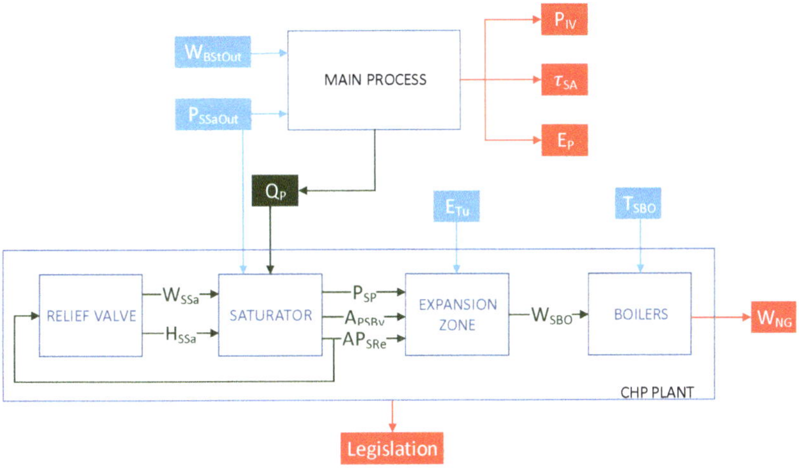

Figure 9. Dependency scheme of the model.

3.10. European Legislation

Following Directive 2012/27/UE [27], in order to consider the cogeneration system as efficient, an index called Primary Energy Savings (PES) is defined. This compares the performance of the CHP plant studied to some expected heat and power efficiency references given for each type of cogeneration system. Cogeneration systems like the one treated in this work, which has a capacity of more than 1 MWe, will be considered efficient if its PES index is greater than 0.10.

$$PES(t) = 1 - \frac{1}{\frac{\mu Q_{CHP}(t)}{\mu Q_{Ref}} + \frac{\mu E_{CHP}(t)}{\mu E_{Ref}}} \tag{27}$$

In order to obtain the efficiency reference values, the European regulation 2015/2402 must be consulted [28]. The actual heat and power efficiencies are defined, respectively, as the useful heat and power energy obtained divided by the energy obtained from fuel combustion:

$$\mu Q_{CHP}(t) = \frac{Q_{CHP}(t)}{F_{CHP}(t)} \tag{28}$$

$$\mu E_{CHP}(t) = \frac{E_{CHP}(t)}{F_{CHP}(t)} \tag{29}$$

Useful heat in this problem has been obtained as the difference between the steam enthalpy at the input of the saturator, and the water enthalpy at the input of the boiler. This difference is the energy given by the fuel. The heat lost in purging the boilers has been neglected.

$$Q_{CHP}(t) = W_{SSAT}(t) \cdot (H_{SSAT}(t) - H_{WBo}) \tag{30}$$

According to the European law, energy can be generated in "cogeneration mode" or not. While the useful heat generated in cogeneration systems is always considered as generated in cogeneration mode, the same cannot be applied to the useful electricity (E_{CHP}) and the fuel used (F_{CHP}). This is done to avoid trickery, as electricity may be generated only for selling it to the external grid, and not for the main process operation. These variables will be considered the same if and only if the global efficiency of the plant for one year is considered greater than or equal to 0.75. This limit has been obtained from Directive 2012/27/UE, considering a cogeneration system with counter pressure turbines and without condensation. The global efficiency is a percentage of how much of the energy generated from burning fuel in the boilers has been used and not wasted.

$$\mu_G(t) = \frac{E_{Plant}(t) + Q_{CHP}(t)}{F_{Plant}} \tag{31}$$

In this expression, E_{plant} is the actual electricity being generated in the turbines at instant t ($E_{Tu}(t)$), while F_{plant} is the energy contained in the fuel being burned at instant t. For the process considered, the fuel used to obtained steam is natural gas.

$$F_{plant}(t) = W_G(t) \cdot PCI_{NG} \tag{32}$$

Should the global efficiency of the plant be below 0.75, the useful electricity and the fuel energy would have to be divided into two different parts, cogeneration and non-cogeneration mode. For the case study considered, the global efficiency is always above 0.75, using restrictions in the optimization problem. Therefore, more equations are not necessary to compute the PES of the system.

Table 1 shows the principal features of the model developed and compares them with the model used for simulation.

Table 1. Comparison between the model used for simulation and the model obtained for optimization.

	Simulator	Optimization Model
Number of equations	6485	52
- Static	6036	38
- Dynamic	449	14
Parameters	2131	23
Variables	6456	48
Input variables	29	4
Non-linear algebraic loops	8	3

4. Validation

In this section, the validity of the model developed in the previous section is analyzed. In order to do so, the model will be simulated in EcosimPro® [41], an object oriented modeling and simulation software. Several experiments, which are shown in Figure 10, have been carried out to test the response of the model. These tests correspond to different typical operational points.

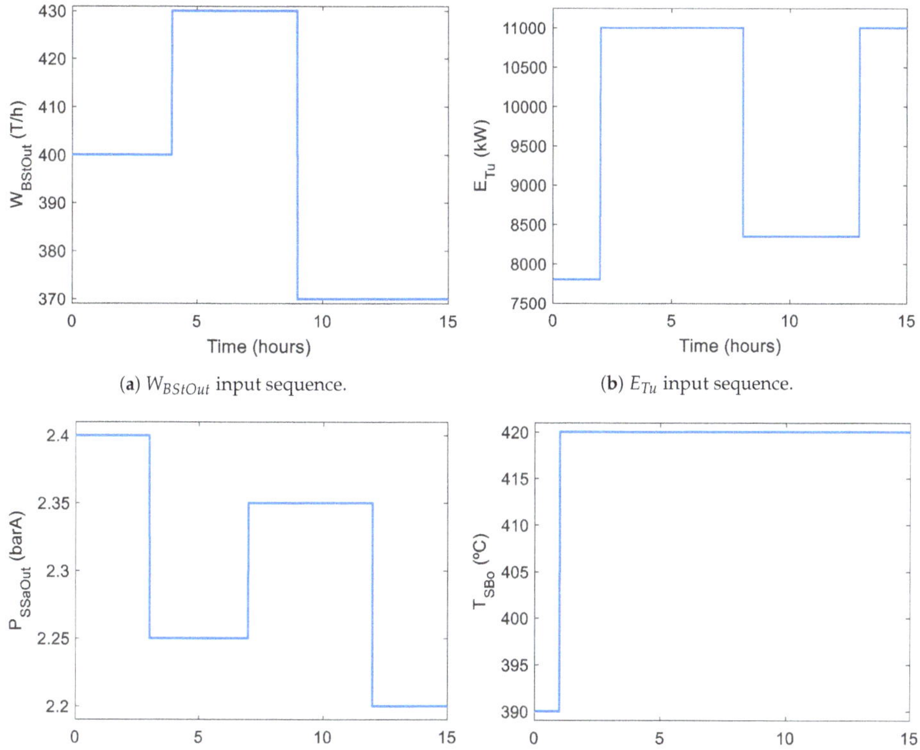

(**a**) W_{BStOut} input sequence. (**b**) E_{Tu} input sequence.

(**c**) P_{SSaOut} input sequence. (**d**) T_{SBo} input sequence.

Figure 10. Inputs for validation.

As the methodology for validation, the same inputs have been introduced to both the simulator and the optimization model. Their response is analyzed in two different ways: on the one hand, both responses can be compared graphically in Figure 11; while on the other, the Root Mean Squared Error (RMSE) has been computed to show the error of the model numerically. This index is one of the most used in the literature and calculates the error of the model with respect to the measurements, weighing the farthest predictions.

$$RMSE = \sqrt{\frac{1}{N}\sum_{i=1}^{N}(y_i - \hat{y}_i)^2} \quad (33)$$

Discussion of the Results

In this subsection, the accuracy of the model is discussed. This is a highly relevant topic in optimization problems, since a bad model may lead the optimization to suboptimal or even infeasible results when the responses are sent to the plant. If the accuracy of the model is not good enough, caused for example by uncertainty in process measurements, this will have to be considered explicitly in the optimization problem. The way to do it is using a different approach that considers this kind of uncertainty explicitly. Several methodologies can be applied for this topic, like the two-step approach [42], modifier adaptation [43], or extremum seeking control [44], among others [45].

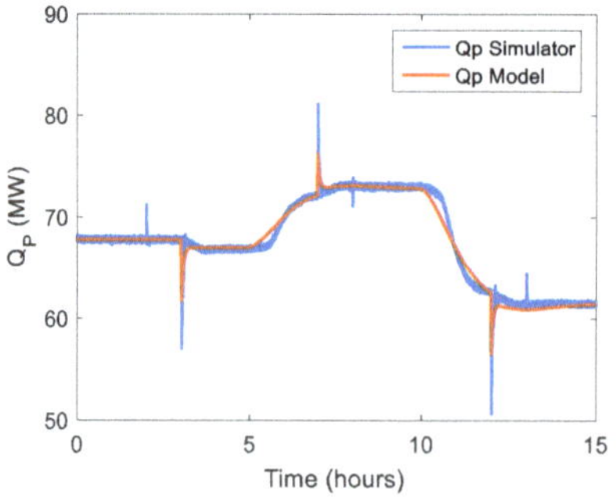

(**a**) Q_p results from validation.

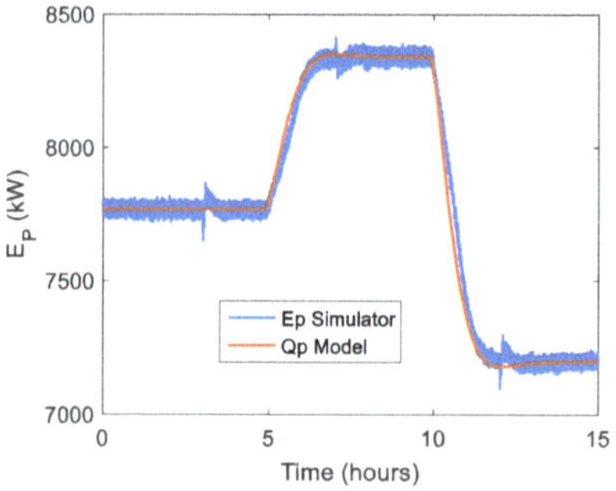

(**b**) E_p results from validation.

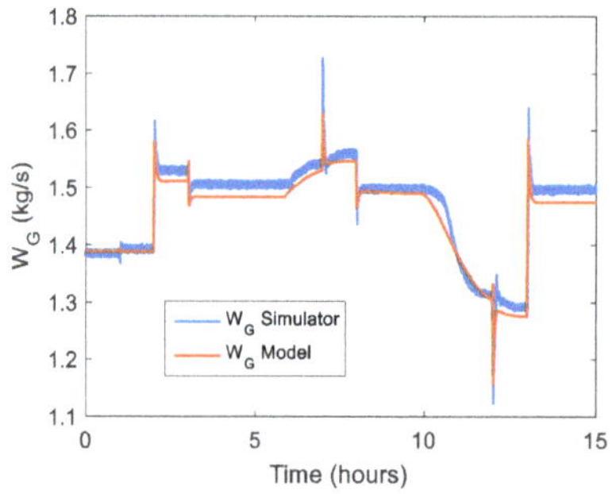

(**c**) W_G results from validation.

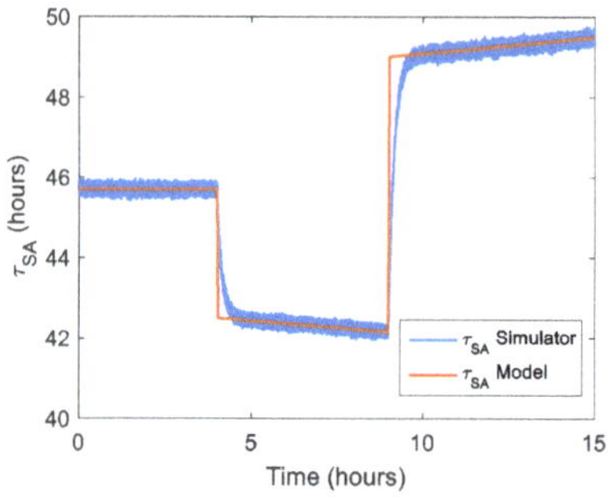

(**d**) τ_{SA} results from validation.

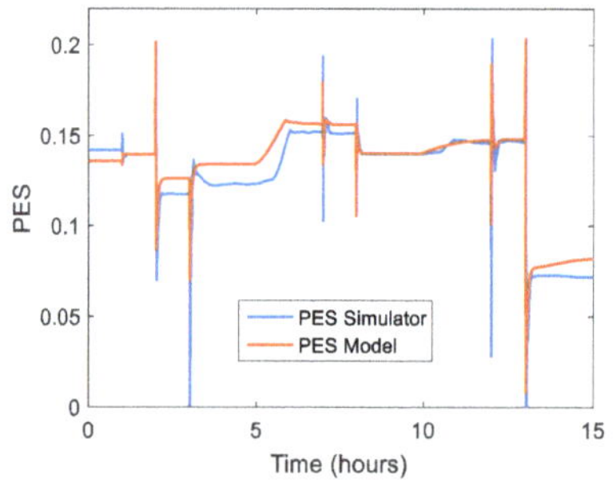

(**e**) PES results from validation.

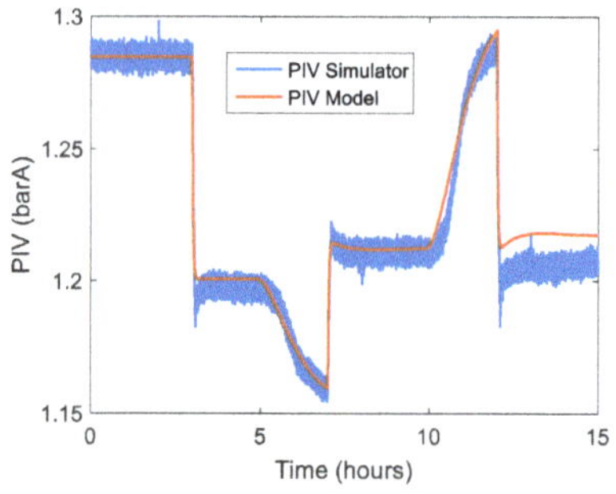

(**f**) P_{IV} results from validation.

Figure 11. Outputs from data validation.

Looking at Figure 10 three typical operational points, corresponding to three different beet production rates, can be discerned. The energy generated in turbines is moved in order to sell more or less electricity and the steam boilers temperature and evaporation working pressure are modified to try to maintain the P_{IV} and PES between limits.

Figure 11 shows that, for almost all the time, the output of the optimization model follows the shape of the output of the simulator. For every output variable, it can be seen how the stationary state is captured reasonably well. The transient state is clearly affected by the linear assumptions made when obtaining the model. Here, the behaviour of the model during slow and quick transients can be differentiated. For slow transients, caused by changes in production, the dynamic of the real plant is approximated by a straight line. This can be observed, for example, in W_G from 10 to 13 h approximately. Although this response is not perfect, for the purpose of the model, it is sufficiently close to the real value of the plant. On the other hand, when quick transients are considered, in most cases, it can be seen that the model is able to predict these changes, but that fails when computing the amplitude. This is especially relevant in the prediction of the PES. However, since those transients are too short and this index is evaluated yearly, it is enough for the desired purpose.

The graphic results can be corroborated in Table 2. Here, for every variable except PES, the relative error is below 2%. This means that the response of the model is very close to the response of the plant. In the case of the PES, the quick transients, and a bias from hour 3 to hour 7, make the relative error move to 6.31. Traditionally, an error below 5% is considered sufficiently good. In our case, this value is slightly exceeded because of the reasons stated above, so it can be concluded that it is acceptable, but additional caution will have to be taken during optimization.

With these results, it can be concluded that the model accuracy is good enough for the operational range considered, and therefore an optimization approach that deals with model mismatch will not be needed.

Table 2. Indexes for the output variables of the model.

Output Variable	RMSE [Ud]	Output Mean [Ud]	Relative Error [%]
Q_p	566.03 kW	67,614.00 kW	0.84
E_p	36.72 kW	7793.89 kW	0.47
W_G	0.018 kg/s	1.47 kg/s	1.24
τ_{St}	0.52 h	45.96 h	1.12
PES	0.01	0.13	6.31
P_{IV}	0.01 barA	1.22 barA	0.52

5. Extension to Other Case Studies

In this section, the methodology used for obtaining the model is presented in a schematic way. Thus, it can be used for other case studies.

1. Establishment of the simulation objectives.
2. Set model inputs and outputs.
3. Search for relations between the inputs and outputs.
    3.1. Look for control volumes to disaggregate the model.
    3.2. Select where to apply first principles or experimental models.
4. Experimental modeling.
    4.1. Analyze the response of the selected outputs, when steps are performed in the inputs.
    4.2. Check linearity between the inputs and outputs selected.
    4.3. If the relation is linear, apply linear identification techniques. Otherwise, look for another approach like symbolic regression [46], Neural Networks or Fuzzy models [47].
5. Parametrization of the model.
6. Symbolic manipulation of the gray-box model obtained. To do so, we recommend the use of a software like EcosimPro [41] as modeling and simulation environment.
    6.1. Perform a degree of freedom analysis.
    6.2. Look for algebraic loops.
    6.3. Check the existence of high-index problems.
7. Initialization of the model. Use a stationary point obtained by the real system.
8. Validation of the model.

8.1. Qualitative validation. Check the transient response and robustness of the model for different inputs.
8.2. Quantitative validation. Compare the output of the obtained model with the real process. This has been done in a graphical and analytical way.

If the CHP scheme is different from the one proposed, all this steps must be performed in order to obtain a new model. However, if it is maintained, several points can be avoided. This way, the first step would be to change the inputs and outputs of the model, since they are process dependent (point 2). Next, the control volumes could be kept, and the experimental models would have to be obtained for the new process (Point 4). The parametrization of the system is also required (point 5), but the model would not need to be manipulated again. Finally, the initialization and validation of the model would be required for the specific case study (points 7 and 8).

6. Conclusions

In this paper, the methodology to obtain a gray-box model for optimizing the production rate of an industrial process, along with the management of an on-site CHP plant, has been reported. This model is especially designed for cases where cogeneration is coupled to the main process and electricity generation depends on the production rate. This is the case of the sugar factory presented as case study. A model of this process was obtained in a previous work; however, due to its size, it cannot be used for optimization. Nevertheless, thanks to its performance and response, which has already been validated, it has been used as a real plant, making the development of the optimization model faster. The resulting model is dynamic and non-linear, mixing first principle equations with linear identification. Thus, if the CHP structure is maintained, it can be extended to other industries simply by identifying their own response, and using their own parameters, which are easy to obtain. It has been shown to be able to predict the behavior of the system reasonably well, simultaneously computing the European legislation indexes for CHP efficiency with only 52 equations, which makes it suitable for optimization.

In future work, this model will be used in the optimization, according to the electricity Day-Ahead market, of the process and CHP management of the simulated sugar factory.

Author Contributions: Conceptualization, C.P., A.M., L.F.A.; Methodology, C.P.; Software, C.P.; validation, C.P.; Investigation, C.P., A.M., L.F.A.; Resources, C.P., A.M., L.F.A.; writing—original draft preparation, C.P.; writing—review and editing, C.P.; visualization, C.P.; supervision, A.M., L.F.A.; project administration, A.M., L.F.A.

Funding: This research was funded by the Spanish MICINN and FSE under FPI grant BES-2016-079388. C.Pablos also acknowledges support by the regional government of Castilla y León and the EU-FEDER (CLU 2017-09).

Acknowledgments: The authors wish to express their gratitude to the personnel from "ACOR" for the aid provided for this work.

Conflicts of Interest: The authors declare no conflict of interest.

Appendix A. Notation of the Model

The notation of the model variables can be divided into five different groups:

1. Streams.
2. Zone/Equipment.
3. Split Range PI controller.
4. Legislation.
5. Identification models.

The notation of the variables related to streams is composed by up to four indexes that states: first the magnitude of the variable, second the element, third the equipment directly related to the variable, and lastly the direction of the current with respect to that equipment. For example, $WSTuIn$ refers to the mass flow of steam that goes into the turbines. The use of all indexes is not necessary for most of the variables, and the least possible amount is used to avoid confusion. The rest of the variables and parameters are fully explained in Tables A1 and A2.

Table A1. Model variables notation.

VARIABLES		
Name	**Description**	**Units**
	1. STREAMS	
Magnitude		
H	Specific Enthalpy	kJ/kg
P	Pressure	barA
Q	Heat Flow	kJ/s
T	Temperature	°C
W	Mass Flow	kg/s
Element		
B	Beet	
G	Natural Gas	
S	Steam	
W	Water	
Equipment		
Bo	BStOutilers	
By	Bypass Valve	
Re	Relief Valve	
Sa	Stturator	
St	Storage zone	
Tu	Turbines	
Direction		
In	Flow that goes into the specified zone	
Out	Flow that leaves the specified zone	
	2. PROCESS	
E_p	Power consumed by the main process	kW
E_{Tu}	Power energy generated in turbines	kW
m_{St}	Accumulated mass beet in the storage zone	T
P_{IV}	Pressure in the fourth effect of the evaporation	barA
Q_p	Heat consumed by the main process	kW
τ_{St}	Beet residence time in storage zone	h
	3. SPLIT RANGE PI CONTROLLER	
Ap_{By}	Opening of the bypass valve	%
Ap_{Re}	Opening of the relief valve	%
e	Error	barA
$P_{SSaOutRef}$	Pressure set-point	barA
v	Output signal	%
v_i	Integral action	%
	4. LEGISLATION	
E_{CHP}	Power generated in cogeneration mode	kW
E_{plant}	Power generated in turbines	kW
F_{CHP}	Energy obtained from fuel in cogeneration mode	kJ/s
F_{plant}	Energy obtained from fuel	kJ/s
μE_{CHP}	Power efficiency of the CHP	%
μ_G	Global efficiency	%
μQ_{CHP}	Heat efficiency of the CHP	%
PES	Primary Saving Energy index	
Q_{CHP}	Heat generated in cogeneration mode	kJ/s
	5. IDENTIFICATION MODELS	
ΔVar	Increase in the value of the variable with respect to the equilibrium point	
Var_{xn}	nth internal state of the variable	

Table A2. Parameters value and notation.

PARAMETERS			
Name	**Description**	**Value**	**Units**
$E_{p_{eq}}$	Value of E_p at the identification point	7769.26	kW
H_{WBo}	Specific enthalpy of the water used in the boilers	550.52	kJ/kg
H_{WSa}	Specific enthalpy of the water used in the saturator	125.80	kJ/kg
k	Polytropic index	1.20	
kp	Proportional gain	140	%/bar
Kv_{By}	Bypass rated valve coefficient	0.50	kg/(s·bar)
Kv_{Re}	Relief rated valve coefficient	5.00	kg/(s·bar)
K_{Tu}	Turbines experimental parameter	23.35	(kg/s·°C)/ bar
μE_{Ref}	Reference cogeneration power efficiency	0.53	
μQ_{Ref}	Reference cogeneration heat efficiency	0.87	
μ_{Tu}	Efficiency of the steam turbine	0.95	
P_{atm}	Atmospheric pressure	1.00	barA
$P_{IV_{eq}}$	Value of P_{IV} at the identification point	1.28	barA
P_{SBo}	Steam Pressure at the output of the boilers	37.30	barA
$P_{SSaOut_{eq}}$	Value of P_{SSaOut} at the identification point	2.40	barA
PCI_G	Natural Gas Lower Heating Value (LHV)	48,130.09	kJ/kg
PCS_G	Natural Gas Higher Heating Value (HHV)	52,200	kJ/kg
Pr_G	Natural Gas Price		€/kWh
Pr_E	Electricity Price		€/kWh
ρ_G	Natural Gas density for the input conditions	3.65	kg/m^3
$Q_{p_{eq}}$	Value of Q_p at the identification point	67,779.55	kW
$Q_{SBo_{eq}}$	Value of Q_{SBo} at the identification point	75,590.83	kW
Ti	Integral gain	10.00	s
T_N	Predicted horizon time		s
v_{max}	Maximum output signal of the splite range controller	100.00	%
$W_{BStOut_{eq}}$	Value of W_{BStOut} at the identification point	400.00	T/h
W_{BStIn}	Arrival of beet to the storage zone	400.00	T/h
$W_{NG_{eq}}$	Value of W_G at the identification point	1.39	kg/s

References

1. Kehlhofer, R.; Hannemann, F.; Rukes, B.; Stirnimann, F. *Combined-Cycle Gas & Steam Turbine Power Plants*; Pennwell Books: Tulsa, OK, USA, 2009.
2. Ashok, S.; Banerjee, R. Optimal operation of industrial cogeneration for load management. *IEEE Trans. Power Syst.* **2003**, *18*, 931–937. [CrossRef]
3. Bindlish, R. Power scheduling and real-time optimization of industrial cogeneration plants. *Comput. Chem. Eng.* **2016**, *87*, 257–266. [CrossRef]
4. Kim, J.S.; Edgar, T.F. Optimal scheduling of combined heat and power plants using mixed-integer nonlinear programming. *Energy* **2014**, *77*, 675–690. [CrossRef]
5. Mitra, S.; Sun, L.; Grossmann, I.E. Optimal scheduling of industrial combined heat and power plants under time-sensitive electricity prices. *Energy* **2013**, *54*, 194–211. [CrossRef]
6. Tina, G.; Passarello, G. Short-term scheduling of industrial cogeneration systems for annual revenue maximisation. *Energy* **2012**, *42*, 46–56. [CrossRef]
7. Salgado, F.; Pedrero, P. Short-term operation planning on cogeneration systems: A survey. *Electr. Power Syst. Res.* **2008**, *78*, 835–848. [CrossRef]
8. Yusta, J.; De Oliveira-De Jesus, P.; Khodr, H. Optimal energy exchange of an industrial cogeneration in a day-ahead electricity market. *Electr. Power Syst. Res.* **2008**, *78*, 1764–1772. [CrossRef]
9. Dvořák, M.; Havel, P. Combined heat and power production planning under liberalized market conditions. *Appl. Therm. Eng.* **2012**, *43*, 163–173. [CrossRef]
10. Agha, M.H.; Thery, R.; Hetreux, G.; Hait, A.; Le Lann, J.M. Integrated production and utility system approach for optimizing industrial unit operations. *Energy* **2010**, *35*, 611–627. [CrossRef]

11. Zhang, Q.; Grossmann, I.E. Planning and scheduling for industrial demand side management: Advances and challenges. In *Alternative Energy Sources and Technologies*; Springer: Berlin, Germany, 2016; pp. 383–414.
12. Hadera, H.; Harjunkoski, I.; Sand, G.; Grossmann, I.E.; Engell, S. Optimization of steel production scheduling with complex time-sensitive electricity cost. *Comput. Chem. Eng.* **2015**, *76*, 117–136. [CrossRef]
13. Leo, E.; Engell, S. Integrated day-ahead energy procurement and production scheduling. *at-Automatisierungstechnik* **2018**, *66*, 950–963. [CrossRef]
14. Biegler, L.T. *Nonlinear Programming: Concepts, Algorithms, and Applications To Chemical Processes*; Siam: Philadelphia, PA, USA, 2010; Volume 10.
15. Lanz, T.; Epple, B.; Heinze, C.; Mertens, N.; Alobaid, F.; Starkloff, R. Progress in dynamic simulation of thermal power plants. *Prog. Energy Combust. Sci.* **2016**, *59*, 79–162.
16. Ringkjøb, H.K.; Haugan, P.M.; Solbrekke, I.M. A review of modelling tools for energy and electricity systems with large shares of variable renewables. *Renew. Sustain. Energy Rev.* **2018**, *96*, 440–459. [CrossRef]
17. Deneux, O.; El Hafni, B.; Péchiné, B.; Di Penta, E.; Antonucci, G.; Nuccio, P. Establishment of a model for a combined heat and power plant with ThermosysPro library. *Procedia Comput. Sci.* **2013**, *19*, 746–753. [CrossRef]
18. Merino, A.; Mazaeda, R.; Alves, R.; Rueda, A.; Acebes, L.; de Prada, C. Sugar factory simulator for operators training. *IFAC Proc. Vol.* **2006**, *39*, 259–264. [CrossRef]
19. Palacios-Bereche, R.; Nebra, S.A. Thermodynamic modeling of a cogeneration system for a sugarcane mill using Aspen-Plus, difficulties and challenges. In Proceedings of the 20th International Congress of Mechanical Engineering, Gramado-RS, Brazil, 15–20 November 2009; pp. 1–9.
20. Keshavarz, M.; Barkhordari Yazdi, M.; Jahed-Motlagh, M.R. Piecewise affine modeling and control of a boiler-turbine unit. *Appl. Therm. Eng.* **2010**, *30*, 781–791. [CrossRef]
21. Sørensen, K.; Karstensen, C.M.S.; Condra, T.; Houbak, N. Modelling and simulating fire tube boiler performance. In *Proceedings of SIMS 2003*; Malardalen University: Malardalen, Sweden, 2003; ISBN 91-631-4716-5.
22. Sunil, P.U.; Barve, J.; Nataraj, P.S. Boiler Model and Simulation for Control Design and Validation. In Proceedings of the 3rd International Conference on Advances in Control and Optimization of Dynamical Systems ACODS 2014, Kanpur, India, 13–15 March 2014; pp. 936–940.
23. Dulau, M.; Bica, D. Mathematical Modelling and Simulation of the Behaviour of the Steam Turbine. *Procedia Technol.* **2014**, *12*, 723–729. [CrossRef]
24. Merino, A. Librería de Modelos del Cuarto de Remolacha de una Industria Azucarera para un Simulador de Entrenamiento de Operarios. Ph.D. Thesis, Universidad de Valladolid, Valladolid, Spain, 2008.
25. Merino, A.; Alves, R.; Acebes, L. A training simulator for the evaporation section of a beet sugar production process. In Proceedings of the 2005 European Simulation and Modelling Conference, University of Porto, Porto, Portugal, 24–26 October 2005; pp. 24–26.
26. Pablos, C.; Felipe Acebes, L.; Merino, A. Sugar plant simulator for energy management purposes. In Proceedings of the 29th European Modeling and Simulation Symposium, Barcelona, Spain, 18–20 September 2017; pp. 370–379.
27. Directive 2012/27/EU of the European Parliament and of the Council of 25 October 2012 on Energy Efficiency. 2012. Available online: http://data.europa.eu/eli/dir/2012/27/oj (accessed on 25 February 2019).
28. Commission Delegated Regulation (EU) 2015/2402 of 12 October 2015 Reviewing Harmonised Efficiency Reference Values for Separate Production of Electricity and Heat. 2015. Available online: http://data.europa.eu/eli/regdel/2015/2402/oj (accessed on 25 February 2019).
29. Van der Poel, P. *Sugar Technology. Beet and Cane Sugar Manufacture*; van der Poel, H., Schiweck, T., Eds.; Verlag Dr. Albert Vartens KG: Berlin, Germany, 1998.
30. Asadi, M. *Beet-Sugar Handbook*; John Wiley & Sons: Hoboken, NJ, USA, 2006.
31. Starzak, M.; Davis, S.B. MATLAB® modelling of a sugar mill: Model development and validation. In Proceedings of the 89th Annual Congress of the South African Sugar Technologists' Association (SASTA 2016), Durban, South Africa, 16–18 August 2016; South African Sugar Technologists' Association: Mount Edgecombe, South Africa, 2017; pp. 517–536
32. Yarnal, G.S.; Puranik, V.S. Energy management in cogeneration system of sugar industry using system dynamics modeling. *Cogener. Distrib. Gener. J.* **2009**, *24*, 7–22. [CrossRef]
33. Chantasiriwan, S.; Charoenvai, S. Improving Cogeneration in Sugar Factories by Superheated Steam Drying of Bagasse *Int. J. Appl. Eng. Res.* **2018**, *13*, 8700–8707.

34. Ljung, L. *System Identification: Theory for the User*; Prentice-Hall: Upper Saddle River, NJ, USA, 1987.
35. Isermann, R.; Münchhof, M. *Identification of Dynamic Systems: An Introduction with Applications*; Springer Science & Business Media: Berlin, Germany, 2010.
36. Urbaniec, K. (Ed.) *Modern Energy Economy in Beet Sugar Factories*; Sugar Series; Elsevier: Amsterdam, The Netherlands, 1989; Volume 10.
37. MATLAB 8.6.0.267246 (R2015b). The MathWorks Inc. 2015. Available online: https://www.mathworks.com/products/matlab.html (accessed on 11 February 2019).
38. Frankenfeld, T.; Voss, C. Electrical power consumption-an European benchmarking-exercise-Part I: Survey. *Zuckerindustrie* **2004**, *129*, 407–414.
39. Perry, R.H.; Green, D.W.; Maloney, J.O. *Perry's Chemical Engineers' Handbook*; McGraw-Hill: New York, NY, USA, 2015.
40. Chaibakhsh, A.; Ghaffari, A. Steam turbine model. *Simul. Model. Pract. Theory* **2008**, *16*, 1145–1162. [CrossRef]
41. EcosimPro 5.4.19. Empresarios Agrupados. 2018. Available online: https://www.ecosimpro.com/ (accessed on 4 February 2019).
42. Chen, C.Y.; Joseph, B. On-line optimization using a two-phase approach: An application study. *Ind. Eng. Chem. Res.* **1987**, *26*, 1924–1930. [CrossRef]
43. Rodríguez-Blanco, T.; Sarabia, D.; Pitarch, J.; De Prada, C. Modifier Adaptation methodology based on transient and static measurements for RTO to cope with structural uncertainty. *Comput. Chem. Eng.* **2017**, *106*, 480–500. [CrossRef]
44. Krstic, M.; Wang, H.H. Stability of extremum seeking feedback for general nonlinear dynamic systems. *Automatica* **2000**, *36*, 595–601. [CrossRef]
45. Navia López, D.A. Handling Uncertainties in Process Optimization. Ph.D. Thesis, Universidad de Valladolid, Valladolid, Spain, 2013.
46. Cozad, A.; Sahinidis, N.V. A global MINLP approach to symbolic regression. *Math. Program.* **2018**, *170*, 97–119. [CrossRef]
47. Nelles, O. *Nonlinear System Identification: From Classical Approaches to Neural Networks and Fuzzy Models*; Springer Science & Business Media: Berlin, Germany, 2013.

processes

MDPI

Article

Mathematical Modelling Forecast on the Idling Transient Characteristic of Reactor Coolant Pump

Xiuli Wang [1], Yajie Xie [1], Yonggang Lu [1,2], Rongsheng Zhu [1,*], Qiang Fu [1,*], Zheng Cai [1] and Ce An [1]

[1] National Research Center of Pumps, Jiangsu University, Zhenjiang 212013, China
[2] School of Mechanical & Aerospace Engineering, Nanyang Technological University, Singapore 639798, Singapore
* Correspondence: ujs_zrs@163.com (R.Z.); ujsfq@sina.com (Q.F.); Tel.: +86-186-0511-0959 (R.Z.); +86-157-5101-0752 (Q.F.)

Received: 29 April 2019; Accepted: 11 July 2019; Published: 15 July 2019

Abstract: The idling behavior of the reactor coolant pump is referred to as an important indicator of the safe operation of the nuclear power system, while the idling transition process under the power failure accident condition is developed as a transient flow process. In this process, the parameters such as the flow rate, speed, and head of the reactor coolant pump are all nonlinear changes. In order to ensure the optimal idling behavior of the reactor coolant pump under the power cutoff accident condition, this manuscript takes the guide vanes of the AP1000 reactor coolant pump as the subject of this study. In this paper, the mathematical model of idling speed and flow characteristic curve of reactor coolant pump under the power failure condition were proposed, while the hydraulic modeling database of different vane structure parameters was modeled based on the orthogonal optimization schemes. Furthermore, based on the mathematical modeling framework of multiple linear regressions, the mathematical relationship of the hydraulic performance of each guide vane in different parameters was predicted. The derived model was verified with the idling test data.

Keywords: reactor coolant pump; vane; costing stopping; mathematical model; idling test

1. Introduction

In the case of a power outage, the reactor coolant pump will be forced to suspend its current operation due to the loss of power. Resulting from a consequent sudden reduction of coolant flow through the core of the reactor, the potential threat for the safety of the reactor is introduced. To ensure the nuclear safety of the reactor nuclear reaction boiling state [1], it is usually required that the core of the primary cooling loop shall be kept running by the unit for a certain period driven by rotational inertia. Therefore, the reactor coolant pump can generate enough flow to absorb the heat generated by the reactor in a short time after the power interruption. The capability of the reactor coolant pump to maintain operation by rotor inertia is called idling characteristics.

The reactor coolant pump coasting under power outage is commonly considered as a transient process. During the transition process of the reactor coolant pump stoppage, all the hydraulic parameters are forced to be continuously varied against times. The traditional velocity distribution, pressure distribution or streamlined diagram can be only applied to identify the overall quality of the flow field or to analyze the flow performance inside the impeller from a macro level. Many studies conducted by domestic and foreign scholars [1–4] can be referred to on the transient changes law after the power failure or shut-down of the reactor coolant pump. Zhang Senru [5] proposed a flow calculation model for the transient process of each loop of the nuclear power plant by using the coolant momentum conservation equation and the torque balance relation of the reactor coolant pump. Moreover, it was also found that the calculation model of the system flow characteristic curve was proposed by scholars like Guo Yujun [6] with the four-quadrant characteristic curve according to the torque balance relation of the reactor coolant pump. The computational model has proven to be important for analyzing the taxiing behavior of reactor coolant pumps. However, the coasting transient condition of the reactor coolant pump of Qinshan Nuclear Power Plant Phase II was calculated by Deng Shaowen [7] through the international conventional transient calculation method. The calculated flow curve of the reactor coolant pump was further compared with that given by the Framatome Atomic Energy Company, and the calculation results were consistent with the actual condition. However, it has also been found that scholars, such as Xu Yiming [8], simplify the calculation of the idling speed model of the reactor coolant pump after power failure through the torque balance relationship. Zhu Rongsheng studied the pressure pulsation characteristics of the reactor coolant pump at low flow conditions, and found that unstable pressure pulsation in the small flow condition was caused by the backflow in the impeller vane flow runner, the backflow mainly existed in the impeller and guide blade import and export [9]. Long Yun [10] and Feng Xiaodong [11] studied the vibration characteristics of the bearing seat in the coasting transition process of the reactor coolant pump in the power failure accidents and the internal flow characteristics of the pump under the condition of small flow. Alatrash, Y. [12] conducted an experimental study of the inertial pumping capability during the coastdown period to confirm whether the coastdown half time requirement given by safety analyses is being satisfied. In the modular modeling system(MMS) simulation model, all of the design data that affect the pump coastdown behavior are reflected. The experimental dataset is well predicted by the MMS model and is confirmed to be valid and consistent. For the study of reactor coolant pump, Brady, D. R. [13] proposed a method of upgrading a 1500 rpm reactor coolant pump motor having a vertically oriented rotor shaft supported by a lower guide bearing disposed of in an oil reservoir. Bang, S. Y. [14] mainly focuses on the performance requirements of the APR1400 and implements the safety, reliability and adaptability goals of the APR1400 system design, and describes the details of the development process, improved design features, and type test results. Metzroth, K. [15] used dynamic probabilistic risk assessment (PRA) methods combined with the MELCOR system code to check the behavior of RCP sealed loss of coolant accident (LOCA) and its impact on the evolution of SBO accident. It was found that the dynamic event tree (DET) analysis produced results were less conservative than those obtained in NUREG-1150, which had predicted larger leak rates earlier in the accident. Lu Y. et al. [16–18] studied the third and fourth generation reactor coolant pumps, main including gas-liquid two-phase flow and the transient characteristics of pump under extreme operating conditions, and finding in different gas fraction conditions, the homogeneous distribution of gas phase component in the fluid area was mainly related to the operating condition of pump, and when the volume flow rate deviates from the designed operating point, the fluid medium's gas phase and liquid phase will separate to some extent.

On the premise of previous researches, the idling speed model of the reactor coolant pump to a reasonable extent by the momentum conservation equation was simplified by this paper, and the calculation model of coasting condition was further derived. Secondly, the hydraulic modeling database of different vane structure parameters was modeled under the orthogonal optimization schemes, and the mathematical relationship of the hydraulic performance of each guide vane in different parameter was predicted on mathematical modeling framework of multiple linear regression, by which the idling mathematical model of main geometrical parameters was derived. At last, the model was verified by the existing power outage test data, and the design criteria of the reactor coolant pump were further obtained from the modeling. Finally, the design parameters of the reactor coolant pump were calculated and verified.

2. Materials and Methods

2.1. Basic Theory of the Idling Condition of the Reactor Coolant Pump

The hydraulic performance and its idling characteristics of the reactor coolant pump are affected by different geometric parameters of guide vanes to a certain extent. The influence of different parameters of guide vane on the performance is not only determined by the size of a single geometric parameter, but also indirectly influenced by the mutual combination of the parameters due to their coupling relationship. Referring to the current research, the study of the idling characteristics of the reactor coolant pump is mainly focused on the idle half-flow time, in which the moment of inertia of the rotor component is much larger than the moment of inertia of the coolant inertia in the loop. Such influence can be ignored in idling mathematical modeling. Therefore, the water moment M_h and the friction torque M_f are both proportional to the square of the angular velocity ω. According to the torque balance principle of reactor coolant pump, a balance equation can be established as follows:

$$-I\frac{d\omega}{dt} = M_h + M_f = C\omega^2. \tag{1}$$

Refereed by the initial conditions $t = 0$, $\omega = \omega_0$, the solution of Formula (1) shall be:

$$\begin{aligned} \omega &= \frac{\omega_0}{1+t/t_p} \\ t_p &= \frac{I}{C\omega_0} \end{aligned} . \tag{2}$$

If the influence of the loop flow inertia on the idler performance is ignored, according to the theoretical formula of the pump, it is found that:

$$P = \frac{g\rho QH}{3600\eta} = \left(M_h + M_f\right)\omega. \tag{3}$$

Therefore, by combining formula (1) and formula (3), it can be obtained that:

$$C\omega^3 = \frac{g\rho QH}{3600\eta} \tag{4}$$

When combining Equation (2), formula (4), and the formula $n = \omega/2\pi$, the formula for the rotational speed of the idling condition is achieved:

$$N(t) = \frac{n_0}{1 + \frac{g\rho Q_e H_e}{4\pi^2 n_e{}^2 I\eta_e} t}. \tag{5}$$

where, the hydraulic performance of the reactor coolant pump under rated conditions is mainly related to the six main geometric parameters, including the inlet placement angle of the guide vane α_3, the outlet placement angle α_4, the blade wrap angle φ, blade thickness δ, outlet width b_4, and clearance between guide vane and impeller R_t. Therefore, formula (6) can be expressed as follow:

$$N(t) = \frac{n_0}{1 + \frac{g\rho Q_0 H(\alpha_3,\alpha_4,\delta,R_t,b_4)}{4\pi^2 n_e{}^2 I\eta(\alpha_3,\alpha_4,\delta,R_t,b_4)} t} \tag{6}$$

In the formula, I am the unit moment of inertia, kg·m^2; ω is rotational speed, r/min; t is time, s; M_h is the water moment; M_f is the friction torque. C is set as a coefficient related to the torque; P is the motor power, W; g = 9.81 m^2/s; ρ is the liquid density; Q is the flow, m^3/s; H is the head of the pump; η is the efficiency of pump; ω_0 is the rated speed, t_p is the time at half speed. Pe is the electrical power, W; Qe is the rated flow, m^3/s; He is the rated head; ηe is the efficiency at rated flow; n_0 is the initial rotation speed, r/min; ne is the rated speed, r/min; $H(\alpha3, \alpha4, \varphi, \delta, R_t, b4)$ is the mathematical relationship between each parameter and the head; and η $(\alpha3, \alpha4, \varphi, \delta, R_t, b4)$ is the mathematical relationship between each parameter and the efficiency.

2.2. Hydraulic Modeling Database Construction on an Orthogonal Optimization Scheme

In order to obtain the relationship between the idling properties and structure parameters of the reactor coolant pump, the main geometrical parameters of the guide vanes were evaluated by taking the efficiency and head as indicators. The inlet placement angle α_3, the outlet placement angle α_4, the blade wrap angle φ, the blade thickness δ, the outlet width b_4, and the clearance between guide vane and impeller R_t of each factor were recorded at three levels as shown in Table 1. As referring to the design parameters of the reactor coolant pump, the hydraulic calculation was carried out for 18 different guide vane models with a different combination of factors. The main geometric parameters and structural diagram of the guide vane are shown in Figure 1.

Table 1. Level of different factors.

Level	Factor					
	$\alpha_3/(°)$	$\alpha_4/(°)$	$\varphi/(°)$	δ/mm	R_t/mm	b_4/mm
1	22	20	70	10	5	280
2	26	25	75	20	15	290
3	30	30	80	30	25	300

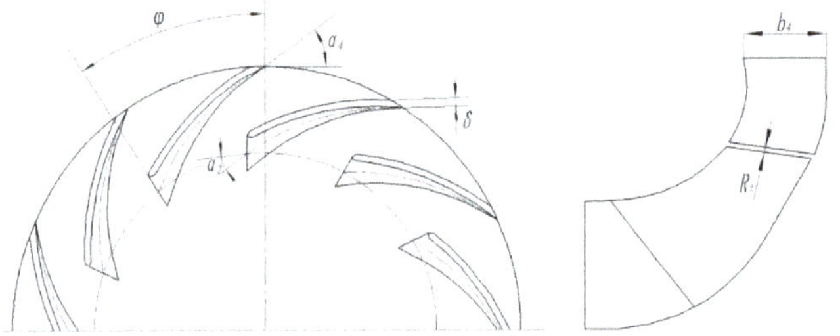

Figure 1. Main structural parameters of the guide vane.

Software Pro/E (4.0, PTC, Boston, MA, USA, 2007) was used to model the water model of the impeller, inlet section, volute, and guide vanes with different structural parameters. The fluid domain grid was divided by ICEM and the number of grids was tested for independence. Considering the computational resources and ensuring the better convergence of the computational model, hexahedral structure grid is adopted in the fluid domain and the boundary layer and interface region on the blade surface are locally encrypted. After grid independence check, it is found that when the grid number of the model was higher than 2.5 million, the head change was less than 0.5%. Taking the calculated resources and calculation accuracy into consideration, the total number of grid dividing units in the fluid domain is about 2.561 million, among which the number of grid units in the inlet section, impeller water body, guide vane water body and volute water body is 341 thousand, 528 thousand, 915 thousand and 771 thousand, respectively. The water models and grid diagram of the reactor coolant pump are shown in Figure 2. Figure 2f shows the grid number independence test diagram of reactor coolant pump. Then, with CFX, the steady calculation of the water body of the reactor coolant pump was carried out as follows: the turbulence model was set to the standard *k-e* model, the boundary conditions are set as the total pressure inlet and mass flow outlet, the discrete scheme was set to the first-order upwind formula, and the numerical calculation adopted the SIMPLE algorithm. The convergence accuracy was set to 10^{-4}. The following 18 sets of reactor coolant pump calculation models for different vanes were applied for the steady computation, and the calculation results are shown in Table 2 below.

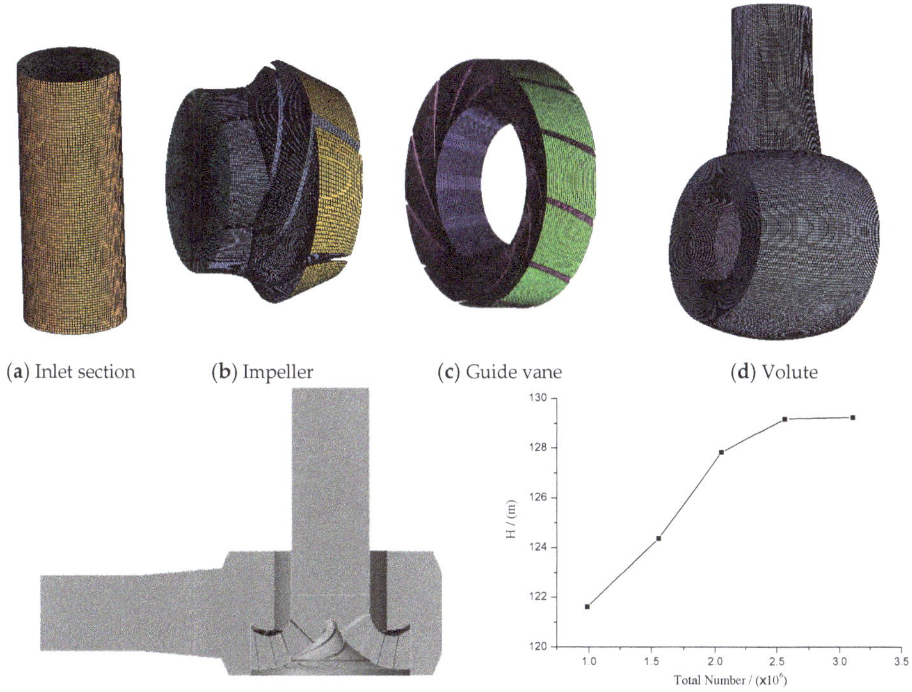

(a) Inlet section (b) Impeller (c) Guide vane (d) Volute

(e) Water model of reactor coolant pump (f) Water model of reactor coolant pump

Figure 2. The water model and grid diagram of the reactor coolant pump.

Table 2. Numerical calculation results of the orthogonal test.

Serial Number	$\alpha_3/°$	$\alpha_4/°$	$\varphi/°$	δ/mm	R_t/mm	b_4/mm	Indicator	
							$\eta/\%$	H/m
1	22	18	70	15	5	280	79.59	131.532
2	22	20	75	20	10	290	82.30	130.936
3	22	22	80	25	15	300	82.98	125.46
4	26	18	70	20	15	300	80.65	124.954
5	26	20	75	25	5	280	82.23	131.721
6	26	22	80	15	10	290	83.12	128.98
7	30	18	75	25	10	300	82.50	128.234
8	30	20	80	15	15	280	82.80	125.961
9	30	22	70	20	5	290	78.20	131.295
10	22	18	80	20	10	280	83.20	130.161
11	22	20	70	25	15	290	76.41	124.076
12	22	22	75	15	5	300	82.52	134.112
13	26	18	75	15	15	290	82.67	126.207
14	26	20	80	20	5	300	83.52	132.032
15	26	22	70	25	10	280	77.36	126.807
16	30	18	80	25	5	290	83.43	136.537
17	30	20	70	15	10	300	79.71	127.18
18	30	22	75	20	15	280	81.72	123.568

2.3. Mathematics Modeling Under Multiple Linear Regression Frameworks

The variables $x1$, $x2$, $x3$, $x4$, $x5$, and $x6$ were assigned to the inlet placement angle α_3, the outlet placement angle α_4, the blade wrap angle φ, the blade thickness δ, the outlet width b_4 and the clearance between guide vane and impeller R_t respectively, and the variable y represents efficiency and head. Assuming that the variable y has a linear regression relationship with the variables $x1$, $x2$, $x3$, $x4$, $x5$ and $x6$, the mathematical model of multivariate linear regression was concluded as follows:

$$\begin{aligned} y_1 &= \beta_0 + \beta_1 x_{11} + \beta_2 x_{12} + \beta_3 x_{13} + \beta_4 x_{14} + \beta_5 x_{15} + \beta_6 x_{16} + \varepsilon_1, \\ y_2 &= \beta_0 + \beta_1 x_{21} + \beta_2 x_{22} + \beta_3 x_{23} + \beta_4 x_{24} + \beta_5 x_{25} + \beta_6 x_{26} + \varepsilon_2, \\ y_N &= \beta_0 + \beta_1 x_{N1} + \beta_2 x_{N2} + \beta_3 x_{N3} + \beta_4 x_{N4} + \beta_5 x_{N5} + \beta_6 x_{N6} + \varepsilon_N. \end{aligned} \tag{7}$$

To convert to matrix form, it was set as follow:

$$X = \begin{bmatrix} 1 & x_{11} & x_{12} & \cdots & x_{16} \\ 1 & x_{12} & x_{22} & \cdots & x_{26} \\ & & \vdots & & \\ 1 & x_{n1} & x_{n2} & \cdots & x_{n6} \end{bmatrix} \tag{8}$$

$$Y = (y_1, y_2, \ldots y_6) \tag{9}$$

$$\beta = [\beta_0, \beta_1, \beta_2 \ldots \beta_6]' \tag{10}$$

$$\varepsilon = [\varepsilon_0, \varepsilon_1, \varepsilon_2 \ldots \varepsilon_6]' \tag{11}$$

Then the multiple linear matrices is:

$$Y = X\beta + \varepsilon. \tag{12}$$

In which, $\beta_0, \beta_1, \beta_2, \ldots \beta_6$ is defined as n + 1 regression estimation parameter, while ε_0, ε_1, ε_2, ... , ε_N are defined as the random variables that are independent of each other and subject to the same normal distribution $N(0, \sigma^2)$. When the least-squares estimation method is adopted to estimate the parameters $\beta_0, \beta_1, \beta_2, \ldots \beta_6, b_0, b_1, b_2, \ldots b_6$ is assumed to represent the least-squares regression coefficients of parameter $\beta_0, \beta_1, \beta_2, \ldots \beta_6$ respectively. The multiple linear regression equation is obtained as follows:

$$\hat{y} = b_0 + b_1 x_1 + b_2 x_2 + \ldots + b_6 x_6 \tag{13}$$

For each set of observations $x_{i1}, x_{i2}, x_{i3}, \ldots x_{i6}$, a regression value could be determined by the formula and the sum of the deviations of all regression values of $\hat{y}_i$ and y_i was obtained:

$$\hat{y}_i = b_0 + b_1 x_{i1} + b_2 x_{i2} + \ldots + b_6 x_{i6} \tag{14}$$

$$Q(b_0, b_1, b_2, \ldots b_6) = \sum_{i=1}^{N} (y_i - \hat{y}_i)^2 \tag{15}$$

According to the principle of least squares, b_0, b_1, b_2, ... , should be minimized by $Q(b_0, b_1, b_2, \ldots, b_6)$. Since the quadratic Formula (15) is a non-negative function, there always should be a minimum value. The least- squares regression model could be achieved to estimate coefficients of b_0, b_1, b_2, ... , b_6 by solving the extreme value theorem of differential calculus.

3. Results

3.1. Solution of a Mathematical Model of Hydraulic Performance Based on MATLAB

A multivariate regression model was constructed to solve the optimization results of the orthogonal test of all parameters of the guide vane based on MATLAB (R2010a, MathWorks, Natick, MA, USA, 2010), and the mathematical relationship between all parameters of the guide vane and the hydraulic performance was predicted. The main processes are summarized as follows:

(1) The relationships between different geometrical parameter combinations of guide vane and efficiency and head of the reactor coolant pump were sorted and recorded as *txt* files, respectively. The test serial number was retained in the first column and the last seven columns were the six geometric parameter values and efficiency index or head index of different guide vane schemes, respectively. Each column was further separated by a comma and preserved as the *efficiency.txt* and the *head.txt*, and then they were deposited into the MATLAB workbench for easy access.

(2) The following structure is used to analyze the efficiency relationship due to a similar process for solving multiple linear regression models of head and efficiency. For the purpose of content simplification, the order for solving the head regression model was not repeated hereof. The following command lines were entered in the window of the command line of the work interface (content after % was defined as the interpretation of the current step) after the MATLAB was opened. The definition of the analysis data is completed as follows:

Load ('efficiency.txt');% read the work table file of *efficiency.txt;*
a = load ('efficiency.txt');% assign the value of *efficiency.txt* to the contents *a* matrix;
x1=a (:, 2);% assigns the value of the second column of *a* matrix to *x1;*
x2=a (:, 3);% assigns the value of the third column of *a* matrix to *x2;*
x3=a (:, 4);% assigns the value of the fourth column of *a* matrix to *x3;*
x4=a (:, 5);% assigns the value of the Fifth column of *a* matrix to *x1;*
x5=a (:, 6);% assigns the value of the sixth column of *a* matrix to *x4;*
x6=a (:, 7);% assigns the value of the seventh column of *a* matrix to *x5;*
y=a (:, 8);% assigns the value of the eighth column of *a* matrix to *x6;*
X = *[ones (length (y), 1), x1, x2, x3, x4, x5, x6];*

The parentheses were merged by %d into the new matrix X, in which the command of *ones (length (y), 1)* was designed to create a column matrix that is equal to the *y* row number with the value of 1;

(3) Furthermore, the regression equation model as defined by the above data block *(y,X)* was constructed based on the least-squares estimation method. Its regression coefficient b_i *(i = 0,1,2,3,4,5,6)*, regression coefficient interval *bint*, residual *r*, confidence interval *rint*, and regression model test coefficient *stats* were output. Correlation coefficient r^2, significance test statistic value *F* and probability *p* corresponding to *F* value were incl uded. The command line was defined as follows:

$$[b, bint, r, rint, stats] = regress\ (y, X). \tag{16}$$

The data block (*Y*, *X*) is applied with regression analysis by % *d*, and then *b*, *bint*, *r*, *rint*, *stats*, and related data were output.

The values of the regression coefficient for b_0, b_1, b_2, b_3, b_4, b_5, and b_6, as well as the regression coefficient interval *bint*, were output after the completion of the command line. The output values of the regression model test coefficient *stats* were further read, including the correlation coefficient $r^2 = 0.8659$, the significance test statistic $F = 9.4896$, and the corresponding significance level $p = 0.008$. The test values of the independent and dependent variables *t*, as well as the significant level *p*, are indicated in Table 3. Moderate correlation due to correlation coefficient ($r^2 = 0.8891 < 0.9$) and significant level distribution test $p = 0.008 < 0.05$, significance difference distribution test *t* was further performed on the coefficients, where the significant level and efficiency of x6 is less than 0.001, this is considered negligible, i.e., no significant impact on the efficiency is posted by x_6. Therefore, the re-regression analysis was performed for the remaining parameters after x_6 was removed from the regression model above. A new regression model along with its coefficients b_0, b_1, b_2, b_3, b_4, b_5 and regression coefficient interval *bint* were obtained after repeating the data definition and command input process of model construction above. See Table 4.

Table 3. Regression coefficient interval.

	Value	*bint*	*t*	*p*
b_0	42.0264	[17.3924,66.6604]	3.5802	0.005
b_1	0.0283	[−0.1473,0.2040]	0.3385	0.742
b_2	−0.2558	[−0.6072,0.0955]	−1.5280	0.1575
b_3	0.4522	[0.3116,0.5927]	6.7517	0.501
b_4	−0.0917	[−0.2322,0.0489]	−1.3688	0.201
b_5	−0.0377	[−0.1782,0.1029]	−0.5624	0.5862
b_6	0.0415	[−0.0288,0.1118]	1.2394	0.0008

Table 4. New regression coefficient interval.

b_i	*Value*	*bint*
b_0	54.061	[39.9724,68.1503]
b_1	0.0273	[−0.1505,0.2072]
b_2	−0.256	[−0.6135,0.1018]
b_3	0.441	[0.3091,0.5952]
b_4	−0.092	[−0.2347,0.0514]
b_5	−0.038	[−0.1807,0.1054]

The output value in the regression model test coefficient *stats* was successfully read, including the correlation coefficient $r^2 = 0.9159$, the significant test statistics $F = 3.4491$, and the significant level of $p = 0.053$ corresponding to value F. It was indicated that the correlation coefficient increased to a strong correlation with a revised significance level ($p > 0.05$). Therefore, the regression model shows a good significance and a good reference for predicting pump rated efficiency of different geometrical parameters of guide vanes. Furthermore, a regression model formula for efficiency can be obtained:

$$H = 54.061 + 0.0273\alpha_3 - 0.256\alpha_4 + 0.441\varphi - 0.092\delta - 0.038R_t. \quad (17)$$

Since the difference test met the requirements, the fitting degree of the regression model was further analyzed by drawing the curve between the fitting value and the actual value as well as the residual confidence interval distribution graph. It can be seen from Figure 3 that the fitted value curve and the actual value curve have a high degree of coincidence in most of the intervals. The deviation of the two curves was very small, and only a few parameter combination points have large deviations. The fitting degree between the two curves was satisfied within the allowable error range for the orthogonal experimental model. It can be seen from Figure 4 that all residual values are within the upper and lower limits of the confidence interval, indicating that the regression model is normal.

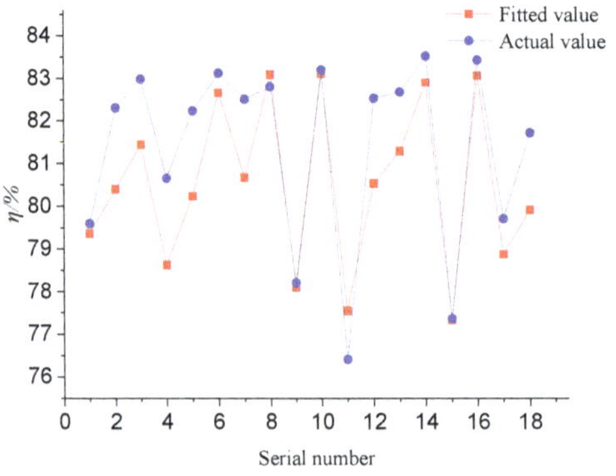

Figure 3. Comparison of fitted value and actual value for efficiency.

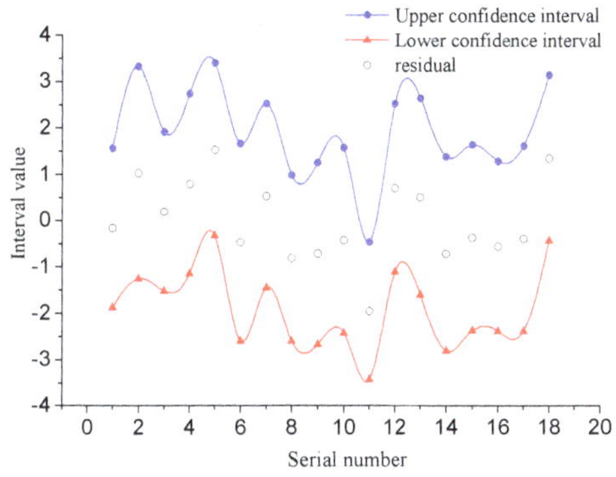

Figure 4. Residual confidence interval.

According to the above operation, the corresponding command line was input to solve the mathematical regression model of the head and the geometric parameters of the different guide. The final output is as follows: the regression coefficients obtained are b_0 = 123.176, b_1 = −0.073, b_2 = −0.3085, b_3 = 0.221, b_4 = −0.0189, b_5 = −0.783, b_6 = 0.0185, r^2 = 0.9131, significant test statistics F = 19.2651, and the probability p = 0.064 corresponding to F value. It was indicated that the correlation coefficient ($r^2 > 0.9$) shows a strong correlation with a significant level ($p > 0.05$). Therefore, the regression model shows a positive significance. The regression coefficient range is shown in Table 5.

Table 5. Regression coefficient Interval.

b_i		*bint*
b_0	123.176	[93.5186,152.8324]
b_1	−0.073	[−0.2845,0.1385]
b_2	−0.3085	[−0.7315,0.1145]
b_3	0.221	[0.0523,0.3906]
b_4	−0.0189	[−0.1881,0.1502]
b_5	−0.783	[−0.9523,-0.6142]
b_6	0.0185	[−0.0661,0.1031]

The regression model formula for the head of the reactor coolant pump is as follows:

$$H = 123.176 - 0.073\alpha_3 - 0.3085\alpha_4 + 0.221\varphi - 0.0189\delta - 0.783R_t + 0.0185b_4. \quad (18)$$

It can be seen from the comparison between the fitting value and the actual value curve in Figure 5 that the coincidence degree between the fitting value curve (red line) and the actual value curve (blue line) is very high. The deviation value between the two is less than 1%, by which a very good fitting degree of the regression model was inculcated. It can also be seen from Figure 6 that the residual values are all within the confidence interval, indicating a normal regression model.

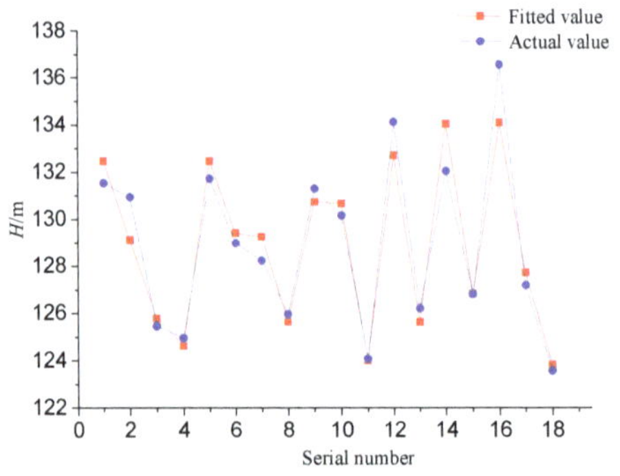

Figure 5. Comparison of fitted value and actual value for the head.

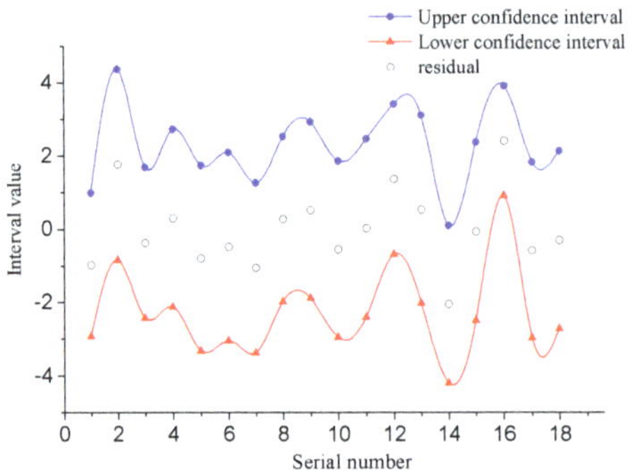

Figure 6. Residual confidence interval.

According to the above formula, the relationship between the geometric parameters of the guide vane and the idling speed and idling flow under the idling operation condition was derived. The established mathematical model of the idling speed is shown in Formula (19):

$$\begin{cases} N(t) = \dfrac{n_0}{1+\frac{g\rho Q_0 H(\alpha_3,\alpha_4,\delta,R_t,b_4)}{4\pi^2 {n_e}^2 I\eta(\alpha_3,\alpha_4,\delta,R_t,b_4)}t} \\ \eta = 54.061 + 0.0273a_3 - 0.256a_4 + 0.441f - 0.092\delta - 0.038R_t \\ H = 123.176 - 0.073a_3 - 0.3085a_4 + 0.221f - 0.0189\delta - 0.783R_t + 0.0185b_4 \\ \\ 22 \le a_3 \le 30 \\ 18 \le a_4 \le 22 \\ 70 \le \varphi \le 80 \\ 15 \le \delta \le 25 \\ 5 \le R_t \le 15 \\ 280 \le b_4 \le 300 \end{cases} \quad (19)$$

3.2. Test System of Reactor Coolant Pump Coasting

In order to verify the accuracy of the mathematical model of inertia, the open experiment table of the reactor coolant pump was built to complete the coasting test of the reactor coolant pump shut-down. The clear water was applied as the conveying medium of the experiment table system. The hydraulic characteristic, as well as the inertia-turn characteristic of the reactor coolant pump, was tested. The experiment table was designed to collect the instantaneous flow rate, the transient inlet, the outlet pressure and the instantaneous speed of the test pump in real-time. The experiment table was set as an open system test bench consisting of an idler wheel, motor, model pump, regulating valve, booster pump, pressure-stabilizing tank and measuring and collecting equipment (torque transducer, press transmitter, flow meter), and connecting pipes. Those were all listed in Figure 7, titled a test on the effect of model pump coasting behavior. The sectional flywheel was installed at the end of the shaft by means of key links. The control valve was adjusted to the maximum output. The model pump was opened to a certain period of stable operation and the control valve was adjusted to make the pump run stably under the condition of $1.0Q_0$. The power supply was later turned off, and the variations of speed, flow, and head of the pump idleness transition process transmitted were observed by measuring and collecting equipment through the console computer for three tests.

Figure 7. Site of the experiment table system.

In order to verify the universality of the model, the model of a group of guide vanes was established for experimental verification. Its main geometric parameters of guide vanes are concluded in Table 6. Meanwhile, the moment of inertia is I = 931 kg·m^2, and the parameters are substituted into Equation (19) to obtain the variation curve Equation (20) of the speed during the idling transition process under the corresponding guide vane parameters. The hydraulic performance test results of the reactor coolant pump are indicated in Figure 8.

$$N(t) = \frac{74}{3 + 1.2178t}(rad/s) \tag{20}$$

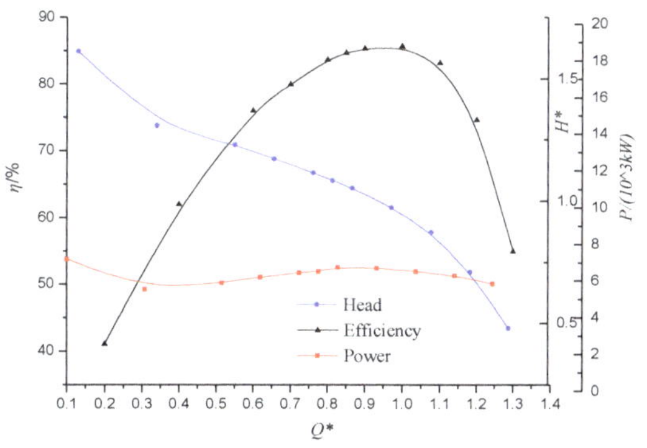

Figure 8. Numerical results of the reactor coolant pump.

Table 6. Geometry parameter size of guide vanes.

Parameter	$\alpha_3/^\circ$	$\alpha_4/^\circ$	$\varphi/^\circ$	δ/mm	R_t/mm	b_4/mm
Value	24	18	78	22	6	296

3.3. Verification of Mathematical Models

Based on the full characteristic curve of the pump combined with the test piping setup, the following factors are defined in the mathematical model of the test system: the water tank part pressure is represented by P_0; the pipe outlet to the model pump inlet section is represented by $L1$; the model pump outlet to the inlet of the water tank is represented by $L2$. The Q–H curve, Q–P curve, and the Q–η curve are fitted as polynomial equations, which are set as the system variables of the experimental mathematical model. The equation of fluid unsteady flow in the pressurized pipeline is combined with the principle of control process and rigid theory Equation (21) to establish the dynamic mathematical models (Equation (22)). As shown in Equation (22), the transient process of hydraulic change of the pump could be better reflected by this model when the pump was changed in a quasi-steady state under different working conditions. The model pump real-time flow Q and head H are defined in real-time by the input *flow-head* curve signal $H(Q)$, while the input torque of the motor M_d is defined in real-time by the input flow-power curve signal $P(Q)$. The resistance moment Mf is defined by both the input flow-torque curve signal $P(Q)$ and the flow-efficiency curve signal slave $\eta(Q)$. The mathematical model of pump speed is concluded in Formula (23):

$$\frac{1}{g}\frac{dQ}{dt}\int_0^L \frac{dx}{A} + \frac{v^2 - v_1^2}{2g} + H - H_1 + h_f = 0 \tag{21}$$

$$\begin{aligned} P_1 &= P_0 \rho g H_0 - \rho \tfrac{L_1}{S_1}\tfrac{dQ}{dt} - \left(\lambda_1 \tfrac{L_1}{d_1} + \xi_1\right)\tfrac{\rho}{2S_1^2}Q^2 - \rho g L_1 \\ P_2 &= P_0 \rho g H_0 - \rho \tfrac{L_1}{S_1}\tfrac{dQ}{dt} - \left(\lambda_2 \tfrac{L_2}{d_2} + \xi_2\right)\tfrac{\rho}{2S_2^2}Q^2 - \rho g L_1 + \tfrac{\rho}{C_F^2}Q^2 \\ H &= \tfrac{P_2 - P_1}{\rho g} \end{aligned} \tag{22}$$

$$\frac{d_n}{d_t} = \frac{30}{\pi J} \times \left(M_d - M_f\right) \tag{23}$$

where, P_1 and P_2 are defined as the inlet pressure and outlet pressure of the model pump respectively, Pa; P_0 is defined as the liquid level pressure of the tank, Pa; H_0 is defined as the level of the water tank, m; L_1 and L_2 are defined as length of inlet pipe and length of outlet pipe, respectively, m; S_1 and S_2 are defined as cross-sectional areas of the inlet pipe and outlet pipe, respectively, m^2. d_1 and d_2 are defined as pipe diameters of inlet pipe and outlet pipe, respectively, m; λ_1 and λ_2 are defined as the resistance coefficients of the inlet pipe and outlet pipe, respectively; ρ is defined as the fluid density, 1000 kg/m^3; g is defined as the acceleration of gravity; Q is defined as the real-time flow rate of the model pump, m^3/h; H is defined as the real-time head of the model pump, m; C_F is defined as the valve resistance coefficient; n is defined as the pump speed, r/min; J is defined as the moment of inertia of rotor parts, kg·m^2; M_d is defined as the input torque of the motor, kg·m; M_f is defined as the resistance torque of the motor, kg·m; t is defined as the simulation process time, s.

In Formulas (22) and (23), the inertia of the experimental unit is indicated by J = 931 kg·m^2, while the pipe diameter and length are indicated by $d_1 = d_2$ = 0.76 m, L_1 = 20 m and L_2 = 30 m, respectively. The water tank liquid level pressure was indicated by $P_0 = 1\ atm$ and the water tank level is indicated by H_0 = 0.8 m. The starting speed of the motor is set as 1480 r/min and the mathematical model based on MATLAB was applied for the data output. The coasting output value of dQ/dt and dn/dt was obtained as 60 s while the flow rate and the speed curve of the corresponding time point were achieved by the integral operation. The comparison between the test output coasting speed changes and the calculation results of the mathematical model coasting speed are indicated in Figure 9. It was shown that the trend of the rotational speed curve is similar to that of the 25 s while the error in the t = 10 s is approximately

2.7%, which is in the acceptable range. Therefore, the coasting rotational speed mathematical model is reliable for predicting the rotational speed of the coasting transition process.

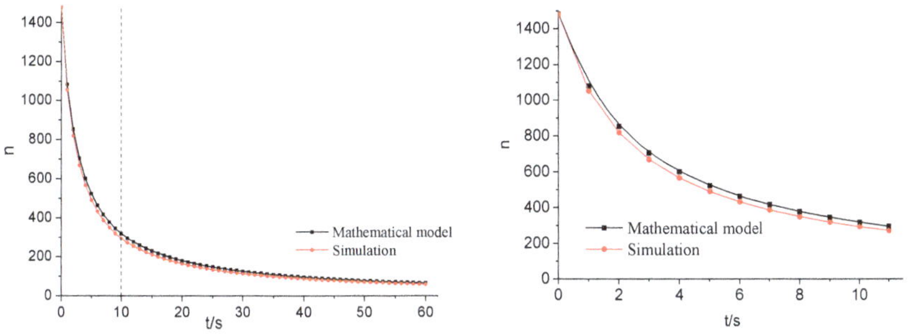

(**a**)The whole process (**b**)The first 10s of process

Figure 9. Chart of simulation test and mathematical model calculation results.

4. Conclusions

(1) Benefiting from the mathematical derivation of the equilibrium equation of the nonlinear inertia transient process, the mathematical relationship between the rotational speed of the inertia transition process and the fixed head and the efficiency of the reactor coolant pump were obtained.

(2) The hydraulic model of 18 sets of different guide vane structure parameters was established under the orthogonal optimization scheme. Based on the multiple linear regression theory, the orthogonal test optimization results of the parameters of the guide vane were calculated by multiple regression analysis. At last, the mathematical relationship of the hydraulic performance of the guide vane parameters was predicted while the mathematical model on the main geometrical parameters of the given guide vane was deduced by the simultaneous mathematical relationship among the rotational speed, the fixed head and the efficiency of the reactor coolant pump.

(3) The mathematical model of coasting was verified by the test results and the speed curve of the idling transition process and the speed curve derived from the mathematical model were evidenced with a good coincidence degree, indicating that the coasting transition process can be well predicted by the constructed coasting model.

Author Contributions: X.W.: Experiments and simulations; Y.X.: The writing and revision of the paper; R.Z.: Ideas and fund support of the paper; Y.L., Q.F., Z.C. and C.A.: Experimental and simulated data processing.

Funding: National Youth Natural Science Foundation of China (51509112), Natural Science Foundation of Jiangsu Province of China (BK20171302); Key R & D programs of Jiangsu Province of China (BE2015129, BE2016160,337 BE2017140); Foshan Science and Technology Innovation Project(2016AG101575),Prospective joint research project of Jiangsu Province (BY2016072-02).

Conflicts of Interest: The authors declare no conflict of interest.

References

1. Sun, H. *Third Generation Nuclear Power Technology AP1000*, 2nd ed.; China Electric Power Press: Beijing, China, 2016.
2. Yanjun, W.; Wenhong, L.; Jun, D.; Ling, G. Comparison of past and present the Chernobyl and the Fukushima nuclear accident and elicit thinking. *Chin. J. Radiol. Health* **2016**, *25*, 459–462.
3. Dien, L.; Diep, D. Verification of VVER-1200 NPP simulator in normal operation and reactor coolant pump coast-down transient. *World J. Eng. Technol.* **2017**, *5*, 507–519. [CrossRef]
4. Turnbull, B. RELAP5/MOD3.3 analysis of the reactor coolant pump trip event at NPP Krško for different transient scenarios. In Proceedings of the International Conference Nuclear Energy for New Europe 2005, Bled, Slovenia, 5–8 September 2005.

5. Senru, Z. Calculation of transient characteristics for main circulating pump. *Nuclear Power Eng.* **1993**, *2*, 183–190.
6. Yujun, G.; Jinling, Z.; Huizheng, Q.; Guanghui, S.; Dounan, J.; Zhenwan, Y. Calculating model of coolant pump flow characteristics for reactor system. *Nucl. Sci. Eng.* **1995**, *3*, 220–225, 231.
7. Shaowen, D. Transient calculation of main pump in Qinshan Nuclear Power Plant Phase Two Project. *Nucl. Power Eng.* **2001**, *6*, 494–496, 507.
8. Xu, Y. *Numerical Simulation of Interior Flow Field of Reactor Coolant Pump Under Station Blackout Accident*; Dalian University of Technology: Dalian, China, 2011.
9. Zhu, R.; Long, Y.; Fu, Q.; Yuan, S.; Wang, X. Pressure fluctuation characteristics of nuclear main pump under low flow conditions. *J. Vibr. Shock* **2014**, *33*, 143–149.
10. Yun, L.; Zhu, R.; Fu, Q.; Yuan, S.; Xi, Y. Numerical analysis on unstable flow of reactor coolant pump under small flow rate condition. *J. Drain. Irrig. Eng.* **2014**, *32*, 290–295.
11. Feng, X.; Wu, D.; Yang, L.; Jia, Y. CNP1000 shaft seal reactor coolant pump technology. *J. Drain. Irrig. Mach. Eng.* **2016**, *34*, 553–560.
12. Alatrash, Y.; Kang, H.O.; Yoon, H.G.; Seo, K.; Chi, D.Y.; Yoon, J. Experimental and analytical investigations of primary coolant pump coastdown phenomena for the Jordan Research and Training Reactor. *Nucl. Eng. Des.* **2015**, *286*, 60–66. [CrossRef]
13. Brady, D.R.; Loebig, T.G. Reactor Coolant Pump Motor Load-Bearing Assembly Configuration. U.S. Patent No. 9,273,694, 1 March 2016.
14. Bang, S.Y.; Chu, S.M.; Chang, J.Y. Development of Reactor Coolant Pump for APR1400. In Proceedings of the KNS 2015 Fall Meeting, Kyungju, Korea, 28–30 October 2015.
15. Metzroth, K.; Denning, R.; Aldemir, T. Dynamic Event Tree Modeling of a Reactor Coolant Pump Seal LOCA. *Adv. Conc. Nucl. Energy Risk Assess. Manag.* **2018**, *1*, 305.
16. Lu, Y.; Zhu, R.; Fu, Q.; Fu, Q.; An, C.; Chen, J. Research on the structure design of the LBE reactor coolant pump in the lead base heap. *Nucl. Eng. Technol.* **2019**, *51*, 546–555. [CrossRef]
17. Lu, Y.; Zhu, R.; Wang, X.; Yang, W.; Qiang, F.; Daoxing, Y. Study on the complete rotational characteristic of coolant pump in the gas-liquid two-phase operating condition. *Ann. Nucl. Energy* **2019**, *123*, 180–189.
18. Lu, Y.; Zhu, R.; Wang, X.; An, C.; Zhao, Y.; Fu, Q. Experimental study on transient performance in the coasting transition process of shutdown for reactor coolant pump. *Nucl. Eng. Des.* **2019**, *346*, 192–199. [CrossRef]

processes

Article

Numerical Simulation of Water Absorption and Swelling in Dehulled Barley Grains during Canned Porridge Cooking

Lei Wang [1], Mengting Wang [1], Mingming Guo [1,2,3], Xingqian Ye [1,2,3], Tian Ding [1,2,3] and Donghong Liu [1,2,3,*]

1 College of Biosystems Engineering and Food Science, Zhejiang University, Hangzhou 310058, China; leiwang94@zju.edu.cn (L.W.); mtwang@zju.edu.cn (M.W.); mingguo@zju.edu.cn (M.G.); psu@zju.edu.cn (X.Y.); tding@zju.edu.cn (T.D.)
2 Fuli Institute of Food Science, Zhejiang University, Hangzhou 310058, China
3 Zhejiang Key Laboratory for Agri-Food Processing, National Engineering Laboratory of Intelligent Food Technology and Equipment, Hangzhou 310058, China
* Correspondence: dhliu@zju.edu.cn; Tel.: +86-0571-8898-2169; Fax: +86-0571-8898-2169

Received: 18 October 2018; Accepted: 13 November 2018; Published: 20 November 2018

Abstract: Understanding the hydration behavior of cereals during cooking is industrially important in order to optimize processing conditions. In this study, barley porridge was cooked in a sealed tin can at 100, 115, and 121 °C, respectively, and changes in water uptake and hygroscopic swelling in dehulled barley grains were measured during the cooking of canned porridge. In order to describe and better understand the hydration behaviors of barley grains during the cooking process, a three-dimensional (3D) numerical model was developed and validated. The proposed model was found to be adequate for representing the moisture absorption characteristics with a mean relative deviation modulus (P) ranging from 4.325% to 5.058%. The analysis of the 3D simulation of hygroscopic swelling was satisfactory for describing the expansion in the geometry of barley. Given that the model represented the experimental values adequately, it can be applied to the simulation and design of cooking processes of cereals grains, allowing for saving in both time and costs.

Keywords: barley; simulation; hydration; swelling; cooking; porridge

---

1. Introduction

Barley (*Hordeum vulgare* L.) is an ancient and widely adapted grain. It ranks fourth among grains in terms of quantity produced (142 M mt, 2014–2017 mean), behind corn (*Zea mays* L., 1027 M mt), wheat (*Triticum aestivum* L., 744 M mt), and rice (*Oryza sativa* L., 482 M mt), and ahead of sorghum (*Sorghum bicolor*, 63 M mt), oat (*Avena sativa*, 23 M mt), and rye (*Secale cereale* L., 13 M mt) [1]. In recent times, because of the high dietary fiber content of barley and the effectiveness of barley β-glucan in lowering cholesterol, the interest in barley food products—such as tea, soup, beverage, snacks, and porridge—is increasing worldwide [2,3].

Barley porridge, similarly to other cereal porridge, is a traditional food in eastern countries. Conventionally, whole or pearled barley grain is boiled in water to gelatinize starch in barley and fully expand it [3], but the cooked barley porridge can only be stored for a few days at ambient temperature. Additionally, ready-to-eat (RTE) porridge has attracted a great deal of attention in many countries due to its excellent storage stability [4]. However, most commercial instant porridge needs to be mixed with hot water before being consumed. Hence, canned barley porridge is more attractive because it could be consumed without any preparation [5].

The cooking of barley porridge in industry is a hydrothermal process to provide the desired attributes in the final product, which are strongly affected by cooking conditions, such as temperature, processing time, and cooking media [6]. Changes in the moisture content and volume of barley grains are the two main phenomena during cooking. Controlling the moisture content in barley grains during cooking is of great importance as water molecules play many roles in food reactions and food quality [7]. In order to predict the optimum processing conditions, the water content variations with the grains are needed as quantitative information [8]. Therefore, it is essential to understand the hydration kinetics of kernels during cooking, and the effect of cooking conditions on the assimilation of moisture [9]. What is more, the swelling of kernels during cooking affects the moisture absorption rate [10]. Hence, the instantaneous kernel volume is a key parameter for better understanding the water absorption process [11]. Many empirical models have been applied to predict the hydration behavior of grains, such as the Expansional, Peleg, and Weibull models. However, considering that these models are just fitted to experimental values, the fitting results are restricted to the test conditions used. Consequently, the phenomenological models are incrementally selected to better describe the phenomena included in the hydration process [12].

Recently, Bakalis et al. [13] used COMSOL Multiphysics®software to simulate the diffusion of moisture in the parboiled grain during cooking. However, their study was limited to starch gelatinization at 70 °C, while cooking is usually done at or above 100 °C. Balbinoti et al. [14] also used COMSOL Multiphysics®software to simulate moisture transfer in the parboiling process step in two- and three-dimensional space under four different temperatures ranging from 35 to 60 °C. Perez et al. [15] built a comparative 3D simulation on water absorption and hygroscopic swelling in japonica rice grains for a soaking temperature of 25, 35, 45, and 55 °C. Montanuci et al. [9] developed a three-dimensional model to describe the hydration process of barley grains under various temperatures, from 10 to 25 °C. To our best knowledge, no research has simulated the hydration curve and large deformation of barley kernels during cooking when the temperature was 100 °C and above.

Therefore, the phenomena of mass transfer, heat transfer, and deformation involved in the cooking period of canned barley porridge are coupled at 100, 115, and 121 °C in this context, in order to evaluate the effect of processing time and temperature on the moisture absorption and volume expansion. Furthermore, the mathematical model developed in this study is also intended to be used for conditions different from those tested in the present study, generating substantial savings in time, energy, and costs by reducing the experimental tests required.

2. Model Development

In this section, a numerical model is presented to describe the distribution and amount of moisture content, as well as the volumetric expansion of a dehulled barley grain undergoing hygroscopic swelling at three different temperature cooking conditions. In order to simplify the complexity conditions, the following assumptions are applied: the barley grain was considered as continuous, homogenous and isotropic; the initial moisture content is considered as homogeneous; and the initial surface temperature of the barley grain is equal to the water temperature.

2.1. Diffusion

The transient model applied to describe the phenomena of mass transfer and heat transfer during the cooking of canned porridge was developed based on Fick's second Law and Fourier's Law, respectively, according to Equations (1) and (2).

$$\frac{\partial c_i}{\partial t} + \nabla.(-D_i \nabla c_i) + u.\nabla c_i = R_i \quad (1)$$

$$\rho C_p \frac{\partial T}{\partial t} + \rho C_p u \nabla T = \nabla(k \nabla T) + Q \quad (2)$$

where c_i, D_i and u are the water concentration (mol/m^3), water diffusion coefficient (m^2/s), and velocity field (m/s), respectively; R_i is the mass generation (kg/m^3), ρ is the barley density (kg/m^3),

C_p is specific heat (J/kg·K), k is the thermal conductivity (W/m·K), T is the grain temperature (K), and Q is the heat production (W·m).

2.2. Hygroscopic Swelling

Dehulled barley grains are predominantly composed of starch (about 65–68%) [2]. During cooking, the starch granules contained in dehulled barley kernel absorb water and swell due to its gelatinization. The hygroscopic strain (i.e., moisture-induced strain) is caused by the swelling of the starch molecules after absorbing water, which can be expressed as an equation in related to the hygroscopic expansion coefficient (β_h) and the moisture content gradient (Δc) [16]:

$$\epsilon_{hs} = \beta_h \Delta c \quad (3)$$

2.3. Boundary and Initial Conditions

The boundary conditions for the governing equations are shown in Figure 1 and described in detail as below. The initial conditions are also listed in Table 1 with the input parameters.

2.3.1. Solid Mechanics

The displacement of side B parallel to the symmetry planes (side C) is set to zero while side A is free to deform (Figure 1).

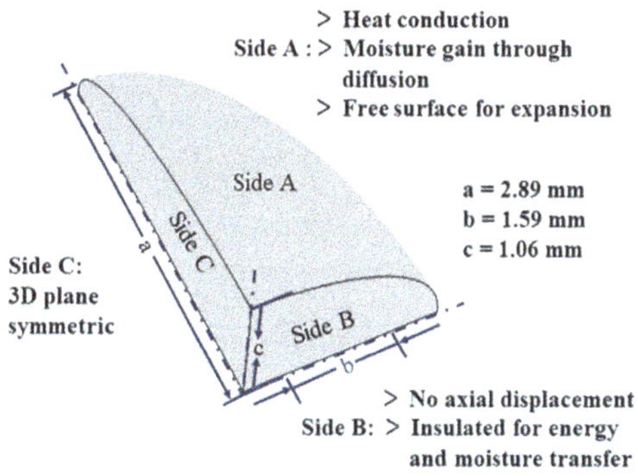

Figure 1. Schematic showing the barley geometry used for simulations and boundary conditions for coupled solid mechanics–heat and mass transport model.

2.3.2. Heat and Mass Transfer

The initial and boundary conditions established for heat transfer in the barley grain throughout the process are: (a) known initial temperature in the barley grain (Equation (4)); (b) no heat flux across the symmetry region (Equation (5)); and (c) convection occurs at the solid-fluid interface (Equation (6)). For mass transfer, the conditions are: (a) the initial moisture content in the grain is uniform and known (Equation (7)); (b) no mass flow in the symmetry region (Equation (8)); (c) the convective boundary conditions on the grain surface are known (Equation (9)).

$$T = T_0 \text{ for } t = 0 \quad (4)$$

$$\frac{\partial T}{\partial t} = 0 \text{ for } r = 0 \quad (5)$$

$$\frac{\partial T}{\partial t} = k(T_w - T) \text{ for t} > 0 \quad (6)$$

$$\text{c} = c_0 \text{ for t} = 0 \quad (7)$$

$$\frac{\partial c_i}{\partial t} = 0 \text{ for r} = 0 \tag{8}$$

$$\frac{\partial c_i}{\partial t} = D(c_e - c) \text{ for t} > 0 \tag{9}$$

where: T_0 is the temperature of uncooked barley (K), T_w is the water temperature (K), k and D are the heat transfer coefficient (W/m·K) and the mass transfer coefficient (m/s), respectively, t is the cooking time (s), and c_0 and c_e are the initial moisture content (kg/kg) and the equilibrium moisture content (kg/kg) of the barley grains, respectively.

2.4. Input Parameters

The input parameters used to simulate the barley cooking process are shown in Table 1. The diffusion coefficient is discussed here in detail.

2.4.1. Diffusion Coefficient

Previous studies have proved the temperature dependency of the diffusion coefficient of grains under hydrothermal conditions, following an Arrhenius type relationship [14,15]. As a consequence of this equation (Equation (10)), the diffusion coefficient increases with temperature.

$$D_t = D_0 \exp\left(\frac{-E_a}{RT}\right) = 1.203 \times 10^{-5} \exp\left(\frac{-4147.7}{T}\right) \tag{10}$$

where D_t is the effective coefficient of the mass transfer (m/s), D_0 is a constant, E_a is the activation energy (J/mol), and R and T are the universal rate constant and absolute temperature (K), respectively. D_0 and E_a are found to be equivalent to 1.203×10^{-5} and 4147.7 J/mol, respectively [9].

Table 1. Input parameters used in the simulations for barley porridge cooking.

Parameter	Value	Units	Source
Dimensions			
Major axis, a	2.89	mm	This study
Major axis, b	1.59	mm	This study
Major axis, c	1.06	mm	This study
Density			
Water, ρ_w	998	Kg/m^3	[17]
Barley, ρ_b	1304	Kg/m^3	[9]
Thermal conductivity			
Water, k_w	$0.57109 + 0.0017625 - 6.7306 \times 10^{-6}T^2$	W/m·K	[17]
Barley, k_b	0.1590	W/m·K	[9]
Specific heat capacity			
Water, C_{pw}	$4176.20 - 0.0909(T - 273) + 5.4731 \times 10^{-3}(T - 273)^2$	J/kg·K	[17]
Barley, C_{pb}	1800	J/kg·K	[9]
Equilibrium concentration of water, c_e	47,222	Mol/m^3	This study
Diffusion coefficient, D	1.203×10^{-5} exp(−4147.7/T)	m^2/s	[9]
Young's modulus, E	Equation (11)	Pa	[18]
Poisson's ratio, V_r	Equation (12)	-	[18]
Hygroscopic expansion coefficient of water, β	1.35×10^{-3}	M^3/kg	This study
Molecular weight of water, M_{mw}	0.0180	Kg/mol	
Initial conditions			
Water concentration, C_0	9598	Mol/m^3	This study
System temperature, T_0	298.15	K	This study

2.4.2. Mechanical Properties

Mechanical properties, such as the elastic modulus (E) and Poisson's ratio (ν) of barley, are required as functions of phase transition temperature (T_g). Barley starch granules are initially glassy, and transform to a rubbery state when the temperature is higher than T_g. For barley starch in a glassy state, E_g and ν_g have been reported as 500 MPa and 0.28, respectively [19]. In the rubbery state, the elastic modulus of starch (E_r) is expected to be of the order of 1 kPa [17]. Poisson's ratio in a rubbery state (ν_r) has been estimated to be about 0.5. In order to avoid singularity and help with convergence of the numerical scheme, a value of 0.49 was adopted during computations. The temperature dependency of the elastic modulus and Poisson's ratio, in consideration of T_g, were approximated using the following functions [20]:

$$\mathrm{E(T)} = \frac{1}{2}(E_g + E_r) - \frac{1}{2}(E_g - E_r)\tanh\frac{T - T_g}{\beta} \quad (11)$$

$$v(\mathrm{T}) = \frac{1}{2}(v_g + v_r) - \frac{1}{2}(v_g - v_r)\tanh\frac{T - T_g}{\beta} \quad (12)$$

Here, β is a parameter related to the temperature range across which phase transition occurs. A value of β = 5 °C was assumed based on values reported for other glassy polymers [18].

2.5. Solution Methodology Geometry, Mesh, and Implementation

A 3D geometry was used in this study and, owing to symmetry, a one-eighth ellipsoid was created (Figure 1). A tetrahedron mesh consisting of 9943 elements was used (Figure 2a), and the quality of the mesh was evaluated according to the color gradient (Figure 2b); the colors varied from red (low quality: 0) to green (high quality: 1).

A commercially available finite element software, COMSOL Multiphysics 5.3a (Comsol Inc., Burlington, MA, USA), was used to solve the governing equations described in Sections 2.1 and 2.2. The modules used were Transport of Diluted Species, Heat Transfer in Solids and Solid Mechanics for the mass transfer, heat transfer, and deformation phenomena, respectively. The simulation was calculated using a 2.93 GHz 6-core Intel Xeon Workstation with 24 GB RAM.

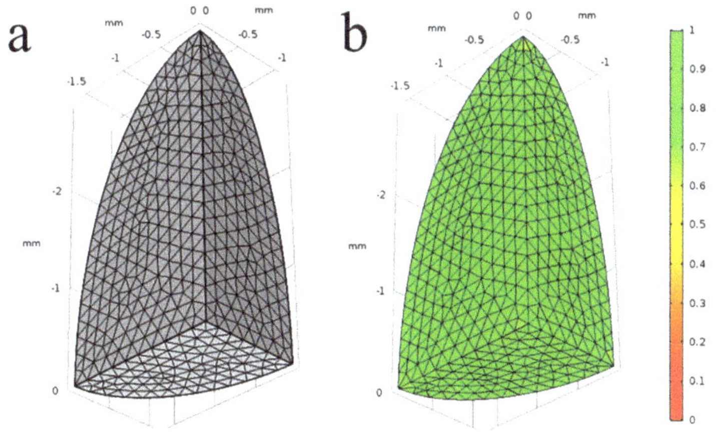

Figure 2. Meshed geometry with tetrahedron mesh elements (**a**); distribution of the evaluated quality of the mesh (**b**).

3. Experimental Methodology

3.1. Materials

Dehulled barley grains were obtained from Wal-Mart (China) Investment Co., Ltd. (Shenzhen, China). The barley grains were selected carefully so that the initial dimensions were almost the same. The initial dimensions of the barley grains (a = 2.89 mm, b = 1.59 mm, c = 1.06 mm) were determined using a vernier caliper (precision: 0.01 mm).

3.2. Cooking Process

Barley porridge was prepared according to the method of Kwang et al. [21], with some modifications. Dehulled barley grains (15 g) were washed and rinsed three times with tap water. Water was added to the barley with the ratio of 8:1 (v/w) in a commercial tin can (ΦA = 73 mm, H = 59.8 mm) (ORG Packaging Co. Ltd., Beijing, China), which was then sealed using a hand seamer (YJ-C200, Zhangjiagang Yijie Automation Equipment Co., Ltd., Suzhou, China). The sealed can was put into an autoclave (Beijing Fanwen Trade Co. Ltd., Beijing, China) pre-heated to 60 °C. Then, the autoclave was further heated up to 100, 115, and 121 °C, maintained for 0–98.5, 0–86, and 0–79 min, respectively, and cooled down quickly using manual exhaust. In order to minimize the impact of the cooling stage on the changes in moisture and volume during porridge cooking, as soon as the vapor temperature dropped to around 100 °C, the can was transferred from the autoclave to cold water (25 °C) for further cooling until the water (inside the tin can) reached room temperature. The temperature changes of the vapor (heating medium, inside the autoclave chamber and outside the can) and water (heat transfer medium, inside the can) during cooking were monitored using MPIII Temperature Data Loggers (M4T12396, Mesa Laboratories, Inc., Lakewood, CA, USA). During the cooking periods at 100 °C and 121 °C, the vapor temperature changes and theoretical and actual time points for sample collection during cooking are shown in Figure 3.

Specifically, the vapor temperatures reached 100, 115, and 121 °C after 8.5, 16, and 19 min, respectively. In the heating-up period, the grains samples were collected at 0, 3, 6, and 8.5 min when the cooking temperature was 100 °C, 0, 3, 6, 8.5, and 16 min at 115 °C, and 0, 3, 6, 8.5, 16, and 19 min at 121 °C. At the stage of preservation, the collection interval was 5 min in the first 30 min, and thereafter, samples were collected every 10 min up to the equilibration of the hydration.

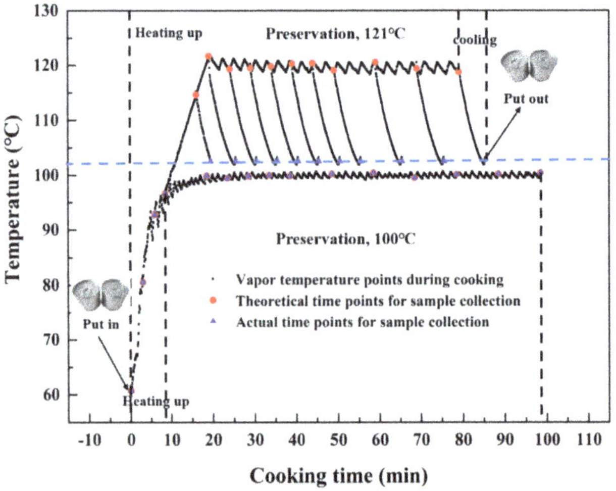

Figure 3. Temperature history and time points for sample collection (theoretical and actual) during cooking at 100 °C and 121 °C.

3.3. Measurement of Moisture Content and Volume Variation

After the removal of the surface water, the moisture content, volume, and expansion ratio of the cooked barley were analyzed. The moisture content (g·g^{-1}, wet basis) of the samples at each time step was obtained based on the increase in sample mass at the corresponding times [22]. The volume variation (mm^3) and expansion ratio (mm^3·mm^{-3}) were determined using water displacement in a measuring cylinder following the method of Fracasso et al. [23], with some modification. All the grains of the cooked barley were placed inside a 250 mL measuring cylinder, which originally contained

100 mL water. The total volume of 15 g cooked barley grains was equal to the volume increment of water. The expansion ratio of the cooked barley grains at different time points for collection was calculated by dividing the volume expansion at corresponding times with the volume of uncooked kernels. All the experiments described above were conducted in three replicates.

4. Results and Discussion

The model developed was validated by comparing the moisture content and expansion ratio change histories of the cooked barley. The performances of the models were determined according to their coefficient of determination (R^2), the root mean square error (RMSE, %), and the mean relative deviation modulus (P). Transient changes in the distribution of moisture content and the shape of the barley are also discussed next.

4.1. Moisture Absorption Characteristic

The moisture content changes of barley grains during the cooking process at 100 °C, 115 °C, and 121 °C are monitored and modeled. The plots of moisture content–cooking time are shown in Figure 4, and here the moisture content is represented by the increment of mass gain of barley grains. The experimental data fitted the simulated values well (Equations (1) and (2)), with the R^2 ranging from 0.993 to 0.997, RMSE ranging from 0.046 to 0.068 $g \cdot g^{-1}$, and P values below 5.058%, as shown in Table 2. According to Jideani and Mpotokwana [22], a P value of less than 10% indicates a good fit for practical purposes. However, the RMSE values in our present study are comparatively larger than the results reported by Perez et al. [16], wherein the RMSE values for the simulated water content of rice grains ranged from 0.0066 to 0.0252 $g \cdot g^{-1}$, in four hydration conditions. This is probably due to the different processing conditions used and the varieties of the kernels. Figure 4 also shows the hydration behavior of dehulled barley during different cooking conditions. In the heating-up period (I), the diffusion coefficient increases with temperature according to exponential law (Equation (10)). Hence, the hydration rate became increasingly fast although the water concentration gradient between the surface and the inside of the barley kernel was reduced simultaneously. At the stage of preservation (II), the curve exhibited the characteristic progression whereby an initial high rate of water gain is followed by slower absorption in a later stage. As cooking proceeds, the amount of moisture absorbed approaches an equilibrium value (about 3 $g \cdot g^{-1}$).

Table 2. Statistical analysis of the fitting of the numerical models to different moisture content and expansion ratio data of dehulled barley grains (100–121 °C) during canned porridge cooking.

T (°C)	Moisture Content ($g \cdot g^{-1}$)			Expansion Ratio		
	R^2	P (%)	RMSE ($g \cdot g^{-1}$)	R^2	P (%)	RMSE
100	0.993	4.325	0.053	0.978	7.230	0.250
115	0.997	5.058	0.068	0.982	6.418	0.230
121	0.997	4.581	0.046	0.990	5.207	0.174

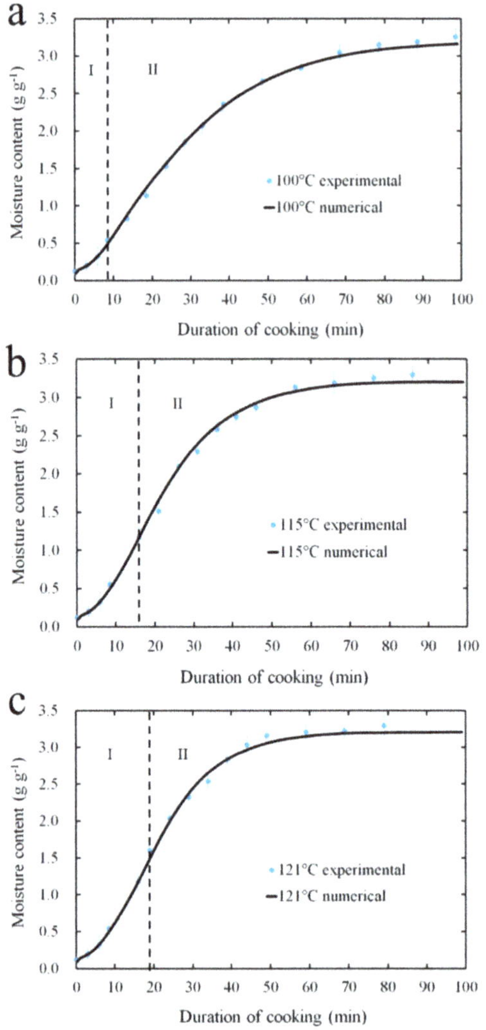

Figure 4. Changes in predicted and experimentally observed moisture content (wet basis) of barley kernels during cooking (I: heating-up stage, II: preservation) of canned porridge at 100 °C (**a**), 115 °C (**b**), and 121 °C (**c**). Vapor temperatures reached 100, 115, and 121 °C at 8.5, 16, and 19 min, respectively.

Figure 5 shows the evident change and the distribution of moisture in the dehulled barley during the cooking. Although the simulation of hydration was performed for all cooking temperatures, only the results for 121 °C are presented because these are representative of the other conditions explored. The different colors represent the different values of the moisture field. It can be observed that the surface layer of the grain is hydrated in the beginning of the cooking process, leaving the central core of the kernel dry. At 24 min of cooking at 121 °C, the average water content in barley is 1.989 $g \cdot g^{-1}$, varying from to 1.2630 to 2.2957 $g \cdot g^{-1}$ according to the position inside the kernel (Figure 5). After 49 min of cooking, the average moisture content is 3.205 $g \cdot g^{-1}$ (Figure 4), uniformly distributed across the kernel. This indicated that the equilibrium value was reached. A similar moisture diffusion performance was reported by Montanuci et al. [9], when the barley grain was soaked at a temperature of 10–25 °C.

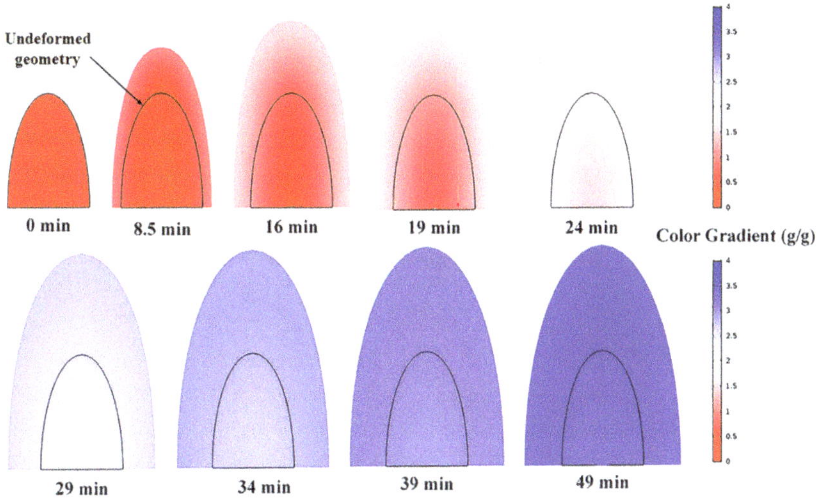

Figure 5. Simulation of moisture uptake in the cooking process at 121 °C.

4.2. Volume Change During Hygroscopic Swelling

The volumetric changes of barley kernels cooked at three different temperatures in sealed cans are presented in Figure 6. The barley kernels swelled obviously throughout the geometry as soon as the grains began to absorb water. The volumes of the barley samples expanded faster at a higher temperature. It took about 49 min to reach the maximum expansion at 121 °C, whereas 78.5 min was required at 100 °C. This observation was consistent with the results reported by Amogha et al. [24], wherein the equilibrium length of rice grain was achieved in 60, 30, and 20 min, respectively, for pre-soaked rice, and 70, 40, and 35 min respectively for un-soaked rice when the temperature was 80, 90, and 97 °C. This could be explained by the fact that water diffusion occurs more readily at higher temperatures [25]. The binding of starch molecules and water induces the volume expansion of barley grains [26].

The experimental data relating to the volume of the grain can also be reasonably fitted using the simulated values (Equation (3)). In our study of volume simulation, the R^2 for the simulated volumes of dehulled barley were 0.978, 0.982, and 0.990, the P values were 7.230%, 6.418%, and 5.207%, and the RMSE values were 0.250, 0.230, and 0.174, respectively, for 100, 115, and 121 °C (Table 2). The RMSE values are relatively close to those in the report of Perez et al. [15], which range from 0.184 to 0.794. Therefore, the numerical model can be used in the prediction of the swelling of barley. The results of the simulation of the hygroscopic swelling in barley grains using the fixed value of the hygroscopic expansion coefficient were a little bit off at 115 °C and 121 °C (Figure 6).

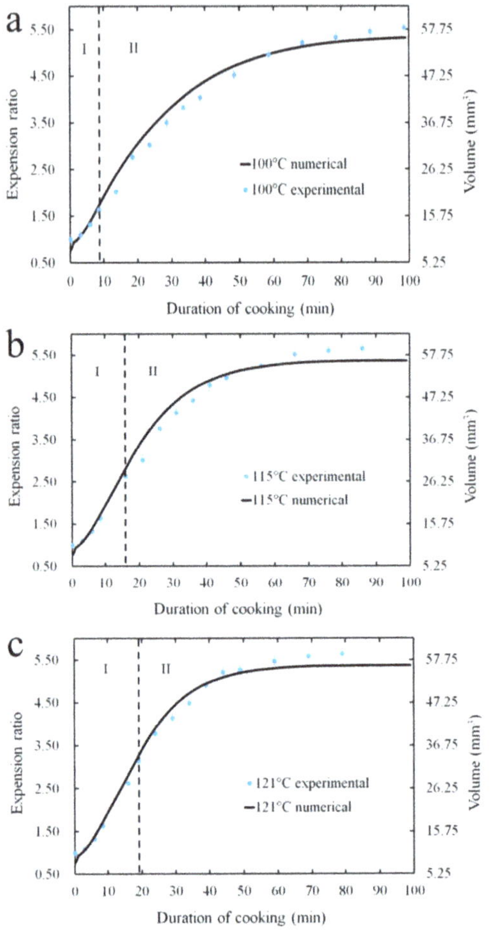

Figure 6. Computed and measured changes in volume and expansion ratio of barley kernels during cooking (I: heating-up stage, II: preservation) of canned porridge at 100 °C (**a**), 115 °C (**b**) and 121 °C (**c**). The vapor temperatures reached 100, 115, and 121 °C at 8.5, 16, and 19 min, respectively.

As mentioned in Section 4.1, the results at 121 °C are representative of other conditions. Both the experiment and the simulation dynamics of the swelling characteristics of barley at 121 °C are shown in Figure 7. The variation in the shade of the kernel shows obvious volumetric changes in barley. The volumetric expansion of the barley begins soon after the water has been absorbed throughout the kernels. As is shown in Figures 6 and 7, the swelling is more pronounced after 8.5 min of cooking (the expansion ratio is only 1.64 at 8.5 min, but 2.62 at 16 min), and becomes less obvious after 49 min (the volume increment is only 3.9 mm when the cooking time increases from 49 min to 79 min). It is obvious that the swelling of the kernels was higher at the tip than along the sides of the kernels during cooking. The major reason for this observation is the geometry of the barley kernels. On account of the elliptical shape of the kernel, more starch molecules are located along the sides of the kernel than at its tip [16]. The swelling of the starch granules along the side section resulted in greater compressive stress. This force pushes more starch molecules towards the tip of the kernel, causing more starch granules to further elongate.

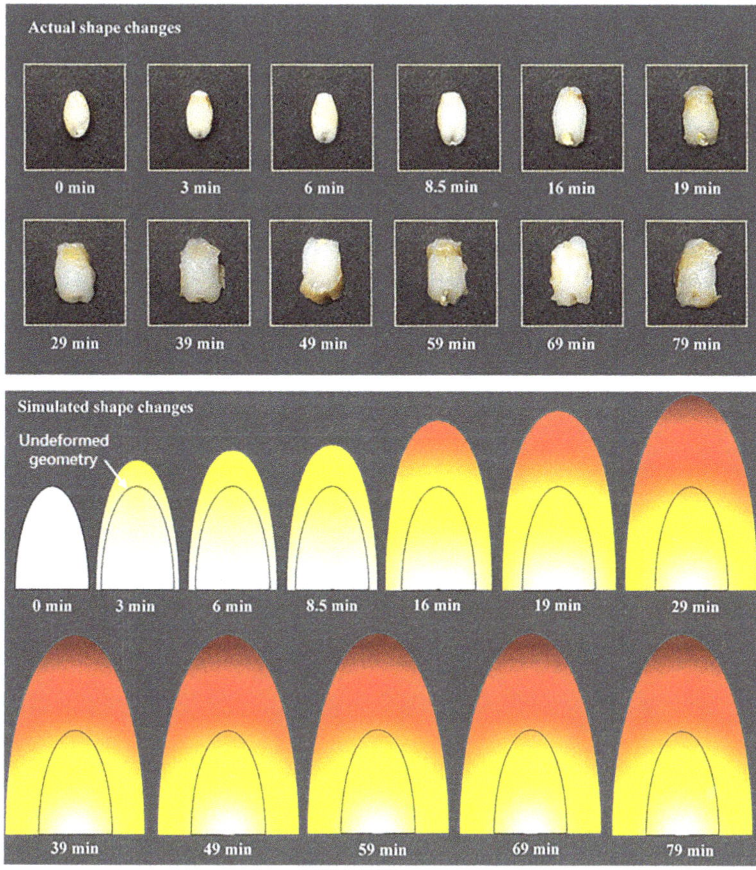

Figure 7. Computed and experimentally observed shape changes of barley kernels at different cooking times (121 °C).

5. Conclusions

Cooking canned barley porridge is a complicated process, which involves several processes, such as moisture uptake, gelatinization of starch, and swelling taking place simultaneously. Computational techniques and numerical modeling can be used to study and understand these complex changes occurring during the cooking of canned porridge, which helps to optimize the process with consequences for the curve of the moisture content and its distribution.

In the current study, a numerical simulation was developed using the finite element method making it possible to predict the amount and distribution of moisture content, as well as the volume of barley during the cooking of canned porridge. The phenomenological model, verified by experiments, can be applied to predict and optimize the hydrothermal processes of barley and other cereals, even in conditions not tested experimentally. The rate of hydration and swelling increased evidently with temperature increments, leading to a decrement of about 50% of the time to reach the equilibrium moisture content and volume of barley grains by increasing the cooking temperature from 100 to 121 °C. However, temperature does not noticeably affect the equilibrium value of the average moisture content.

Author Contributions: D.L. put forward the idea of this work, L.W. conducted the simulation and wrote this paper, M.W. contributed to the results analysis and post-processing, M.G. revised this paper, T.D. and X.Y. supervised the process.

Acknowledgments: This research was funded by the National Major R & D Program of China (grant no. 2016YFD0400301).

Conflicts of Interest: The authors declare that no conflicts of interest exist.

References

1. USDA–United States Department of Agriculture. Available online: https://www.fas.usda.gov/data/grain-world-markets-and-trade (accessed on 12 June 2018).
2. Baik, B.; Ullrich, S.E. Barley for food: Characteristics, improvement, and renewed interest. *J. Cereal Sci.* **2008**, *48*, 233–242. [CrossRef]
3. Baik, B.K. Current and Potential Barley Grain Food Products. *Cereal Foods World* **2016**, *61*, 188–196. [CrossRef]
4. Mandge, H.M.; Sharma, S.; Dar, B.N. Instant multigrain porridge: effect of cooking treatment on physicochemical and functional properties. *J. Food Sci. Technol.* **2014**, *51*, 97–103. [CrossRef] [PubMed]
5. Teerin, C.; Maradee, P. Effect of sterilizing temperature on physical properties of rice porridge mixed with legumes and job's tear in retortable pouch. *J. Food Process. Preserv.* **2015**, *39*, 2356–2360.
6. Tamura, M.; Nagai, T.; Hidaka, Y.; Ogawa, Y. Changes in Nonwaxy Japonic Rice Grain Textural-Related Properties During Cooking. *J. Food Qual.* **2014**, *37*, 177–184. [CrossRef]
7. Jian, F.; Jayas, D.S.; Fields, P.G.; White, N.D.G. Water sorption and cooking time of red kidney beans (*Phaseolus vulgaris* L.): Part II—mathematical models of water sorption. *Int. J. Food Sci. Technol.* **2017**, *52*, 2412–2421. [CrossRef]
8. Yadav, B.K.; Jindal, V.K. Water uptake and solid loss during cooking of milled rice (*Oryza sativa* L.) in relation to its physicochemical properties. *J. Food Eng.* **2007**, *80*, 46–54. [CrossRef]
9. Montanuci, F.D.; Perussello, C.A.; De Matos Jorge, L.M.; Jorge, R.M.M. Experimental analysis and finite element simulation of the hydration process of barley grains. *J. Food Eng.* **2014**, *131*, 44–49. [CrossRef]
10. Bello, M.; Tolaba, M.P.; Aguerre, R.J.; Suarez, C. Modeling water uptake in a cereal grain during soaking. *J. Food Eng.* **2010**, *97*, 95–100. [CrossRef]
11. Yadav, B.K.; Jindal, V.K. Modeling changes in milled rice (*Oryza sativa* L.) kernel dimensions during soaking by image analysis. *J. Food Eng.* **2007**, *80*, 359–369. [CrossRef]
12. Shanthilal, J.; Anandharamakrishnan, C. Computational and numerical modeling of rice hydration and dehydration: A review. *Trends Food Sci. Technol.* **2013**, *31*, 100–117. [CrossRef]
13. Bakalis, S.; Kyritsi, A.; Karathanos, V.T.; Yanniotis, S. Modeling of rice hydration using finite elements. *J. Food Eng.* **2009**, *94*, 321–325. [CrossRef]
14. Balbinoti, T.C.V.; Jorge, L.M.D.M.; Jorge, R.M.M. Modeling the hydration step of the rice (*Oryza sativa*) parboiling process. *J. Food Eng.* **2018**, *216*, 81–89. [CrossRef]
15. Perez, J.H.; Tanaka, F.; Uchino, T. Comparative 3D simulation on water absorption and hygroscopic swelling in japonica rice grains under various isothermal soaking conditions. *Food Res. Int.* **2011**, *44*, 2615–2623. [CrossRef]
16. Perez, J.H.; Tanaka, F.; Uchino, T. Modeling of mass transfer and initiation of hygroscopically induced cracks in rice grains in a thermally controlled soaking condition: With dependency of diffusion coefficient to moisture content and temperature—A 3D finite element approach. *J. Food Eng.* **2012**, *111*, 519–527. [CrossRef]
17. Gulati, T.; Datta, A.K. Coupled multiphase transport, large deformation and phase transition during rice puffing. *Chem. Eng. Sci.* **2016**, *139*, 75–98. [CrossRef]
18. Srivastava, V.; Chester, S.A.; Ames, N.M.; Anand, L. A thermo-mechanically-coupled large-deformation theory for amorphous polymers in a temperature range which spans their glass transition. *Int. J. Plast.* **2010**, *26*, 1138–1182. [CrossRef]
19. Shitanda, D.; Nishiyama, Y.; Koide, S. Compressive strength properties of rough rice considering variation of contact area. *J. Food Eng.* **2002**, *53*, 53–58. [CrossRef]
20. Dupaix, R.B.; Boyce, M.C. Constitutive modeling of the finite strain behavior of amorphous polymers in and above the glass transition. *Mech. Mater.* **2007**, *39*, 39–52. [CrossRef]
21. Kim, K.O.; Kim, H.Y.L.; Lee, Y.C. Optimization of freeze dried instant rice production for infant foods. *Foods Biotechnol.* **1996**, *5*, 14–20.
22. Jideani, V.A.; Mpotokwana, S.M. Modeling of water absorption of Botswana bambara varieties using Peleg's equation. *J. Food Eng.* **2009**, *92*, 182–188. [CrossRef]

23. Fracasso, A.F.; Perussello, C.A.; Haminiuk, C.W.I.; Jorge, L.M.M.; Jorge, R.M.M. Hydration kinetics of soybeans: Transgenic and conventional cultivars. *J. Cereal Sci.* **2014**, *60*, 584–588. [CrossRef]
24. Amogha, V.; Shinde, Y.H.; Pandit, A.B.; Joshi, J.B. Image analysis based validation and kinetic parameter estimation of rice cooking. *J. Food Process Eng.* **2017**, *40*, e12552. [CrossRef]
25. Shinde, Y.H.; Amogha, V.; Pandit, A.B.; Joshi, J.B. Kinetics of cooking of unsoaked and presoaked split peas (Cajanus cajan). *J. Food Process Eng.* **2017**, *40*, e12527. [CrossRef]
26. Shinde, Y.H.; Gudekar, A.S.; Chavan, P.V.; Pandit, A.B.; Joshi, J.B. Design and development of energy efficient continuous cooking system. *J. Food Eng.* **2016**, *168*, 231–239. [CrossRef]

processes

MDPI

Article

Wave Characteristics of Coagulation Bath in Dry-Jet Wet-Spinning Process for Polyacrylonitrile Fiber Production Using Computational Fluid Dynamics

Son Ich Ngo [1], Young-Il Lim [1,*] and Soo-Chan Kim [2]

[1] Center of Sustainable Process Engineering (CoSPE), Department of Chemical Engineering, Hankyong National University, Jungang-ro 327, Anseong-si 17579, Korea; ngoichson@hknu.ac.kr

[2] Research Center for Applied Human Sciences, Department of Electrical and Electronic Engineering, Hankyong National University, Jungang-ro 327, Anseong-si 17579, Korea; sckim@hknu.ac.kr

* Correspondence: limyi@hknu.ac.kr; Tel.: +82-31-670-5207; Fax: +82-31-670-5209

Received: 30 April 2019; Accepted: 22 May 2019; Published: 25 May 2019

Abstract: In this work, a three-dimensional volume-of-fluid computational fluid dynamics (VOF-CFD) model was developed for a coagulation bath of the dry-jet wet spinning (DJWS) process for the production of polyacrylonitrile (PAN)-based carbon fiber under long-term operating conditions. The PAN-fiber was assumed to be a deformable porous zone with variations in moving speed, porosity, and permeability. The Froude number, interpreted as the wave-making resistance on the liquid surface, was analyzed according to the PAN-fiber wind-up speed (v_{PAN}). The effect of the PAN speed on the reflection and wake flow formed by drag between a moving object and fluid is presented. A method for tracking the wave amplitude with time is proposed based on the iso-surface of the liquid volume fraction of 0.95. The wave signal for 30 min was divided into the initial and resonance states that were distinguished at 8 min. The maximum wave amplitude was less than 0.5 mm around the PAN-fiber inlet nozzle for v_{PAN} = 0.1–0.5 m/s in the resonance state. The VOF-CFD model is useful in determining the maximum v_{PAN} under an allowable air gap of the DJWS process.

Keywords: polyacrylonitrile-based carbon fiber; coagulation bath; dry-jet wet spinning process; computational fluid dynamics; wave resonance; maximum wave amplitude

---

1. Introduction

Carbon fiber (CF) has attracted attention owing to the increase of its uses in the fields of aerospace, automotives, sporting goods, biomedicine, building, and infrastructure [1–4]. The need for high-quality and high-productivity CF has emerged due to a need for economic efficiency [5,6]. Polyacrylonitrile (PAN)-based carbon fibers constitute an overwhelming share, i.e., more than 90%, of the world's CF production [7]; they are advantageous in terms of their high tensile strength, low density, and reasonable cost [8].

Several spinning processes such as wet, dry, dry-jet wet, and melt spinning are used to fabricate PAN-fiber [7,9,10]. The dry-jet wet spinning (DJWS) process possesses the advantages of high-speed fiber formation, high concentration of dope, high degree of jet stretch, and control capability of coagulation kinetics [9,11]. The DJWS process is characterized by the fact that the fiber solution is extruded into air or a gaseous environment, and is pulled inside a coagulation bath [11,12].

The gap between the spinneret and the coagulation bath surface, called the air gap, varies with the type of polymer and technology being used. The air gap does not only facilitate jet-stretch, but also provides resistance to counter diffusion when the dope is immersed inside the coagulation bath; this is a crucial factor for preventing the void structure of the fiber product [13]. By increasing the air gap height, the maximum attainable spinning speed decreases sharply [11]. A large air gap results in a yarn of low tenacity [11] and high elongation stress [13]. A small air gap can lead to fiber breakage by contact between the spinneret nozzle and the coagulation bath surface. It has been reported that the molecular orientation of fibers induced by shear stress within the spinneret can relax in an air gap of 1 cm [14]. A high-speed PAN-fiber production has been presented using a solvent-free DJWS process [7].

Computational fluid dynamics (CFD) has been shown as a powerful tool for analyzing single-phase hydrodynamics [2,3] as well as multiphase interactions [15–18]. The volume-of-fluid (VOF) CFD for multiphase flows is widely used for capturing interphase surface characteristics such as tracking the interface between gas and liquid phases in structured packing [19], and free-surface wave flow around a surface-piercing foil [20]. The VOF method can handle highly distorted or breaking interfaces without the need for a grid conformation [18–20]. However, few researchers have addressed the VOF-CFD model the wave resonance on a free surface in the presence of obstacles and moving objects. Moreover, the detection of wave amplitude is required to identify an allowable air gap in the DJWS process.

The purpose of this study was to develop a VOF-CFD model of a coagulation bath in the DJWS process for the identification of wave resonance and amplitude on the liquid surface during a long-term operating condition. The coagulation bath geometry for a DJWS process, CFD mesh structure, and boundary conditions are presented. A method for tracking the wave amplitude with time is proposed based on the iso-surface of the liquid volume fraction. The Froude number, wave speed, and maximum wave amplitude (MWA) are investigated according to the PAN-fiber spinning speed using the VOF-CFD model.

2. Coagulation Bath of DJWS Process

The symmetric coagulation bath of a DJWS process adopted from a patent [21] is shown in Figure 1. The coagulation bath is 1400 mm × 300 mm × 1310 mm (length × width × height). The diameter of the spinneret nozzle is 120 mm (see Figure 1a). The PAN solution is extruded through a commercial spinneret nozzle with 3000 holes of 0.5 mm in each, and subsequently passes through the air gap before being immersed into a dimethyl sulfoxide (DMSO)–water solution in the coagulation bath. The as-spun fiber passes through a static guide roller immersed in the coagulation bath. In this CFD study, the PAN solution uniformly injected from the nozzle is assumed as a porous medium. The non-slip boundary condition was applied to the surface of the roller and the wall of the bath.

During the coagulation process, the PAN solution is spun, stretched, and wound to create the fiber. As shown in Figure 1b,c, the shape of the PAN-bundle (or PAN solution) deforms from a circle to an ellipse, and its size reduces linearly. The ellipse maintains the same form from A to B. The large ellipse shrinks linearly to a small one from B to C. It is known that a short air gap benefits jet-stretch and high production speed [11]. The DJWS process is operated for 30 min in the flow time.

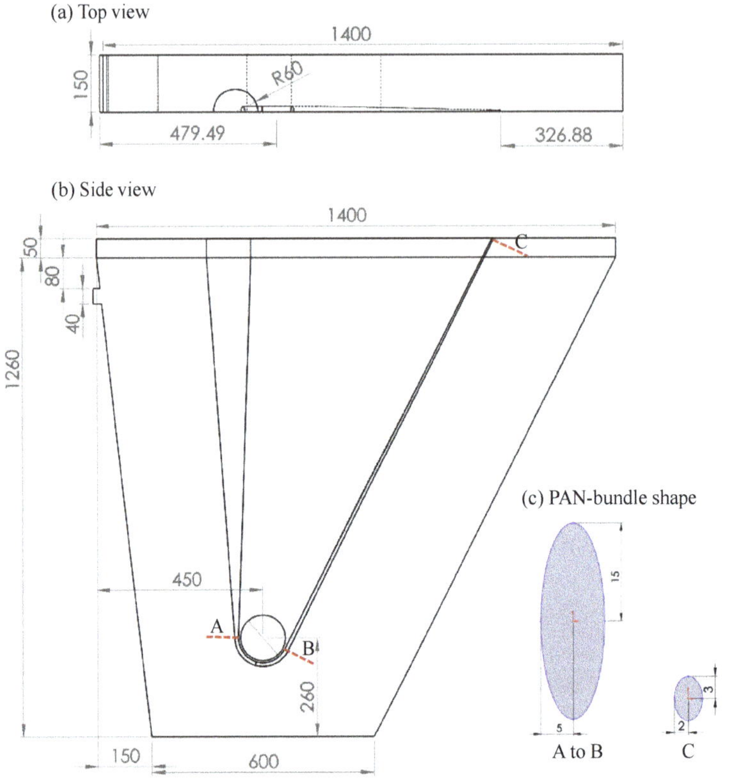

Figure 1. Geometry of coagulation bath and PAN bundle shape in dry-jet wet spinning (DJWS) process.

2.1. Mesh Structure and Material Properties

The boundary and mesh structure of the CFD domain are shown in Figure 2a. The inlet of the DMSO–water solution is located below the liquid outlet at the top of the bath. The liquid overflows from the liquid outlet. The PAN solution is injected 50 mm above the liquid surface. The PAN fiber solidified in the coagulation bath exits at the end of the bath.

The CFD domain has a polyhedral mesh structure of 0.67 million cells. The mesh is concentrated on the gas–liquid interface, PAN fiber, and walls. The near-PAN zone from the PAN inlet to the middle roller where a strong perturbation of liquid can occur is compartmentalized and a relatively dense mesh is used. The area at the top of the coagulation bath is named the freeboard.

Because the diameter and shape of the PAN fiber change and the PAN fiber is stretched through the roller, the PAN fiber is assumed to be a continuous porous zone with variable moving velocity (v), porosity (ϕ), and permeability (K). As shown in Figure 2b, the speed of the PAN fiber is divided into four zones near the middle roller: v^{in}, v_1^{mid}, v_2^{mid}, and v^{out}. The first speed (v^{in}) and the last speed (v^{out}) are regarded as the PAN-fiber spinning speed (v_s) and wind-up speed (v_{PAN}), respectively.

The one-filament diameters of the PAN-bundle are $d_f^{in} = 0.5$ (as the spinneret hole diameter), $d_{f,1}^{mid} = d_{f,2}^{mid} = 0.2$, and $d_f^{out} = 0.1$ mm [7]. The local porosity (ϕ) of the porous zone is calculated from the ratio of the cross-sectional area of all the filaments to that of the PAN bundle. Thus, ϕ is defined as the reciprocal of a linear function of the Y-coordinate, i.e.,

$$\begin{cases} \frac{1}{\phi^{in}} = 0.9479 - 4.79 \times 10^{-2} \times (Y - 0.26) \\ \frac{1}{\phi^{mid}_{1,2}} = 0.9 - 1.61 \times (Y + 0.09449) \\ \frac{1}{\phi^{out}} = 0.6875 \end{cases} \tag{1}$$

where the origin of Y is the bottom of the coagulation bath.

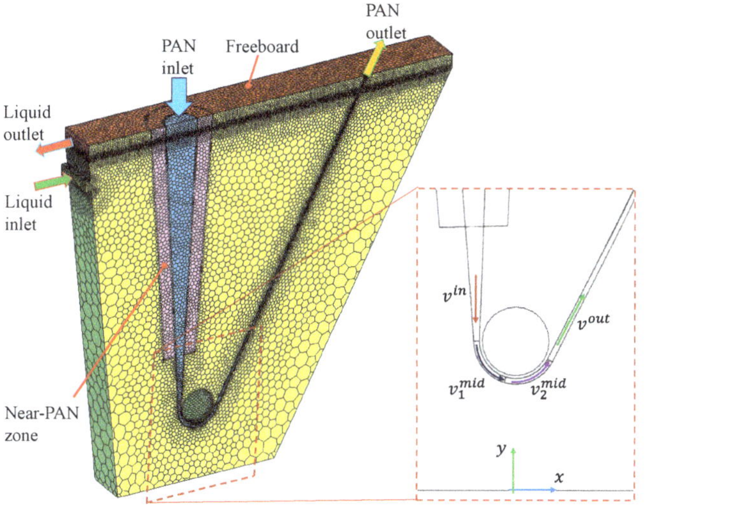

(a) Polyhedral mesh structure (b) PAN-fiber moving velocity (v) around middle roller

Figure 2. Mesh structure and polyacrylonitrile (PAN)-fiber moving velocity.

The transverse and parallel permeabilities ($K_\perp$ and $K_{||}$, respectively) are calculated from [22,23]

$$K_\perp = \frac{16}{9\pi\sqrt{6}}\left(\sqrt{\frac{1-\phi_c}{1-\phi}} - 1\right)^{2.5} R_f^2 \tag{2}$$

$$K_{||} = \frac{8R_f^2}{c}\frac{\phi^3}{(1-\phi)^2} \tag{3}$$

where $\phi_c = 1 - \pi/\left(2\sqrt{3}\right)$ is the critical porosity for hexagonal packing, at which the filaments come into contact, thus preventing flow in the transverse direction and c is the geometrical shape factor equal to 53 for hexagonal packing. R_f is the filament radius (= $d_f/2$). The porous viscous resistance ($1/K$) is the reciprocal of the permeability. $K_\perp$ and $K_{||}$ along the Y-axis are defined as

$$\begin{cases} \frac{1}{K_\perp^{in}} = 3.86 \times 10^6 + 7.15 \times 10^7 (Y - 0.26) \\ \frac{1}{K^{mid}_{1,2,\perp}} = 7.15 \times 10^7 + 3.1 \times 10^{10} (Y + 0.09449) \\ \frac{1}{K_\perp^{out}} = 4.17 \times 10^9 \end{cases} \tag{4}$$

$$\begin{cases} \frac{1}{K_{\|}^{in}} = 3.38E5 + 8.75 \times 10^6 \times (Y - 0.26) \\ \frac{1}{K_{1,2,\|}^{mid}} = 9.09E6 + 5.96 \times 10^9 \times (Y + 0.09449) \\ \frac{1}{K_{\|}^{out}} = 7.96 \times 10^8 \end{cases} \quad (5)$$

In this study, five spinning velocities (v_s) of the dope (polymer solution) at the spinneret nozzle were considered with a jet-stretch ratio (v_{PAN}/v_s) of five [7,24,25]. Table 1 indicates the PAN-bundle velocity (v), porosity (ϕ), and viscous resistance ($1/K$) according to the five cases. The porous zone was divided into four: the inlet, middle roller 1, middle roller 2, and outlet. The five v_{PAN} were 0.1, 0.25, 0.5, 0.75, and 1 m/s. v_{PAN} was five times higher than v_s according to the jet-stretch ratio, implying that the PAN-fiber was elongated by five times during the coagulation. The ranges of porosity and viscous resistance were indicated in each porous zone.

Table 1. Material properties of porous media zones.

Porous Zone	PAN-Bundle Velocity (v, m/s)					Porosity (ϕ)	Viscous Resistance ($1/K$, m^2)
	Case 1	Case 2	Case 3	Case 4	Case 5		
Inlet	0.02	0.05	0.1	0.15	0.2	0.9479–0.9	3.86×10^6–7.54×10^7
Middle roller 1	0.03	0.075	0.15	0.225	0.3	0.9	7.54×10^7
Middle roller 2	0.04	0.1	0.2	0.3	0.4	0.9–0.6875	7.54×10^7–4.17×10^9
Outlet	0.1	0.25	0.5	0.75	1	0.6875	4.17×10^9

2.2. Mesh-Independence Test

A mesh-independence test was performed for Case 2 (v_{PAN} = 0.25 m/s) on coarse, medium, and fine meshes of 0.52, 0.673, and 1.42 million cells, respectively. The time- and volume-averaged velocities in the near-PAN zone and the entire CFD domain are shown in Figure 3 with respect to the cell number. The average velocities of the coarse and medium meshes were significantly different, whereas those on the medium and fine meshes changed slightly. Therefore, the medium mesh was selected for computational efficiency, while maintaining the numerical accuracy.

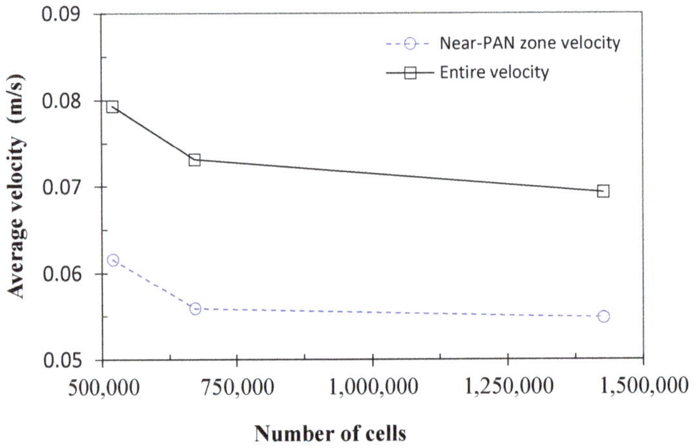

Figure 3. Mesh-independence test of computational fluid dynamics (CFD) domain at v_{PAN} = 0.25 m/s (Case 2).

It is noteworthy that the mesh density at the liquid surface should be sufficiently dense to capture the surface wave. We recommend that the cell size at the liquid surface be smaller than 0.5 mm, according to our experience. The cell size at the liquid surface ranges from 0.38 to 0.45 mm on the selected medium mesh.

3. VOF-CFD Model

The volume-of-fluid computational fluid dynamics model relies on the assumption that two phases do not interpenetrate [2]. An unsteady-state VOF-CFD model was applied to the incompressible gas–liquid system including air and the DMSO–water solution, where the diffusion of the solvent (DMSO) from the PAN solution to the bath was ignored. The standard k–ε model was used for modeling the turbulence, which has been used extensively for industrial applications [19]. Table 2 shows the VOF-CFD model.

Table 2. VOF-CFD model used in this study.

Continuity equation (DMSO): $\frac{\partial \alpha_D}{\partial t} + \vec{\nabla} \cdot (\alpha_D \vec{u}) = 0$	(T1)
(air): $\alpha_D + \alpha_A = 1$	(T2)
Single momentum equation: $\frac{\partial}{\partial t}(\rho \vec{u}) + \vec{\nabla} \cdot (\rho \vec{u} \vec{u}) = -\vec{\nabla} P + \vec{\nabla} \cdot [\eta(\vec{\nabla} \vec{u} + \vec{\nabla} \vec{u}^T)] + \rho \vec{g} + F_{surf}$	(T3)
where the surface tension force (F_{surf}) is $F_{surf} = \vec{\nabla} \cdot \overline{\overline{\tau}}$	(T4)
with the surface stress tensor of $\overline{\overline{\tau}} = \sigma \left(\lvert \vec{\nabla} \alpha_D \rvert \overline{\overline{I}} - \frac{\vec{\nabla} \alpha_D \cdot \left(\vec{\nabla} \alpha_D \right)^T}{\lvert \vec{\nabla} \alpha_D \rvert} \right)$	(T5)
Single properties (density): $\rho = \alpha_A \rho_A + \alpha_D \rho_D$	(T6)
(viscosity): $\eta = \alpha_A \eta_A + \alpha_D \eta_D$	(T7)
k-ε turbulence model: $\frac{\partial}{\partial t}(\rho k) + \vec{\nabla} \cdot \left(\rho k \vec{u} \right) = \vec{\nabla} \cdot \left[\left(\eta + \frac{\eta_t}{D_k^t} \right) \vec{\nabla} k \right] + G_k - \rho \varepsilon$	(T8)
$\frac{\partial}{\partial t}(\rho \varepsilon) + \vec{\nabla} \cdot \left(\rho \varepsilon \vec{u} \right) = \vec{\nabla} \cdot \left[\left(\eta + \frac{\eta_t}{D_\varepsilon^t} \right) \vec{\nabla} \varepsilon \right] + C_{1\varepsilon} \frac{\varepsilon}{k} \left(G_k - C_{2\varepsilon} \rho \frac{\varepsilon^2}{k} \right)$	(T9)
where: $\eta_t = \rho C_\mu \frac{k^2}{\varepsilon}$, $G_k = \eta_t S^2$, and $S = \sqrt{2 S_{ij} Sij}$.	(T10)

The tracking of the interface between the two phases was accomplished by the solution of a continuity equation for the volume fraction (α_D) of the DMSO solution, where the DMSO solution is denoted as the subscript D. Assuming no mass transfer between the two phases and no source term in each phase, the continuity equation of the DMSO solution is expressed in Equation (T1), where $\vec{u}$ is the velocity. The continuity equation of the air phase (denoted as the subscript A) is computed using Equation (T2).

In this study, a special interpolation treatment of α_D to the cells that lie near the interface between the two phases was applied, which was based on the compressive scheme of a second-order slope limiter reconstruction [26]. To specify a cutoff limit for α_D, a cutoff factor of $\alpha_D = 1 \times 10^{-6}$ was used. Thus, all values of α_D below the cutoff value were zero.

In the VOF model, the single momentum equation shown in Equation (T3) was solved throughout the domain, and the resulting velocity field was shared in both phases. The left-hand side of Equation (T3) includes the accumulation and convection of momentum per unit volume. On the right-hand side of Equation (T3), P is the pressure shared in both phases, $\rho \vec{g}$ is the gravity force, $\eta(\vec{\nabla} \vec{u} + \vec{\nabla} \vec{u}^T)$ is the stress tensor of the Newtonian fluid, and F_{surf} is the surface tension force on the interface between the two phases.

The continuum surface stress (CSS) model, expressed in Equation (T4), was used to model F_{surf} conservatively, where the surface stress tensor ($\overline{\overline{\tau}}$) owing to the surface tension is expressed in Equation (T5). The gradient of α_D ($\vec{\nabla} \alpha_D$) is the surface normal vector and $\overline{\overline{I}}$ is the unit tensor. The surface tension coefficient (σ) is assumed as a constant $\sigma = 0.072$ N/m, which is comparable to the air-water system [19].

The single properties of density (ρ) and viscosity (η) are employed to solve the VOF model, as expressed in Equations (T6)–(T7). For the DMSO–water solution, ρ_D = 1048 kg/m^3 and η_D = 0.00212 kg/m/s. For air, ρ_A = 1.225 kg/m^3 and η_A = 1.7894 × 10^{-5} kg/m/s.

The standard k-ε models [27] for turbulence kinetic energy (k) and turbulence energy dissipation (ε) are expressed in Equations (T8) and (T9), respectively. The turbulence viscosity (η_t) is determined by the local values of ρ, k, ε, and a constant C_μ (see Equation (T10)). G_k representing the generation of turbulence kinetic energy is estimated by the turbulence viscosity (η_t) and the modulus of mean rate-of-strain tensor (S). D_k^t and D_ε^t in Equations (T8) and (T9), respectively, represent the diffusion rate of k and ε, respectively [27]. The model constants of the turbulence model were set to $C_{1\varepsilon} = 1.44$, $C_{2\varepsilon} = 1.92$, $C_\mu = 0.09$, $D_k^t = 1.0$, and $D_\varepsilon^t = 1.3$.

3.1. Boundary Conditions of the VOF–CFD Model

Table 3 summarizes the boundary conditions for the present CFD model. The DMSO inlet was set to the mass flow inlet with 0.1 kg/s of the DMSO–water solution. The inlet and outlet of the PAN fiber, the outlet of the DMSO solution, and the freeboard (see Figure 2) were defined as an open channel. The open channel involved a free surface between the flowing fluid and the atmosphere that was often applied for wave propagation (ANSYS Fluent Theory Guide, ANSYS Inc., Washington, PA, USA, 2018).

Table 3. Boundary conditions of coagulation bath CFD simulation. Legend: DMSO, dimethyl sulfoxide.

Boundary Type	Setting	Value	Remarks
DMSO inlet	Mass flow inlet	0.1 kg/s	100% liquid
PAN inlet	Open channel	v_s = 0.02–0.2 m/s	Porous medium
PAN outlet	Open channel	v_{PAN} = 0.1–1.0 m/s	Porous medium
DMSO outlet	Open channel		
Freeboard	Open channel		Gas phase

3.2. Froude Number and Wave Speed

The open-channel flows are characterized by the dimensionless Froude Number (Fr) that is defined as the ratio of the inertial force to the hydrostatic force, i.e.,

$$Fr = \frac{\left|\vec{u}\right|}{\sqrt{gy}} \tag{6}$$

where $\left|\vec{u}\right|$ is the velocity magnitude of the liquid at the surface, g is the gravity, and y is the length scale. Here, y is given by the distance from the bottom of the equipment to the free surface (approximately 1.26 m). The denominator of Equation (6) is the wave speed of the fluid itself.

The wave speed observed by a fixed observer is defined as (ANSYS Fluent Theory Guide, 2018):

$$\left|\vec{u}\right|_w = \left|\vec{u}\right| \pm \sqrt{gy} \tag{7}$$

Based on the Froude number, open-channel flows can be classified into the following three categories. When $Fr < 1$, i.e., $\left|\vec{u}\right| < \sqrt{gy}$, $\left|\vec{u}\right|_w < 0$ or $\left|\vec{u}\right|_w > 0$, and the flow is known to be subcritical where disturbances can travel to the upstream and downstream. In this case, the downstream conditions might influence the upstream flow. When $Fr = 1$, the flow is known to be critical, where waves propagating from the inlet stream remain stationary. When $Fr > 1$, i.e., $\left|\vec{u}\right| > \sqrt{gy}$, $\left|\vec{u}\right|_w > 0$, and the flow is known to be supercritical where disturbances cannot travel to the upstream (ANSYS Fluent Theory Guide, 2018).

4. Detection of Surface Wave

In the VOF multiphase CFD model, the interface between the gas and liquid is located at a liquid volume fraction from 0 to 1 ($0 < \alpha_D < 1$). α_D close to 0 is gas-like while α_D close to 1 is liquid-like. The interface sharpness was highly dependent on the grid resolution around the interface [20]. In this study, $\alpha_{D,min} = 0.95$ as the minimum liquid volume fraction was selected to detect the iso-surface of the liquid. $\alpha_{D,min}$ higher than 0.95 on the medium mesh structure yielded a flatter iso-surface. $\alpha_{D,min}$ lower than 0.95 caused a wave signal with considerable noise. A $\alpha_{D,min}$ lower than 0.95 was reported to capture the interface for the Eulerian multiphase CFD model [28,29].

The detection point of the surface wave is shown in Figure 4. There are five horizontal lines (1–5), six vertical lines (A–F), and six additional points (PA–PF) located near the PAN injection spinneret nozzle. Lines 1 and 5 are located at the symmetry line and the side wall of the coagulation bath, respectively. The distance from Lines 2 to 5 from the symmetry line (Line 1) is 20, 50, 100, and 149 mm, respectively. The vertical lines are positioned in the same interval from the spinneret nozzle (Line B) of 200 mm. Thus, 30 intercepts of the horizontal and vertical lines are created, which are named after the number and letter.

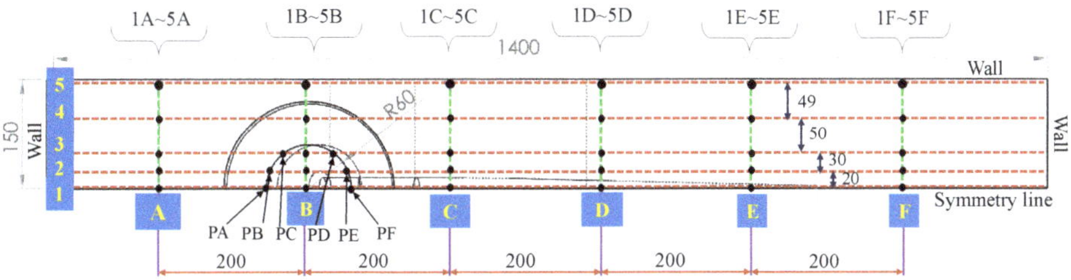

Figure 4. Detection points of surface wave.

At the 36 points, the surface wave signal in the time-series was transformed into the wave height (h_w) by a three-step data processing, as shown in Figure 5. In the first step, the unsteady state CFD simulation results were used to reconstruct the iso-surface of $\alpha_{D,min} = 0.95$. The data sampling interval was set to every one second of flow time, and 1800 CFD data files for 30 min were used. The wave coordinates (Y^i) at a point (P) and a time step (t^i) were obtained from the iso-surface of liquid with $\alpha_D \geq 0.95$. In the second step, the coordinate and time of the iso-surface, $P(Y^i)$, were exported as a new dataset. In the third step, the wave height (h_w) in the time-series at each detecting point was generated. The maximum wave amplitude (MWA) was calculated from the difference between the maximum and minimum values of h_w within a given period.

$$\text{MWA} = \max\,[h_w(t)] - \min\,[h_w(t)] \tag{8}$$

As shown in the third step of Figure 5, the wave height signal is divided into two states: initial wave and resonance wave. The initial wave appears at the first moment of operation, where the wave is unstable due to the inertia force of liquid itself against the moving PAN fiber. For all five cases with different PAN speeds, the initial wave appeared when the flow time was less than 8 min. The resonance wave occurred in the later moment, showing a stable hydrodynamic state with the compromise of inertia and gravity forces, surface tension, and turbulence at the liquid surface.

In this study, the wave speed ($\left|\vec{u}\right|_w$) was calculated from two datasets of the iso-surface of liquid in an interval of 1 s. The distance of two wave peaks (Δl) displacing for one second was used directly as $\left|\vec{u}\right|_w$:

$$\left|\vec{u}\right|_w = \Delta l, \tag{9}$$

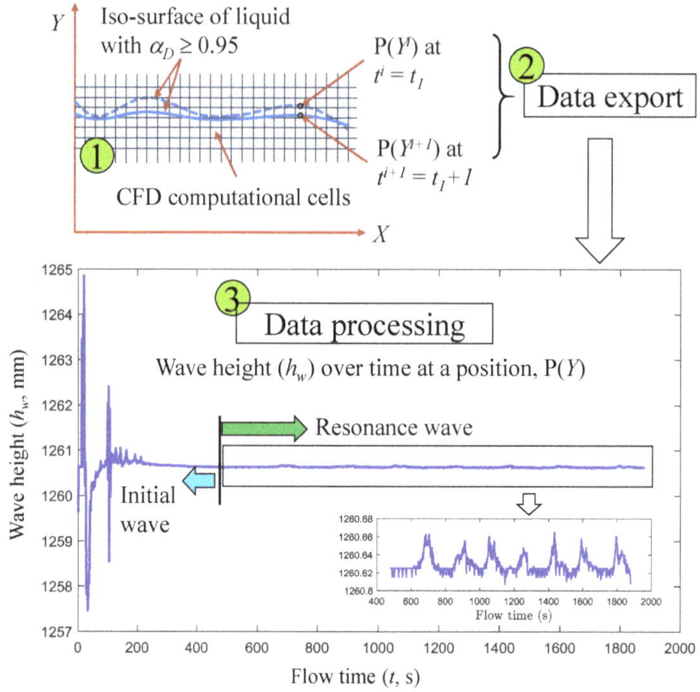

Figure 5. Procedure of detecting wave height from CFD results.

5. Results and Discussion

The three-dimensional (3D) VOF-CFD model was solved using ANSYS Fluent v18.2 (ANSYS Inc., Washington, PA, USA) and a 24-core workstation (Supermicro Inc., San Jose, CA, USA, model: X10DAi, Intel Xeon CPU E5-2670 of 2.3 GHz and 128 GB RAM). The time step of the transient simulation was fixed to 0.01 s. The calculation time was approximately five days for 30 min of the flow time in each case.

The CFD results were analyzed for the five PAN-fiber speeds (v_{PAN}) in terms of the streamlines of liquid flow, Froude number, wave speed on the liquid surface, and wave amplitude. The time-averaged quantities such as velocity and volume fraction were sampled at every 0.1 s for 30 min of flow time.

5.1. Effects of PAN Speed on Flow Streamlines

The side view of the flow streamlines on the symmetrical plane is illustrated in Figure 6 for the five v_{PAN} (0.1, 0.25, 0.5, 0.75, and 1 m/s) at t = 1800 s. The color of the streamlines indicates the velocity magnitude, and the streamlines with the arrow describe the flow history during the residence in the bath. A streamline represents a non-intersected flow. At v_{PAN} = 0.1 m/s, a large recirculation flow is observed in the inner region surrounded by the PAN fiber. One recirculation flow also appears in the top-right of the bath (see Figure 6a). A recirculation flow well-developed beneath the right of the PAN outlet is observed in Figure 6e. It may be attributed from the fact that the drag force between the moving PAN fiber and the liquid promotes the recirculation along the wall of the coagulation bath as v_{PAN} increases. At v_{PAN} = 1.0 m/s, the wake flow that emerges beneath the surface and separates the flow on the surface is observed at the PAN outlet, and in the region between the DMSO outlet and the PAN inlet.

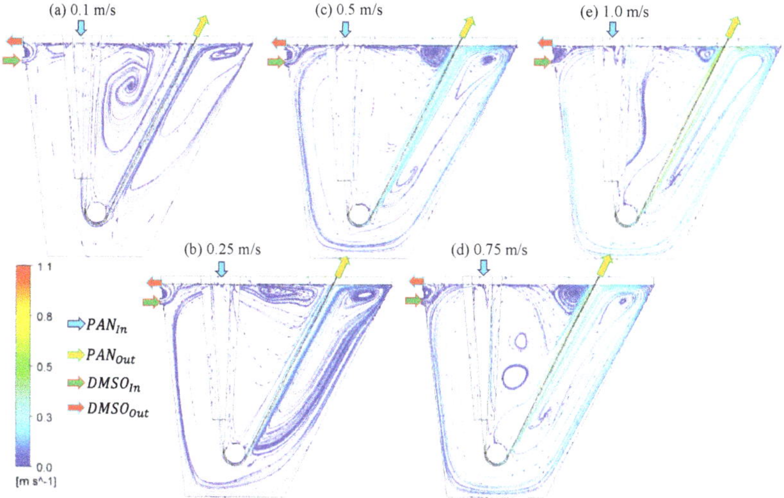

Figure 6. Streamlines of DMSO solution in the side-view at t = 1800 s.

Figure 7 depicts the streamlines from the top view of the liquid surface for the five v_{PAN} at t = 1800 s. The two wake flows are clearly shown for each PAN speed behind the PAN inlet and at the PAN outlet. In the wake flow behind the PAN inlet, the front flow moves directly from the DMSO inlet to the outlet, while the rear flow moves forward to the PAN inlet. The wake points change according to the PAN speed. Because the forward flow from the PAN inlet collides with the backward flow from the PAN outlet, recirculation flows (or vortices) appear between the PAN inlet and PAN outlet.

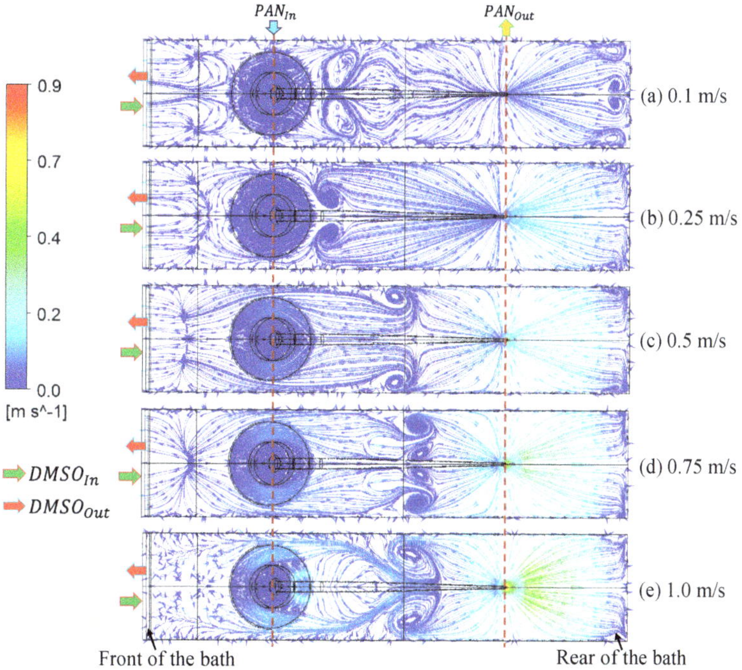

Figure 7. Streamlines of DMSO solution from the top view at t = 1800s.

5.2. Froude Number and Wave Speed

Figure 8 shows the contours of the time-averaged Froude number ($\overline{Fr}$) on the liquid surface at the five v_{PAN}. The maximum value of $\overline{Fr}$ was 0.28 near the PAN outlet at $v_{PAN} = 1$ m/s. However, the scale of $\overline{Fr}$ in Figure 8 was limited from 0 to 0.05 for an easy visualization of the difference in $\overline{Fr}$ between the five PAN speeds. Thus, red implies that $\overline{Fr}$ ranges from 0.05 to 0.28. As $\overline{Fr} < 1$ in this system, the waves can travel to both the front and rear sides of the bath.

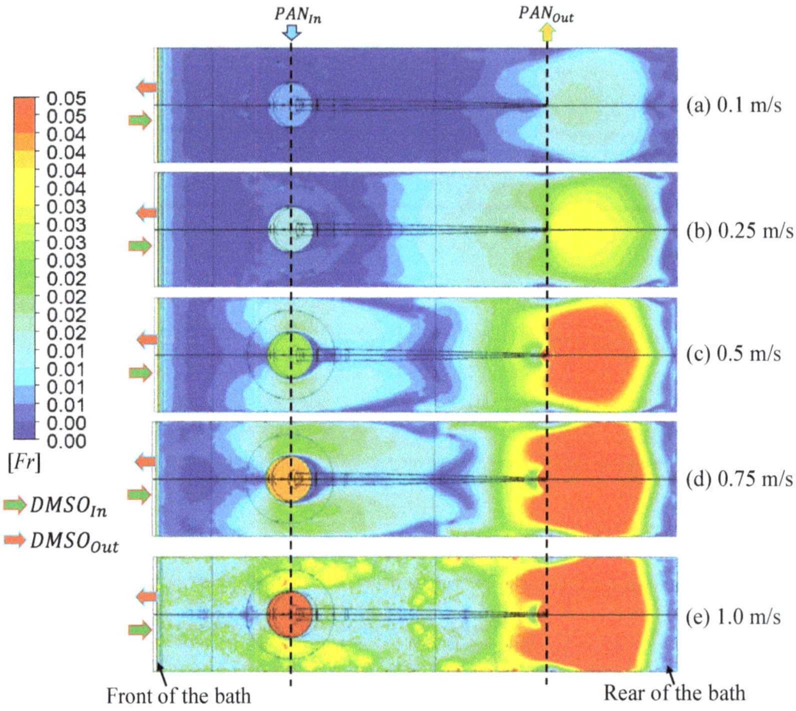

Figure 8. Contours of time-averaged Froude number ($\overline{Fr}$) on the liquid surface.

As explained in Equation (6), a high $\overline{Fr}$ on the liquid surface indicates a high potential for wave formation. The higher the PAN speed (v_{PAN}), the higher the $\overline{Fr}$ becomes. $\overline{Fr}$ is relatively high on the surfaces of the PAN inlet and the rear side of the PAN outlet. $\overline{Fr}$ is relatively low on the surfaces of the right end of the bath, the center of the bath, and behind the PAN inlet. In Figure 8d, two long ellipsoidal shapes are found near the side walls of the PAN inlet, where the wave can be formed. Some waves are observed in the region where the forward flow from the PAN inlet collides with the backward flow from the PAN outlet (see Figure 7), which is coincident with the ripples on the free surface.

Figure 9 shows the wave speed ($|u|_w$) with respect to the liquid velocity ($|u|$). The experimental data of $|u|_w$ were measured in a co-current air–water inclined-pipe flow [30]. Two CFD results were obtained for both the simple inclined-pipe flow and the present coagulation bath at $v_{PAN} = 1.0$ m/s at various positions on the liquid surface. The CFD results for the inclined-pipe flow are comparable to the experimental data, having the same order of the wave speed in magnitude. However, the difference of the wave speed between the simple inclined-pipe flow and the complex coagulation bath flow is approximately two orders of magnitude. The wave speed exhibits a linear relationship with the liquid velocity in the inclined-pipe flow [30], while the present coagulation bath showed a random response.

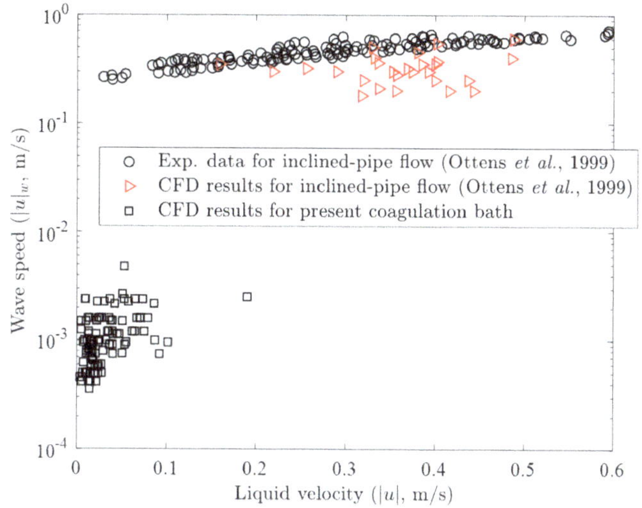

Figure 9. Wave speed ($|u|_w$) with respect to liquid velocity ($|u|$).

The surface wave resonance and breaking are closely related to the hydrodynamic parameters such as density, velocity, surface tension, and viscosity [31,32]. Furthermore, the geometry contributes significantly to the surface wave characteristics owing to numerous reflections from the walls and moving objects. The surface wave resonance appears somewhat chaotic in the coagulation bath, which leads to a random response of the wave speed to the liquid velocity. Therefore, it is difficult to predict the present surface wave behaviors using any previous liquid wave models [32–35].

5.3. Analysis of Surface Wave Signal

As mentioned in Section 4, the surface wave signal can be divided into two states: the initial state for $t \leq 480$ s and the resonance state for $t > 480$ s. Figure 10 depicts the wave height (h_w) at Point PF (see also Figure 4) for v_{PAN} = 0.1, 0.5, 0.75, and 1.0 m/s. Because Point PF is located at the PAN inlet, the baseline of h_w indicated in Figure 10b depends entirely on v_{PAN}. The baseline is lowered as the v_{PAN} increases, as expected. For v_{PAN} = 0.1 and 0.5 m/s, the initial waves are unstable and the resonance waves are stable and periodic (see Figure 10a). As shown in Figure 10b, the resonance wave at v_{PAN} = 0.75 m/s is stable and periodic during the majority of the flow time, but unstable waves are occasionally superpositioned. The wave height at v_{PAN} = 1.0 m/s fluctuates strongly over 480 s.

Figure 11 illustrates the MWA according to v_{PAN}, which was detected at ten points around the PAN inlet. For $t \leq 480$ s, the MWA is relatively high for every detecting point because of the unstable waves in the initial state (see Figure 11a). As shown in Figure 10b, the MWA of v_{PAN} = 1.0 m/s is relatively lower than those of other v_{PAN}. This may be attributed to the fact that the strong injection of the PAN solution protects the invasion of waves into the PAN inlet area at the initial moment.

However, the MWA of v_{PAN} = 1.0 m/s is the highest for $t > 480$ s for most detecting points, as shown in Figure 11b. The MWAs of v_{PAN} = 0.1–0.5 m/s remain low under 0.5 mm, while the MWAs increase considerably for v_{PAN} = 0.75 and 1.0 m/s. The highest value of MWA (= 8 mm) is detected at Point PF, which is located at the axial center of the PAN inlet in the forward direction. The MWA around the PAN inlet provides an air gap that is allowable according to the PAN speed.

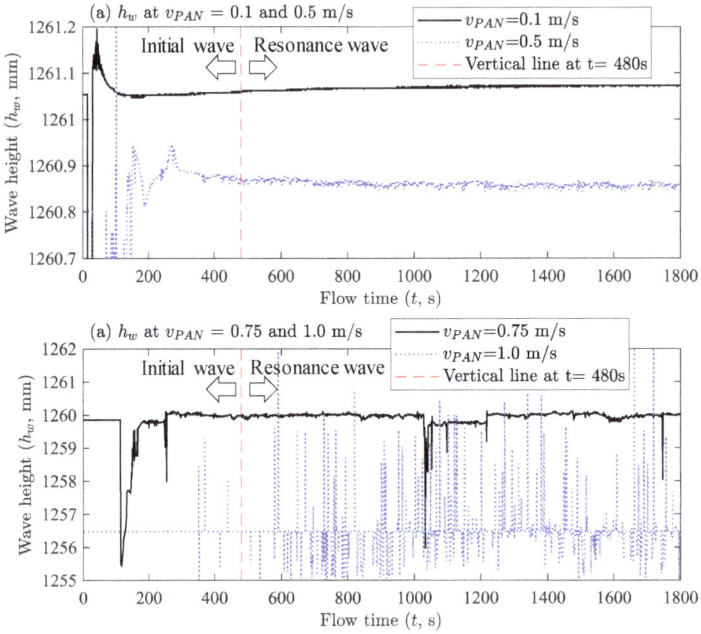

Figure 10. Wave height (h_w) from the bottom of the coagulation bath at Point PF.

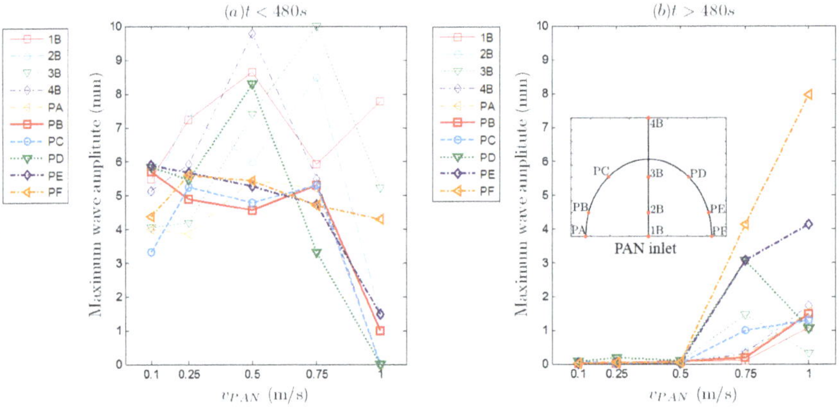

Figure 11. Maximum wave amplitude (MWA) around PAN inlet according to PAN speed (v_{PAN}).

6. Conclusions

In this work, a 3D VOF-CFD model was developed for a coagulation bath for the production of PAN-based CF in the DJWS process. A bundle of 3000 PAN-based CF filaments was assumed as a deformable porous zone with variations in moving velocity, porosity, and permeability along its path through the coagulation bath. The surface wave was detected at 36 points on the free surface by a three-step data processing. The iso-surface having a liquid volume fraction greater than 0.95 was tracked for 30 min of the flow time, and the wave height versus time was obtained.

The streamlines of the liquid, Froude number (*Fr*), wave speed, and wave height were examined for five PAN-fiber speeds (v_{PAN} = 0.1, 0.25, 0.5, 0.75, and 1.0 m/s). The wake flow that emerged beneath the surface and separated the flow on the surface was observed in the liquid streamlines.

The contour of *Fr* indicating the potential for wave formation was coincident with the ripples on the free surface. Due to the reflections from the coagulation bath walls and moving objects, the wave speed of the coagulation bath was much lower than that of an inclined pipe flow without obstacles. The surface wave signal for 30 min was divided into the initial state ($t \leq 480$ s) and the resonance state ($t > 480$ s). The maximum wave amplitude (MWA) around the PAN inlet increased considerably for v_{PAN} = 0.75 and 1.0 m/s in the resonance state. Thus, the maximum PAN production speed of 0.5 m/s was recommended for the DJWS process to maintain a stable resonance for the long-term operating condition.

The MWA can determine the minimum air gap between the spinneret and the bath surface. The maximum PAN production rate can be obtained when the air gap is close to the MWA. The resolution of the MWA depended on the minimum value of the liquid volume fraction used to determine the iso-surface. The minimum value may be confirmed by an experimental validation on the MWA. It would be useful to investigate the effect of solvent diffusion from the PAN solution on the hydrodynamics near the PAN zone to identify the reason for PAN-fiber defects during coagulation.

Author Contributions: Conceptualization, S.I.N. and Y.-I.L.; methodology, S.I.N.; software, S.I.N.; validation, S.I.N., Y-I.L., and S.-C.K.; formal analysis, S.I.N.; investigation, S.I.N.; resources, S.I.N.; data curation, S.I.N.; writing—original draft preparation, S.I.N.; writing—review and editing, Y.-I.L. and S.-C.K.; visualization, S.I.N.; supervision, Y.-I.L. and S.-C.K.; project administration, Y.-I.L.; funding acquisition, Y.-I.L.

Funding: This research was funded by the Ministry of Science, ICT, and Future Planning of Korea, grant number NRF-2016R1A2B4010423.

Acknowledgments: This research was supported by the Basic Science Research Program through the National Research Foundation of Korea (NRF). We acknowledge the Hyosung Research Institute for the internal communications.

Conflicts of Interest: The authors declare no conflict of interest. The funders had no role in the design of the study; in the collection, analyses, or interpretation of data; in the writing of the manuscript, or in the decision to publish the results.

References

1. Chang, I.Y.; Lees, J.K. Recent development in thermoplastic composites: A review of matrix systems and processing methods. *J. Thermoplast. Compos. Mater.* **1988**, *1*, 277–296. [CrossRef]
2. Ngo, S.I.; Lim, Y.-I.; Hahn, M.-H.; Jung, J. Prediction of degree of impregnation in thermoplastic unidirectional carbon fiber prepreg by multi-scale computational fluid dynamics. *Chem. Eng. Sci.* **2018**, *185*, 64–75. [CrossRef]
3. Ngo, S.I.; Lim, Y.-I.; Hahn, M.-H.; Jung, J.; Bang, Y.-H. Multi-scale computational fluid dynamics of impregnation die for thermoplastic carbon fiber prepreg production. *Comput. Chem. Eng.* **2017**, *103*, 58–68. [CrossRef]
4. Rodríguez-Tembleque, L.; Aliabadi, M.H. Numerical simulation of fretting wear in fiber-reinforced composite materials. *Eng. Fract. Mech.* **2016**, *168*, 13–27. [CrossRef]
5. Paul, J.T. Method of Manufacturing Carbon Fiber Using Preliminary Stretch. U.S. Patent 5066433, 19 November 1991.
6. Deng, J.; Xu, L.; Zhang, L.; Peng, J.; Guo, S.; Liu, J.; Koppala, S. Recycling of carbon fibers from CFRP waste by microwave thermolysis. *Processes* **2019**, *7*, 207. [CrossRef]
7. Rahman, M.A.; Ismail, A.F.; Mustafa, A. The effect of residence time on the physical characteristics of PAN-based fibers produced using a solvent-free coagulation process. *Mater. Sci. Eng. A* **2007**, *448*, 275–280. [CrossRef]
8. Mataram, A.; Ismail, A.F.; Mahmod, D.S.A.; Matsuura, T. Characterization and mechanical properties of polyacrylonitrile/silica composite fibers prepared via dry-jet wet spinning process. *Mater. Lett.* **2010**, *64*, 1875–1878. [CrossRef]
9. Bajaj, P.; Paliwal, D.K. Some recent advances in the production of acrylic fibres for specific end uses. *Indian J. Fibre Text. Res.* **1991**, *16*, 89–99.
10. Lee, J.H.; Jin, J.-U.; Park, S.; Choi, D.; You, N.-H.; Chung, Y.; Ku, B.-C.; Yeo, H. Melt processable polyacrylonitrile copolymer precursors for carbon fibers: Rheological, thermal, and mechanical properties. *J. Ind. Eng. Chem.* **2018**. [CrossRef]

11. Baojun, Q.; Ding, P.; Zhenqiou, W. The mechanism and characteristics of dry-jet-wet spinning of acrylic fibers. *Adv. Polym. Technol.* **1986**, *6*, 509–529. [CrossRef]
12. Monsanto, C. Manufacture of Industrial Acrylic Fibers. U.S. Patent 1,193,170, 12 December 1966.
13. Zeng, X.; Hu, J.; Zhao, J.; Zhang, Y.; Pan, D. Investigating the jet stretch in the wet spinning of PAN fiber. *Appl. Polym. Sci.* **2007**, *106*, 2267–2273. [CrossRef]
14. Qin, J.-J.; Gu, J.; Chung, T.-S. Effect of wet and dry-jet wet spinning on the shear-induced orientation during the formation of ultrafiltration hollow fiber membranes. *J. Membr. Sci.* **2001**, *182*, 57–75. [CrossRef]
15. Ngo, S.I.; Lim, Y.-I.; Song, B.-H.; Lee, U.-D.; Lee, J.-W.; Song, J.-H. Effects of fluidization velocity on solid stack volume in a bubbling fluidized-bed with nozzle-type distributor. *Powder Technol.* **2015**, *275*, 188–198. [CrossRef]
16. Ngo, S.I.; Lim, Y.-I.; Song, B.-H.; Lee, U.-D.; Yang, C.-W.; Choi, Y.-T.; Song, J.-H. Hydrodynamics of cold-rig biomass gasifier using semi-dual fluidized-bed. *Powder Technol.* **2013**, *234*, 97–106. [CrossRef]
17. Janssen, J.; Mayer, R. Computational fluid dynamics (CFD)-based droplet size estimates in emulsification equipment. *Processes* **2016**, *4*, 50. [CrossRef]
18. Ramírez-Argáez, M.A.; Dutta, A.; Amaro-Villeda, A.; González-Rivera, C.; Conejo, A.N. A novel multiphase methodology simulating three phase flows in a steel ladle. *Processes* **2019**, *7*, 175. [CrossRef]
19. Hosseini, S.H.; Shojaee, S.; Ahmadi, G.; Zivdar, M. Computational fluid dynamics studies of dry and wet pressure drops in structured packings. *J. Ind. Eng. Chem.* **2012**, *18*, 1465–1473. [CrossRef]
20. Rhee, S.H.; Makarov, B.P.; Krishinan, H.; Ivanov, V. Assessment of the volume of fluid method for free-surface wave flow. *J. Mar. Sci. Technol.* **2005**, *10*, 173–180. [CrossRef]
21. Glawion, E.; Zenker, D. Spinning Bath Vat. U.S. Patent 0141111 A1, 22 May 2014.
22. Tahir, M.W.; Hallström, S.; Åkermo, M. Effect of dual scale porosity on the overall permeability of fibrous structures. *Compos. Sci. Technol.* **2014**, *103*, 56–62. [CrossRef]
23. Gebart, R. Permeability of unidirectional reinforcements for RTM. *J. Compos. Mater.* **1992**, *26*, 1100–1133. [CrossRef]
24. East, G.C.; McIntyre, J.E.; Patel, G.C. The dry-jet wet-spinning of an acrylic-fibre yarn. *J. Text. Inst.* **1984**, *75*, 196–200. [CrossRef]
25. Morris, E.; Weisenberger, M.; Rice, G. Properties of PAN fibers solution spun into a chilled coagulation bath at high solvent compositions. *Fibers* **2015**, *3*, 560–574. [CrossRef]
26. Ubbink, O. Numerical Prediction of Two Fluid Systems with Sharp Interfaces. Ph.D. Thesis, Imperial College of Science, Technology & Medicine, London, UK, 1997.
27. Launder, B.; Spalding, D.B. *Lectures in Mathematical Models of Turbulence*, 1st ed.; Academic Press: London, UK, 1972.
28. Varela, S.; Martínez, M.; Delgado, J.A.; Godard, C.; Curulla-Ferré, D.; Pallares, J.; Vernet, A. Numerical and experimental modelization of the two-phase mixing in a small scale stirred vessel. *J. Ind. Eng. Chem.* **2018**, *60*, 286–296. [CrossRef]
29. Le, T.T.; Ngo, S.I.; Lim, Y.-I.; Park, C.-K.; Lee, B.-D.; Kim, B.-G.; Lim, D.H. Effect of simultaneous three-angular motion on the performance of an air–water–oil separator under offshore operation. *Ocean Eng.* **2019**, *171*, 469–484. [CrossRef]
30. Ottens, M.; Klinkspoor, K.; Hoefsloot, H.C.J.; Hamersma, P.J. Wave characteristics during cocurrent gas-liquid pipe flow. *Exp. Therm Fluid Sci.* **1999**, *19*, 140–150. [CrossRef]
31. Tulin, M.P. Breaking of ocean waves and downshifting. In *Waves and Nonlinear Processes in Hydrodynamics*; Grue, J., Gjevik, B., Weber, J.E., Eds.; Springer: Dordrecht, The Netherlands, 1996; pp. 177–190.
32. Song, C.; Sirviente, A.I. A numerical study of breaking waves. *Phys. Fluids* **2004**, *16*, 2649–2667. [CrossRef]
33. Francois, N.; Xia, H.; Punzmann, H.; Fontana, P.W.; Shats, M. Wave-based liquid-interface metamaterials. *Nat. Commun.* **2017**, *8*, 14325. [CrossRef] [PubMed]
34. Kawahara, M.; Kodama, T.; Kinoshita, M. Finite element method for tsunami wave propagation analysis considering the open boundary condition. *Comput. Math. Appl.* **1988**, *16*, 139–152. [CrossRef]
35. Whitham, G.B. *Linear and Nonlinear Waves*, 1st ed.; Wiley: New York, NY, USA, 2011.

processes

MDPI

Article

Evaluation of the Influences of Scrap Melting and Dissolution during Dynamic Linz–Donawitz (LD) Converter Modelling

Florian Markus Penz [1,*], Johannes Schenk [1,2], Rainer Ammer [3], Gerald Klösch [4] and Krzysztof Pastucha [5]

1 K1-MET GmbH, Stahlstraße 14, A-4020 Linz, Austria; Johannes.Schenk@unileoben.ac.at
2 Chair of Ferrous Metallurgy, Montanuniversität Leoben, Franz Josef Straße 18, A-8700 Leoben, Austria
3 voestalpine Stahl GmbH, voestalpine Straße 3, A-4020 Linz, Austria; Rainer.Ammer@voestalpine.com
4 voestalpine Stahl Donawitz GmbH, Kerpelystraße 199, A-8700 Leoben, Austria; Gerald.Kloesch@voestalpine.com
5 Primetals Technologies Austria GmbH, Turmstraße 44, A-4020 Linz, Austria; Krzysztof.Pastucha@primetals.com
* Correspondence: Florian-Markus.Penz@K1-MET.com; Tel.: +43-3842-402-2244

Received: 22 February 2019; Accepted: 28 March 2019; Published: 31 March 2019

Abstract: The Linz–Donawitz (LD) converter is still the dominant process for converting hot metal into crude steel with the help of technically pure oxygen. Beside hot metal, scrap is the most important charging material which acts as an additional iron source and coolant. Because of the irrevocable importance of the process, there is continued interest in a dynamic simulation of the LD process, especially regarding the savings of material and process costs with optimized process times. Based on a thermodynamic and kinetic Matlab®coded model, the influences of several scrap parameters on its melting and dissolution behavior were determined, with a special focus on establishing the importance of specific factors on the crude steel composition and bath temperature after a defined blowing period to increase the accuracy of the process model. The calculations reported clearly indicate that the dynamic converter model reacts very sensitively to the chemical composition of the scrap as well as the charged scrap mass and size. Those results reflect the importance of experiments for validation on the diffusive scrap melting model in further research work. Based on that, reliable conclusions could be drawn to improve the theoretical and practical description of the dissolution and melting behavior of scrap in dynamic converter modelling.

Keywords: scrap dissolution; scrap melting; thermodynamics; kinetics; dynamic converter modelling

1. Introduction

The Linz–Donawitz (LD) converter steelmaking process was patented in Austria in the early 1950s. The invention of the LD converter enabled the refinement of hot metal to crude steel in short blowing periods of around 20 min, enabling the high productivity of the steelmaking industry today. Technically pure oxygen is blown onto the surface of the liquid melt inside the vessel, which leads to an increase of the reaction surface through the ejection of iron droplets and further to a stronger oxidation of the dissolved elements like carbon, silicon, manganese and phosphorus. These chemical reactions are exothermic, which results in a sharp temperature rise. As a coolant, steel scrap is added at the beginning of each blowing period [1–5].

As process and material costs are getting more important, the modelling of the process is crucial. Several authors gave different approaches for modelling the scrap melting during the LD converter process e.g., by Kruskopf in [6–8], Guo in [9] and Sethi and Shukla et al. in [10–12]. In previous publications by Y. Lytvynyuk et al. [13,14], a single reaction zone model was developed. This paper should point out the sensitivity of various scrap parameters on the simulation results. Therefore, the only parameters which will be changed are scrap parameters.

2. Description of the Dynamic LD Converter Model

Lytvynyuk's model is based on thermodynamic and kinetic calculations [13,14]. It is assumed that in the heterogeneous thermodynamic reaction zone, all components can be conveyed between the slag and metal phases except carbon, which is oxidized to become gaseous carbon monoxide. The chemical oxidation reactions are assumed to be simultaneous at the interfacial surface between the slag and metal phases, whereby the oxidized carbon is removed instantly, and the equilibrium thermodynamics of the post combustion is neglected according to a non-reversible oxidation process. The flowsheet of the LD converter model is presented in Figure 1. Two simulation targets can be attained. During the main calculation loop, the mass and heat balances are calculated. Every single element is considered due to the results of chemical reactions, the consumption of the blown oxygen and the heating and melting of the charged materials. In the heat balance the consumed and generated heats are considered. Additionally, a structure of sub models will be solved during the main calculation loop. In this paper only the sub model of the scrap melting will be explained in detail. A more detailed description of the whole model is given in [4,13,15,16].

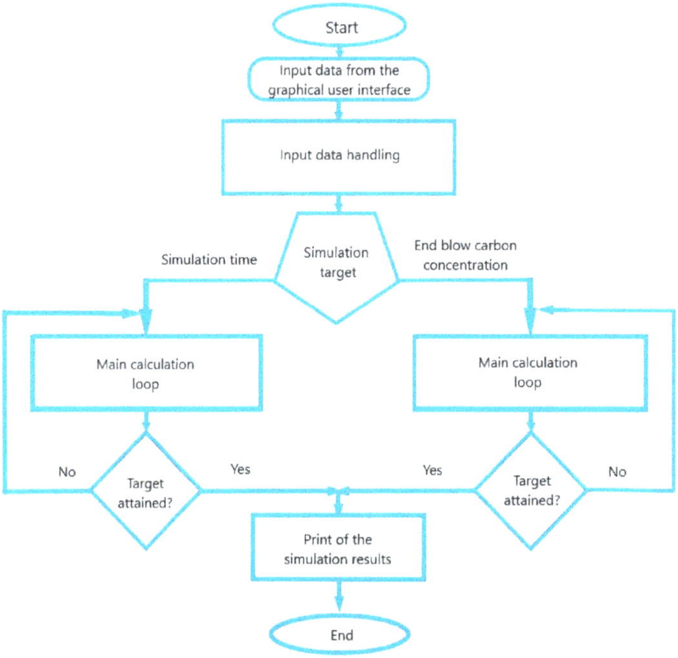

Figure 1. Flow sheet of the Linz–Donawitz (LD) converter model.

During the entire process, the chemical composition of the metal and slag phases changes due to the blowing oxygen consumption theory by V.E. Grum-Grzhimaylo for Bessemer converters [17]. In this theory, it is assumed that only iron is oxidized by blown oxygen and the remaining elements in the metal phase react with the iron oxide. The slag phase is formed by chemical reactions, melting and dissolution of charge materials and iron oxide, which is generated in the hot spot as a result of iron burning. Beside the oxidation, the dissolution and melting behavior of all charged materials influences the melt and slag composition during the entire blowing period. It is considered that the metal phase consists of the charged hot metal. Further, the solid scrap will be dissolved only in the hot metal due to its higher density. The other materials, which able to be charged, like lime, pellets, magnesia or dolomite, are assumed to dissolve only in the slag phase. They can be charged in portions during the heat [13,14].

The coupled reaction model published by Ohguchi et al. in [18] is used to describe the concurrent oxidation–reduction reactions between slag and metal, which is commonly utilized for the determination of the influence of kinetic parameters on chemical reaction rates and dephosphorization processes. [18–25]. In the present model, the simultaneous chemical reactions between the two different phases of the heterogeneous thermodynamic system can be determined by the system of chemical reactions listed in Equation (1). The reactions take place on the interfacial area between the slag and metal phase. Hess´s law is utilized for all calculated reactions. [13] In Equations (1) and (2) following metallurgical convention was used: [], () and {} indicate the metal, slag and gas phases, respectively.

$$\begin{cases} [Si]+2[O] \leftrightarrow (SiO_2) \\ [Mn]+[O] \leftrightarrow (MnO) \\ [P]+2.5[O] \leftrightarrow (PO_{2.5}) \\ [Ti]+2[O] \leftrightarrow (TiO_2) \\ [V]+1.5[O] \leftrightarrow (VO_{1.5}) \\ [Fe]+[O] \leftrightarrow (FeO) \\ [C]+[O] \leftrightarrow \{CO\} \end{cases} \tag{1}$$

For example, the oxidation of phosphorus by iron oxide is expressed by combining two reactions of Equation (1), which will result in Equation (2).

$$\begin{gathered} -\begin{cases} [P]+2.5[O] \leftrightarrow (PO_{2.5}) \\ 2.5[Fe]+[O] \leftrightarrow (FeO) \end{cases} \\ [P]-2.5[Fe]-2.5[O]+2.5[O]=(PO_{2.5})-2.5(FeO) \\ [P]+2.5(FeO)=(PO_{2.5})+2.5\,[Fe] \end{gathered} \tag{2}$$

By the use of the system of the chemical reactions, any change of one component parameter, e.g., the concentration or the activity coefficient, will lead to a change in the whole system of the considered chemical reactions. [13,14,26].

The dynamic LD converter model used is mainly focused on cost reduction due to shorter process times and specified amounts of charged materials. This parameter study on scrap melting and dissolution behavior was performed since alterations in component parameters influence the whole system because of the thermodynamic and kinetic principles of the model and its equation of oxygen balance. For all mathematical and chemical expressions, the following assumptions were taken into account:

- At the interfacial surface all reactions are expeditious and equilibrated at each time step.
- The mass transfer kinetics in the metal and slag phases are the limitation for reaction rates.
- Fluctuations in iron concentration as well as lime concentrations are neglected [13,27].

The fundamental equation to solve the calculation is one algebraic equation, which includes the bulk chemical compositions of the metal and the slag phases as well as thermodynamic and kinetic parameters. The flowchart of the reaction model is given in Figure 2. Further descriptions of the kinetic and thermodynamic calculations and for the melting behavior of slag formers, pellets and FeSi were published by Lytvynyuk et al. [13,26].

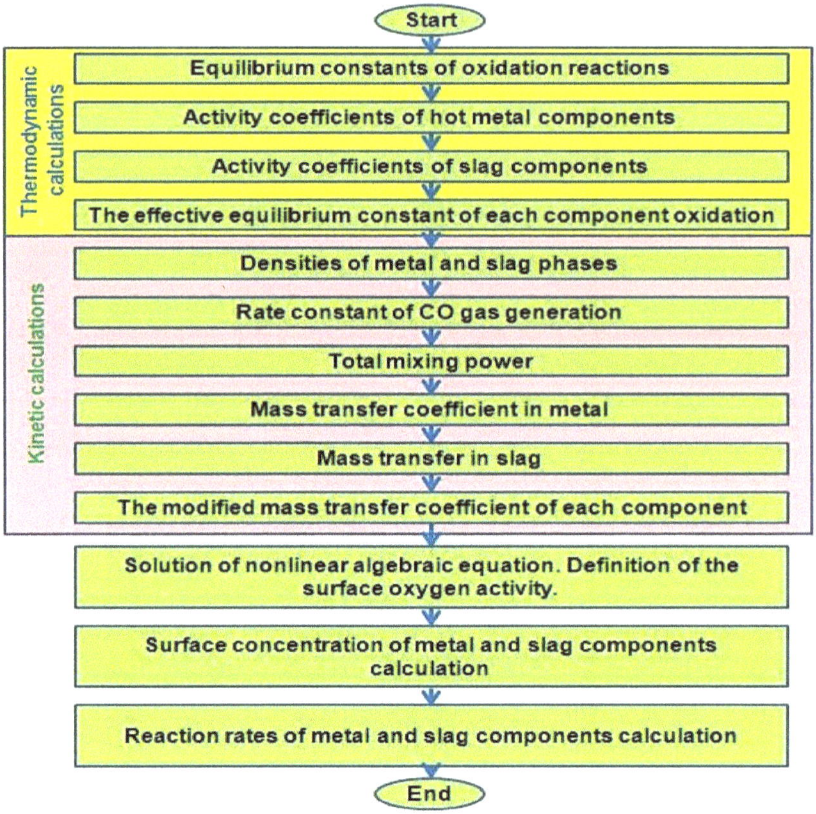

Figure 2. Flowsheet of the reaction model used [14].

To receive the effective equilibrium constant E of the oxidation reaction as a result of the thermodynamic calculation, the equilibrium constants, as well as the activity coefficients of the hot metal components, γ_{HM} and the slag components γ_S have to be determined. [26] Thus will result in Equation (3). The activity coefficients of the slag components are calculated by the collective electron theory described by Grigoryan et al. in [28] or Kolesnikova et al. in [29]. For the determination of the activity coefficients the Wagner–Lupis–Elliot method was chosen, published by Sigworth and Elliott in [30].

$$E = \frac{K \times \gamma_{HM}}{\gamma_S} \tag{3}$$

The equilibrium constant K for example for the oxidation reaction of phosphorus was derived from Equation (4). [28]

$$\log K = \frac{27050}{T} - 14.25 \tag{4}$$

The model of Lytvynyuk et al. was validated based on the output parameters of a commercial 170 t converter, which was published by Lytvynyuk et al. in [31]. The model was in a good agreement with the measured industrial data. The behavior of the temperature, the composition of the metal and slag phases, as well as the melting and dissolution of charge materials requires shutdowns during a converter heat. This kind of research is difficult to realize, due to the high costs incurred by the loss of production. The trends of the temperature or the off gas composition as well as the composition of the metal and slag phases is comparable by information from literature. Lytvynyuk also carried out a validation in this direction and could present in [14] and [31] a good agreement between the model and literature-based information.

3. Mechanisms of Scrap Melting in the LD Model

Two mechanisms describe the scrap melting and dissolution behavior in the LD model used. The scrap is charged into the vessel at the beginning of the process. The scrap geometry is assumed to be spherical. It is also possible to define the scrap to be cylindrical in shape, but in this case a melting only in radial direction can be simulated. Further, a manipulation of the overall surface of the scrap particle can be executed by introducing the form factor sphericity. The sphericity of a particle is defined as the ratio of the surface area of a sphere to the surface area of the particle, whereby the sphere has the same volume as the given particle. [32,33] In this work only a spherical shaped scrap is determined. A study on the differences between cylindrical, spherical and particles manipulated by the use of sphericity was published by the authors in [27] to give a first estimation on scrap melting behavior on the influence of heat balance. For simplification, it is assumed that the surface temperature of the scrap is equal to the hot metal temperature and that the scrap is heated up through thermal conduction. Due to the fact that the solid scrap is denser than hot metal, it assumed that the scrap is covered by liquid hot metal. Therefore, the influence of radiation can be neglected [1,4,34,35].

Forced or convective scrap melting appears if the melt temperature exceeds the melting point of the scrap and diffusive scrap melting occurs at temperatures below the scrap melt point. Forced scrap melting controls the scrap dissolution in the final stage of the LD converter process and the temperature difference between hot metal and scrap acts as the driving force. In this case, heat transfer determines the scrap melting [34,35]. Equation (5) characterizes the model for forced scrap melting of a spherical scrap particle:

$$-\partial r/\partial t = \mathrm{h_{met}} \times \left(\mathrm{T_{HM}} - \mathrm{T_{liq}}\right) / \left(\mathrm{L} + \left(\mathrm{H}\left(\mathrm{T_{scrap}}\right) - \mathrm{H}\left(\mathrm{T_{liq}}\right)\right)\right) \times \rho_{\mathrm{scrap}} \tag{5}$$

The scrap particle´s radius is r in unit (m). The heat transfer coefficient in the metal phase is h_{met} in (W m^{-2} K^{-1}) and the density of the scrap is ρ_{scrap} in (kg m^{-3}). The latent heat of scrap melting is L in (J kg^{-1}). T_{HM} and T_{liq} are the temperatures of the metal phase and the liquidus temperature of the scrap in (K) [13,31]. $H(T_{scrap})$ is the specific enthalpy of scrap at the actual temperature of the scrap surface and $H(T_{liq})$ is the specific enthalpy of the scrap melting point, both in (J kg^{-1}) [4,27].

Equation (6) describes the diffusive melting of a spherical scrap particle, where the driving force is the difference of carbon concentration in the liquid phase and the scrap. It is strongly dependent on the mass transfer coefficient of the system. According to the binary Fe-Fe_3C diagram, low carbon scrap has a higher melting point than hot metal, with around 4.5 wt.% of carbon. [3]

$$-\partial r/\partial t = \mathrm{k_{met}} \times \ln\left(\left(\%\mathrm{C_{HM}} - \%\mathrm{C_{liq}}\right) \times \rho_{\mathrm{liquid}} / \left(\%\mathrm{C_{liq}} - \%\mathrm{C_{scrap}}\right) \times \rho_{\mathrm{scrap}} + 1\right) \tag{6}$$

The mass transfer coefficient in the metal phase is k_{met} in (m s^{-1}). C_{scrap} and C_{HM} are the carbon concentrations in the scrap and hot metal in (wt.%). C_{liq} describes the carbon concentration on the liquidus line. The density of the liquid hot metal is ρ_{liquid} and of the scrap is ρ_{scrap}, both in (kg m^{-3}) [3,27]. The values for the liquidus lines are approximated by a database of Fe-Fe_3C-Si-Mn diagrams, generated by the FactSage™ FSstel database (licensed to Montanuniversität Leoben, Department Metallurgie; Version 7.1, ©Thermfact and GTT-Technologies, Montreal, Canada and Herzogenrath, Germany) [13,36].

The specific mixing power, which is created by bottom stirring and oxygen blowing, provides the basis for the mass transfer coefficient in the metal phase. The mass transfer coefficient is calculated through a function of the total mixing power including the bath depth, the converter geometry, position and geometry of the oxygen lance and the number of bottom-stirring nozzles as well as the flow rate of oxygen and bottom-stirring gas [14,26]. The heat transfer coefficient of the metal phase is defined by the function given in Equation (7), which is dependent on the specific mixing power $\dot{\varepsilon}$ [35]. According to Lytvynyuk et al. the total mixing power is a sum of the mixing power by top-blown oxygen and the mixing power by bottom-blown gas [13].

$$h_{met} = 5000 \times \dot{\varepsilon}^{0.3} \tag{7}$$

4. Simulation Parameters

Based on industrial materials and their chemical composition, the influence of small adjustments in carbon, phosphorus and silicon contents as well as the size and charged mass of scrap were investigated. The variations of the carbon and silicon contents were taken according to their quantity in scrap. The phosphorus content is rather small in common steels; still, through the scrap melting, small quantities of phosphorus are always delivered through to the liquid melt and may influence the final phosphorus content. The aim is to clarify the relevance of the obtained deviation in comparison to the adjustments to maintain the future focus on the detailed description of the melting and dissolution behavior of scrap in an LD converter. In particular, it is necessary to evaluate the sensitivity of the dynamic LD converter model on the adjusted charged scrap for further purposes.

The input parameters are listed in Table 1; Table 2. For modelling, a common rail steel grade was used. It has to be mentioned that the blowing time was fixed at 12.6 min and the amount of blown oxygen was also constant. Hot metal, scrap, solid LD slag and sand are charged at the beginning of the blowing period, whereby for simplification, only lime is charged stepwise during the entire process time. In Table 1, the initial composition, the charged mass and charging temperature of the hot metal and scrap are listed. The charged mass of scrap and the hot metal are constant for all investigations except when the scrap mass was modified.

Table 1. Charging parameters of the hot metal and the standard scrap used.

Definition	Unit	Hot Metal	Standard Scrap
Carbon content	wt.%	4.536	0.737
Silicon content	wt.%	0.410	0.349
Manganese content	wt.%	1.171	1.060
Phosphorus content	wt.%	0.100	0.013
Iron content	wt.%	93.783	97.841
Mass	t	53.60	15.72
Temperature	°C	1318	20

Table 2. Selected chemistry of added slag, sand and lime.

Name	Unit	Initial Slag	Dust Pellets	Sand	Lime
SiO_2 content	wt.%	11.32	-	92.79	0.980
MnO content	wt.%	11.93	2.960	-	-
P_2O_5 content	wt.%	1.330	-	-	-
FeO content	wt.%	29.66	-	-	-
CaO content	wt.%	40.08	7.320	-	92.37
MgO content	wt.%	4.380	4.580	-	3.080
CO_2 content	wt.%	-	-	-	2.400
H_2O content	wt.%	-	-	-	0.170
Fe_2O_3 content	wt.%	-	67.88	-	-
Fe content	wt.%	-	11.09	-	-
Amount of charged material	t	0.001	1.000	0.172	2.800

The chemistry of the initial slag as well as their compositions and the amounts of the charged dust pellets, sand and the added lime are shown in Table 2.

The standard scrap parameters from Table 1 were adjusted to analyze the melting and dissolution behavior of scrap in the BOF process. The phosphorus content in the scrap is very low, at 0.013 wt.%. The adjustment of the phosphorus was therefore set only to a higher content than usual, for example, which can be found in weathering steel. The changed values are listed in Table 3 and their percentage relative adjustment in comparison to the standard scrap is noted. It is assumed that the shape of the scrap is a sphere, whereby the size is defined to be its radius.

Table 3. Variation of the initial parameters of the scrap and their percentage.

Name and Unit	Standard Scrap	Lower Value		Higher Value	
Carbon content (wt.%)	0.7370	0.40	−45.7%	1.00	35.68%
Silicon content (wt.%)	0.3488	0.10	−71.3%	0.70	100.7%
Phosphorus content (wt.%)	0.0130	-	-	0.05	273.1%
Size (m)	0.1	0.08	−20.0%	0.12	20.00%
Mass (t)	15.72	13.0	−17.3%	17.0	%

5. Results and Discussion

This publication displays in the following illustrations of the scrap melting and dissolution behavior during the LD process with the aforementioned parameters. The calculated influence of the adjusted parameters is also shown by the trajectories of carbon, phosphorus and the melt temperature.

5.1. *Influence on the Melting and Dissolution Behavior of Scrap*

In Figures 3 and 4, the dissolution and melting behavior of scrap is pointed out. Between minutes 8 and 10, a kink occurs in all figures, resulting from the change between diffusive and forced scrap melting. At this point, the melt temperature exceeds the melting temperature of the scrap. Under real process conditions, a smooth transition between the two melting mechanisms will take place. While the melting takes place under real conditions in the two-phase area between the solidus and liquidus lines, in this model, the assumption is used that the melting point of scrap is specific to the liquidus line [27].

It is illustrated in Figure 3 that the rate of diffusive melting is faster for higher carbon content in the scrap. In the model, the denominator of the logarithmic term in Equation (4) decreases, since the difference of $\%C_{liq}$ and $\%C_{scrap}$ becomes smaller in comparison to the standard case. A lower carbon content results in the opposite. Also, a higher silicon content in the scrap leads to faster diffusive melting. The effect of the higher silicon content is a decrease of $\%C_{liq}$, which also reduces the value of the denominator of the logarithmic term in Equation (4). Changes in the phosphorus contents do not show a big difference, because they are already too small to influence the melting behavior of scrap.

In Figure 4, it is shown that low-dimension scrap melts a little bit slower in the initial phase of melting because of the higher cooling effect due to a higher specific surface. Once the low-dimension scrap is heated up—between minutes three and four—the melting of low dimension scrap accelerates. The complete melting time of the big dimension scrap with a radius of 0.12 m is higher; however, the time difference is about 1 min in comparison to the low dimension scrap with a radius of 0.08 m. It is obvious that the amount of scrap has an influence on the trajectories of melting behavior. There is a strong increase in the scrap melting of high scrap amounts shortly before the transition point between diffusive and forced scrap melting. This is explainable through Equation (6), where the logarithmic term increases in value. According to the increasing temperatures and decreasing carbon concentrations in the hot metal at this stage of the process, the denominator of the logarithm decreases faster than the numerator.

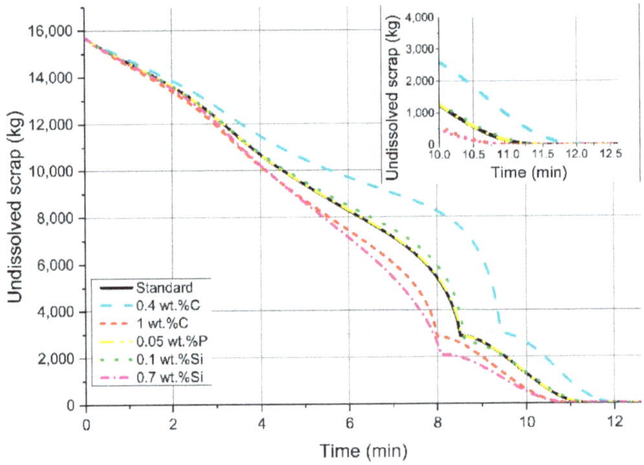

Figure 3. Influence of adjustment on chemical composition on the melting behavior of scrap.

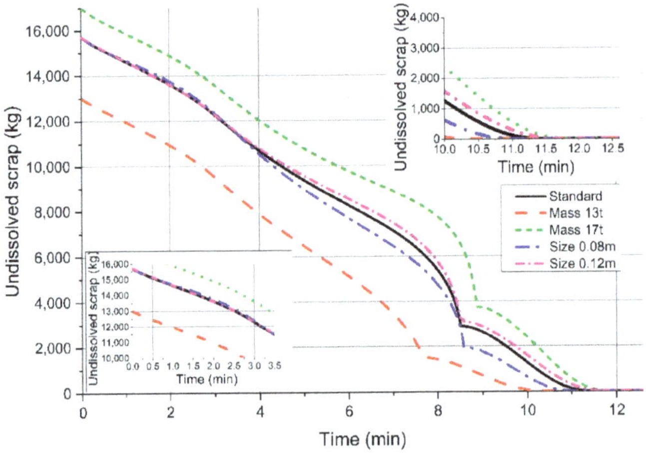

Figure 4. Influence of variable charging scrap amounts and scrap size on the melting behavior.

5.2. Influence on the Final Crude Steel Temperature

Influenced by the varying melting behavior, the calculated final temperature of the liquid crude steel changes slightly. As shown in Figure 5, the highest influence on the final temperature will be reached if the mass of charged scrap is modified. Due to the energy balance, less energy for heating the scrap will be needed with decreasing scrap amounts, which will result in higher tapping temperatures. But in comparison to the relative adjustments used, the influence on the final temperature is still small, whereby 1% means 16 °C.

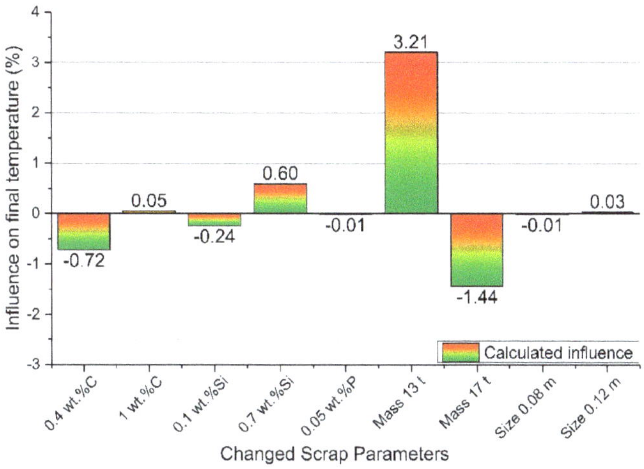

Figure 5. Influence on the final melt temperature; 1% defines a difference of 16 °C.

The influence of the mass and the scrap size can be seen in Figure 6. It is obvious that a lower charged mass of scrap will consume less heat from the system, resulting in a higher final temperature. The variation of the scrap size will only influence the melt temperature in the initial stages of the blowing period. Due to the higher overall surface of small scale scrap (0.08 m in diameter) the exchange area of heat is increasing. This fact results in a higher cooling effect at the beginning of the blowing period.

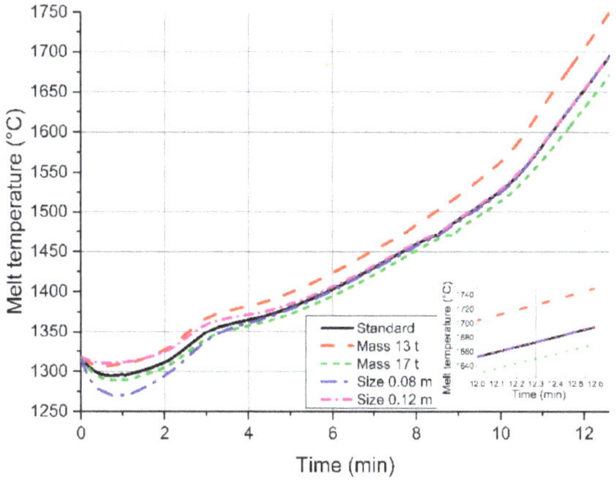

Figure 6. Influence of variable charging scrap amounts and scrap size on the melt temperature.

5.3. *Influence on the Final Carbon Content*

Decarburization is one of the two main tasks of an LD converter. How strongly the calculated final carbon content is influenced by the assumed modifications is shown in Figures 7 and 8. What is interesting to mention is that lower silicon contents in the scrap have a positive effect on decarburization and result in, besides low carbon contents in the scrap, lower final carbon contents in the liquid melt (Figure 7). In Figure 8, it is shown that a lower mass and therefore a lower input of carbon through scrap will also result in lower final values. The same behavior was determined for silicon. There is no large effect of the scrap size and the scrap phosphorus content on the final carbon content visible.

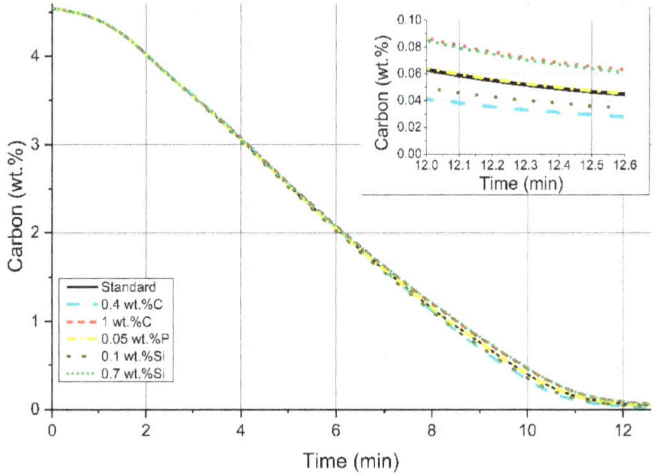

Figure 7. Influence of variable carbon and silicon contents on the carbon content of the liquid crude steel.

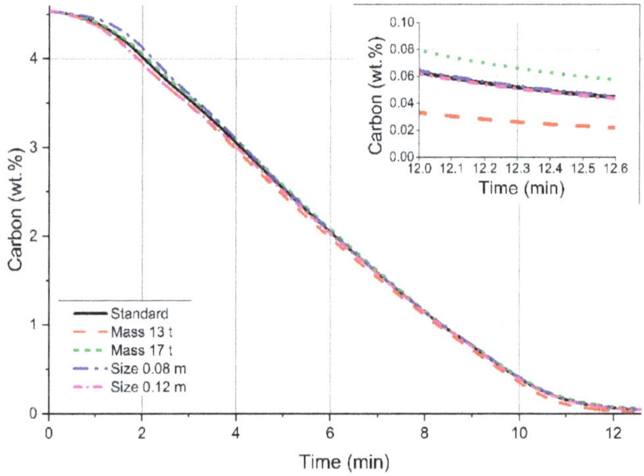

Figure 8. Influence of variable charging scrap amounts and scrap size on the carbon content of the liquid crude steel.

Even though the difference seems to be small in the final carbon content, it has to be carefully analyzed because a discrepancy of around 50% in the scrap carbon composition results in a final deviation of 40% in the final melt carbon composition. Similar values are detected in variations of the silicon content of the scrap and the scrap mass, as shown in Figure 9. A discrepancy in the final carbon composition of 50.67% was detected, if less scrap is charged. This high value results in the chosen rail scrap with a carbon content of 0.737 wt.%. Due to the faster melting of the reduced scrap mass, less carbon is transported into the melt and therefore lower amounts can be reached in the final content. The phosphorus content in the scrap and the particle size show no significant influence.

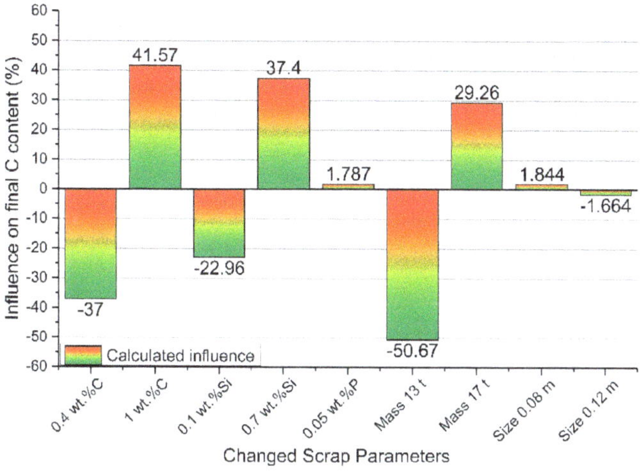

Figure 9. Calculated influence on the final carbon content.

5.4. *Influence on the Final Phosphorus Content*

The second main task of an LD converter is dephosphorization, and the aim of each operator is to reach low phosphorus contents in the tapped crude steel. As shown in Figures 10 and 11, the trajectories of the phosphorus content in the melt are influenced during the entire blowing period through the adjusted parameters. The interaction between an increasing amount of carbon, silicon and phosphorus in the scrap leads to higher final phosphorus contents in the tapped crude steel (Figure 10). Due to the still relatively high carbon activity and the increasing bath temperatures during the process, the stable oxides of manganese and phosphorus will be reduced. In the final stages of the blowing period, low amounts of carbon and silicon lead to an early carbon activity decrease. This results in an earlier resumption of the phosphorus oxidation.

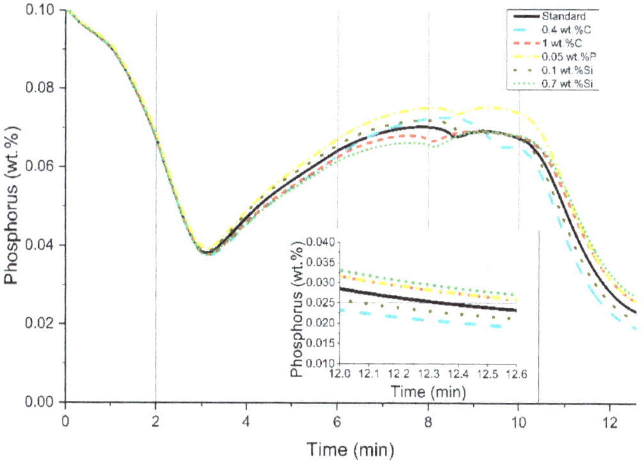

Figure 10. Influence of variable element contents in scrap on the phosphorus content of the liquid crude steel.

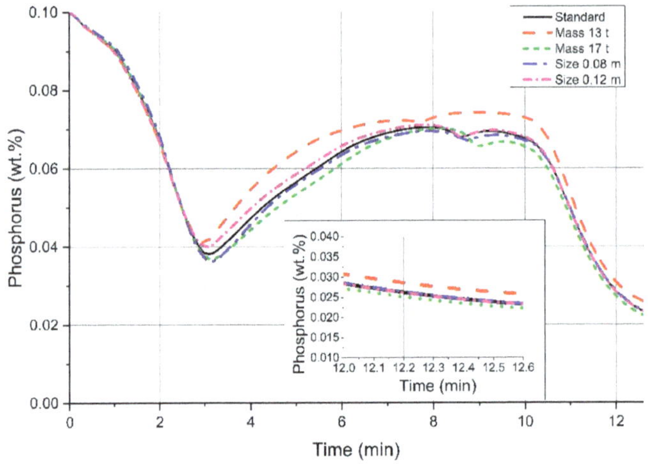

Figure 11. Influence of variable charging scrap amounts and scrap size on the phosphorus content of the liquid crude steel.

Because of the strong influence on the melt temperature, the slag composition and the melting behavior of low-charged scrap amounts lead to higher calculated final phosphorus contents, as pointed out in Figure 11. In the main dephosphorization period at the beginning of the process, the temperature of the melt rises faster if there is less scrap or high volume scrap charged. Therefore, the point of re-phosphorization will start earlier and this causes higher final phosphorus contents in the melt. A possibility to counteract this behavior would be a different charging concept for lime and a modified slag metallurgy.

The percentage of influence on the final phosphorus content in the liquid melt according to the adjusted parameters is shown in Figure 12. It shows that the dynamic LD converter model reacts sensitively to the final phosphorus contents on variations in the chemical scrap composition and charged scrap mass.

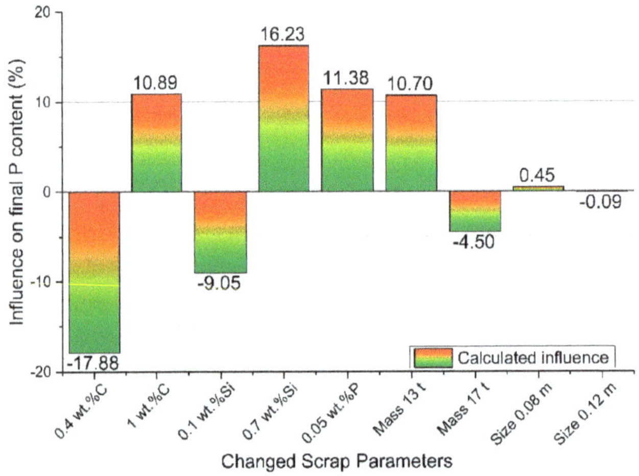

Figure 12. Calculated influence on the final phosphorus content.

6. Conclusions

The present study was done to clarify the relevance of the obtained deviations for the final temperature, composition and scrap melting behavior. Therefore, adjustments in chemical composition, size and mass of charged scrap were considered. The aim was to focus on the detailed description of the melting and dissolution behavior of scrap during the LD converter process for future work. The BOF model used was coded in MatLab® (R2014b, MathWorks Inc., Natick, MA, USA) and describes the behavior of the metal and slag phases during the blowing period of the BOF process using thermodynamic and kinetic equations.

The LD converter model used includes literature-based scrap melting equations. In the literature they are discussed, but insufficient validation reports are available. The model calculations show that around three quarters of the blowing process are dominated by diffusive scrap melting. A faster melting of the scrap could be indicated if the carbon or silicon content of the scrap were increased in comparison to the standard composition. This behavior is based on a lower melting point of scrap due to the higher contents of those elements. In this research, the phosphorus content of the scrap was also investigated. It has to be mentioned that it has no influence on the melting behavior of the scrap, but with increasing phosphorus in the scrap, the final phosphorus content in the liquid crude steel will also increase. It was shown that the amount of charged material has a very strong influence on the melting behavior as well as the final compositions of carbon and phosphorus. The scrap size changes those values solely in a small frame. It is worth pointing out that particularly low contents of carbon and silicon in the scrap also lower the final phosphorus and carbon content in the melt.

To sum up, the calculations reported in this paper clearly indicate that the dynamic BOF model reacts very sensitively to the chemical composition of the scrap as well as the charged scrap mass and size and therefore, the whole melting and dissolution behavior of scrap. Experiments will be necessary for validation of the diffusive scrap melting model. Based on that, reliable conclusions could be drawn to improve the theoretical and practical description of the dissolution and melting behavior of scrap. This description should be as precise as possible if it is necessary to be able to implement a complete dynamic LD converter model for usage in the industry.

Author Contributions: Conceptualization, F.M.P.; Data curation, F.M.P.; Investigation, F.M.P.; Methodology, F.M.P.; Project administration, F.M.P., J.S. and K.P.; Resources, F.M.P.; Software, F.M.P.; Supervision, F.M.P. and J.S.; Validation, J.S., R.A., G.K. and K.P.; Visualization, F.M.P.; Writing—original draft, F.M.P.; Writing—review & editing, F.M.P., J.S., R.A., G.K. and K.P.

Funding: This research project is co-funded by public financial resources from the Austrian Competence Center Programme COMET and by the industrial partners voestalpine Stahl, voestalpine Stahl Donawitz, and Primetals Technologies Austria.

Acknowledgments: The authors gratefully acknowledge the funding support of K1-MET GmbH, metallurgical competence center. The research programme of the K1-MET competence centre is supported by COMET (Competence Centre for Excellent Technologies), the Austrian programme for competence centres. COMET is funded by the Federal Ministry for Transport, Innovation and Technology, the Federal Ministry for Science, Research and Economy, the provinces of Upper Austria, Tyrol and Styria as well as the Styrian Business Promotion Agency (SFG).

Conflicts of Interest: The authors declare no conflict of interest.

References

1. Turkdogan, E.T. *Fundamentals of Steelmaking*; The Institute of Materials: London, UK, 1996; pp. 209–244.
2. Ghosh, A.; Chatterjee, A. *Ironmaking and Steelmaking Theory and Practice*; PHI Learning Private Limited: Delhi, India, 2015; pp. 285–292.
3. Chigwedu, C. Beitrag zur Modellierung des LD-Sauerstoffaufblasverfahrens zur Stahlerzeugung. Ph.D. Thesis, Technische Universität Clausthal, Clausthal-Zellerfeld, Germany, 1997.
4. Penz, F.M.; Bundschuh, P.; Schenk, J.; Panhofer, H.; Pastucha, K.; Paul, A. Effect of Scrap Composition on the Thermodynamics of Kinetic Modelling of BOF Converter. In Proceedings of the 2nd VDEh-ISIJ-JK Symposium, Stockholm, Sweden, 12–13 June 2017; pp. 124–135.

5. Hiebler, H.; Krieger, W. Die Metallurgie des LD-Prozesses. *BHM* **1992**, *137*, 256–262.
6. Kruskopf, A.; Holappa, L. Scrap melting model for steel converter founded on interfacial solid/liquid phenomena. *Metall. Res. Technol.* **2018**, *115*, 201–208. [CrossRef]
7. Kruskopf, A.; Louhenkilpi, S. 1-Dimensional scrap melting model for steel converter (BOF). In Proceedings of the METEC & 2nd ESTAD, Düsseldorf, Germany, 15–19 June 2015; pp. 1–4.
8. Kruskopf, A. A Model for Scrap Melting in Steel Converter. *Metall. Mater. Trans. B* **2015**, *46*, 1195–1206. [CrossRef]
9. Guo, D.; Swickard, D.; Alavanja, M.; Bradley, J. Numerical Simulation of Heavy Scrap Melting in BOF steelmaking. *Iron Steel Technol.* **2013**, *10*, 125–132.
10. Sethi, G.; Shukla, A.K.; Das, P.C.; Chandra, P.; Deo, B. Theoretical Aspects of Scrap Dissolution in Oxygen Steelmaking Converters. In Proceedings of the AISTech, Nashville, TN, USA, 15–17 September 2004; Volume II, pp. 915–926.
11. Shukla, A.K.; Deo, B. Coupled heat and mass transfer approach to simulate the scrap dissolution in steelmaking process. In Proceedings of the International Symposium for Research Scholars on Metallurgy, Materials Science & Engineering, Chennai, India, 18–20 December 2006; pp. 1–14.
12. Shukla, A.K.; Deo, B.; Robertson, D. Scrap Dissolution in Molten Iron Containing Carbon for the Case of Coupled Heat and Mass Transfer Control. *Metall. Mater. Trans. B* **2013**, *44*, 1407–1427. [CrossRef]
13. Lytvynyuk, Y.; Schenk, J.; Hiebler, M.; Sormann, A. Thermodynamic and Kinetic Model of the Converter Steelmaking Process. Part 1: The Description of the BOF Model. *Steel Res. Int.* **2014**, *85*, 537–543. [CrossRef]
14. Lytvynyuk, Y. Thermodynamic and Kinetic Modelling of Metallurgical Processes. Ph.D. Dissertation, Montanuniversität Leoben, Leoben, Austria, 2013.
15. Hirai, M.; Tsujino, R.; Mukai, T.; Harada, T.; Masanao, O. Mechanism of Post Combustion in the Converter. *Trans. ISIJ* **1987**, *27*, 805–813. [CrossRef]
16. Bundschuh, P.; Schenk, J.; Hiebler, M.; Panhofer, H.; Sormann, A. Influence of CaO Dissolution on the Kinetics of Metallurgical Reactions in BOF-process. In Proceedings of the 7th European Oxygen Steelmaking Conference, Trinec, Czech Republic, 9–11 September 2014.
17. Boychenko, B.; Okhotskiy, V.; Kharlashin, P. *The Converter Steelmaking*; Dnipro-VAL: Dnipropetrovsk, Ukraine, 2006; pp. 22–69.
18. Ohguchi, S.; Robertson, D.; Deo, B.; Grieveson, P.; Jeffes, J. Simultaneous dephosphorization and desulphurization of molten pig iron. *Iron Steelmak.* **1984**, *11*, 202–213.
19. Kitamura, S.; Kitamura, T.; Shibata, K.; Mizukami, Y.; Mukawa, S.; Nakagawa, J. Effect of stirring energy, temperature and flux composition on hot metal dephosphorization kinetics. *ISIJ Int.* **1991**, *31*, 1322–1328. [CrossRef]
20. Kitamura, S.; Kitamura, T.; Aida, T.; Sakomura, E.; Koneko, R.; Nuibe, T. Development of analyses and control method for hot metal dephosphorization process by computer simulation. *ISIJ Int.* **1991**, *31*, 1329–1335. [CrossRef]
21. Kitamura, S.; Shibata, H.; Maruoka, N. Kinetic Model of Hot Metal Dephosphorization by Liquid and Solid coexisting slags. *Steel Res. Int.* **2008**, *79*, 586–590. [CrossRef]
22. Pahlevani, F.; Kitamura, S.; Shibata, H.; Maruoka, N. Kinetic Model Dephosphorization in Converter. In Proceedings of the SteelSim, Leoben, Austria, 8–10 September 2009.
23. Mukawa, S.; Mizukami, Y. Effect of stirring energy and rate of oxygen supply on the rate of hot metal dephosphorization. *ISIJ Int.* **1995**, *35*, 1374–1380. [CrossRef]
24. Ishikawa, M. Analysis of hot metal desiliconization behaviour in converter experiments by coupled reaction model. *ISIJ Int.* **2004**, *44*, 316–325. [CrossRef]
25. Higuchi, Y.; Tago, Y.; Takatani, K.; Fukagawa, S. Effect of stirring and slag condition on reoxidation on molten steel. *ISIJ* **1998**, *84*, 13–18.
26. Lytvynyuk, Y.; Schenk, J.; Hiebler, M.; Mizelli, H. Thermodynamic and kinetic modelling of the devanadization process in the steelmaking converter. In Proceedings of the 6th European Oxygen Steelmaking Conference, Stockholm, Sweden, 7–9 September 2011.
27. Penz, F.M.; Bundschuh, P.; Schenk, J.; Panhofer, H.; Pastucha, K.; Maunz, B. Scrap melting in BOF: Influence of particle surface and size during dynamic converter modelling. In Proceedings of the 3rd ABM week, São Paulo, Brazil, 2–6 October 2009.

28. Grigoryan, A.H.; Stomakhin, A.J.; Ponomarenko, A.G. *Physico-Chemical Calculations of the Electric Steel Process*; Metallurgy: Moscow, Russia, 1989.
29. Kolesnikova, K.; Gogunskii, V.; Olekh, T. Calculation of equilibrium in the system metal-slag during steelmaking in electric arc furnace. *Metall. Min. Ind.* **2016**, *6*, 8–13.
30. Sigworth, G.K.; Elliot, J.F. The thermodynamics of liquid dilute iron alloys. *Met. Sci.* **1974**, *8*, 298–310. [CrossRef]
31. Lytvynyuk, Y.; Schenk, J.; Hiebler, M.; Sormann, A. Thermodynamic and Kinetic Model of the Converter Steelmaking Process. Part 2: The Model Validation. *Steel Res. Int.* **2014**, *85*, 544–563. [CrossRef]
32. Wadell, H. Volume, shape, and roundness of quartz particles. *J. Geol.* **1935**, *43*, 250–280. [CrossRef]
33. Penz, F.M. Charakterisierung des Hochofeneinsatzstoffes Sinter, Mittels Optischer 3D-Partikelanalyse. Bachelor's Thesis, Montanuniversitaet Leoben, Leoben, Austria, 2014.
34. Medhibozhskiy, M.Y. *Basis of Thermodynamic and Kinetic of Steelmaking*; Vischa shkola: Kyiv, Ukraine, 1979; p. 229.
35. Isobe, K.; Maede, H.; Ozawa, K.; Umezawa, K.; Saito, C. Analysis of the Scrap Melting Rate in High Carbon Molten Iron. *ISIJ* **1990**, *76*, 2033–2040.
36. Zarl, M. Development and Evaluation of a BOF Pre-Processor Model. Master's Thesis, Montanuniversität Leoben, Leoben, Austria, 2017.

MDPI
St. Alban-Anlage 66
4052 Basel
Switzerland
Tel. +41 61 683 77 34
Fax +41 61 302 89 18
www.mdpi.com

Processes Editorial Office
E-mail: processes@mdpi.com
www.mdpi.com/journal/processes

www.ingramcontent.com/pod-product-compliance
Lightning Source LLC
LaVergne TN
LVHW071938080526
838202LV00064B/6635
* 9 7 8 3 0 3 9 2 1 4 5 5 6 *